calculus
of several
variables

The Houghton Mifflin GCMC Series
under the editorship of A. B. Willcox

Introduction to Calculus 1
A. B. Willcox · R. Creighton Buck · H. G. Jacob · D. W. Bailey

Introduction to Calculus 2
A. B. Willcox · R. Creighton Buck · H. G. Jacob · D. W. Bailey

Introduction to Calculus 1 and 2—combined edition
A. B. Willcox · R. Creighton Buck · H. G. Jacob · D. W. Bailey

Calculus of Several Variables
R. Creighton Buck · A. B. Willcox

Linear Algebra
H. G. Jacob · D. W. Bailey

Introduction to Probability and Statistics
Samuel Goldberg

R. Creighton Buck
University of Wisconsin

Alfred B. Willcox
Executive Director
Mathematical Association of America

Houghton Mifflin Company · Boston

calculus of several variables

New York · Atlanta · Geneva, Illinois · Dallas · Palo Alto

Printed in the USA

Library of Congress Catalog Card Number: 76-137795

ISBN: 0-395-05541-5

To Our Wives

preface

In recent years, many texts have appeared which concentrate on that portion of the calculus dealing with functions of one variable. Bearing titles such as *First Year Calculus*, these books cover what CUPM in its report *GCMC* (*General Curriculum in Mathematics for Colleges*) called Mathematics 1 and Mathematics 2.

As our title suggests, the purpose of the present book is to complete this program by presenting the basic calculus of functions of several variables, covering the content of the *GCMC* course Mathematics 4. The book is designed to follow any one-year introduction to calculus, and the combination will yield a comprehensive calculus course suitable for a science major, arts major, or mathematics major. Not unnaturally, we feel that an ideal text to precede this one is *Introduction to Calculus 1 and 2*, of which we are also authors; however, this preference does not imply that other books cannot be used. All that we have presupposed is a general knowledge of differentiation and integration for functions of one variable; we do not assume that students have been exposed to a rigorous course with emphasis on theory.

The central themes of the book are easily listed: continuity and its consequences, differentiation and its applications, the theory of iterated and multiple integrals, introductory differential equations.

We have tried to strike a balance between rigor and intuition, with the distinction always visible. Much attention has been given to motivation and the role of analogy as a guide. In presenting proofs we have sometimes chosen to give an informal outline first to clarify the pattern of reasoning, and then to complete the proof by a valid detailed argument. We have striven to integrate applications and theory by exploiting the concept of "mathematical model"; we hope that this helps answer the needs of those for whom such motivation is useful. We have chosen to lay stress upon a geometric approach to many of the basic ideas; for example, the concept of "neighborhood" is central in discussions of such topics as continuity. Thus we have tried to avoid concealing the essence of a definition or proof behind a welter of notation.

As with previous books in the Houghton Mifflin *GCMC* series, we wrote this book for students. It is not intended to be a reference treatise, but rather a companion in the classroom; there are many places where the teacher can lead students to explore related ideas or to try a different approach.

We hesitate to suggest a detailed syllabus for use with the book, since instructors differ so widely in their pace and objectives. Our experience suggests that it is possible to cover the book in about 40 lessons. The first two chapters can be completed in about 15 class meetings, perhaps less if some of the 3-space analytical geometry in Chapter 1 is review. Chapters 3 and 4 will also take up about 15 class meetings. We note that Section 3.5, dealing with applications of differentiation, will take several class periods. The same is true for sections 4.4 and 4.5 dealing with the evaluation and existence of double integrals. Section 4.7, on triple integrals, contains many applications which a class may wish to touch only lightly.

Chapter 5 contains a basic introduction to ordinary differential equations and covers most of the common techniques of solution, including the use of power series and numerical methods. Stress has been laid on the relationship between the original physical situation and the differential equation which is its model, so that students will be better prepared for science courses that make use of mathematics. Please note that each of sections 5.5 ("Techniques of Solution") and 5.7 ("Linear Equations") requires two or three class meetings to cover adequately. The last section, on the harmonic oscillator, is a detailed consideration of the phenomena of damping and resonance; it is more fun to teach if one has access to some simple apparatus to demonstrate the behavior of a real system!

Although we have not assumed that students reading this book have completed a course in linear algebra, we recognize that there are aspects of analysis that become more meaningful when viewed against such a background. These aspects are usually met in the study of advanced calculus and are therefore not entirely relevant for a text at the level of the present book; nevertheless, for the benefit of those students with a stronger background or a more professional interest in mathematics, we have included a Postlude which sets forth this relationship to some extent. It may also be helpful in expanding the horizons of all students.

The nature of this book reflects strongly the intimate association of the authors with CUPM and its panels over many years, as well as their experience in teaching calculus for a combined total of at least 45 years. In addition, many persons whose names do not appear on the title page have shared in shaping our ideas and attitudes. Among these, Ellen Buck deserves special mention, since she has once again served as a collaborating author; her experience and skill are responsible for many of the simplifying expositions. Our appreciation also goes to Nancy Buck who criticized the manuscript from the student's viewpoint and who contributed many of the exercises. Portions of Chapter 5 are based on a set of notes prepared by one of the authors in collaboration with John Nohel and Fred Brauer as part of a curriculum-study project at the University of Wisconsin, supported in part by a grant from the National Science Foundation.

Our thanks also go to Mr. Paul Kelly of Houghton Mifflin Company for his constant encouragement, and to Miss Veronica McLoud of Houghton Mifflin Company, whose efficient editing tamed our rather verbose style and clarified our numerous "that's" and "which's."

R.C.B.
A.B.W.

contents

geometry

chapter 1

1.1

Space In studying calculus for functions of one variable, it is hardly ever necessary to step outside the plane. The domain of any function considered is usually an interval of real numbers, which is a portion of the 1-dimensional line $\mathbb{R}$, and its graph is a set of points in the 2-dimensional plane $\mathbb{R}^2$. However, the moment we begin to study functions of two or more variables, we make use of higher dimensional spaces; the graph of a function of two variables will turn out to be a surface in 3-space, and that of a function of three variables can only be thought of as a "surface" in 4-space.

The geometric properties of space (i.e., 3-space) are vastly different from those of a line (1-space) or a plane (2-space). For example, a closed curve in a plane has an inside and an outside, but no such concept applies to a closed curve in space. The geometrical differences between ordinary space and higher dimensional spaces such as 4-space are even more profound, and only a few mathematicians are able to be at home and to develop trustworthy intuitions there. Fortunately, such skills are not needed in working with undergraduate mathematics, and it is enough to be able to "see" in 3-space.

The purpose of this chapter is to cover briefly some of the basic concepts and techniques of analytical geometry of space. As in the case of the plane, we start by introducing a **coordinate system** with three mutually perpendicular axes, whose intersection we call the origin $\boldsymbol{O}$. The position of any point P can then be specified by an ordered triple of numbers (a, b, c) which gives the location of P with respect to the origin and the axes. (See Figure 1–1.) The ordering of the coordinates agrees with a conventional ordering of the axes, as we see indicated in the diagram. Here, the coordinate axes are labeled X, Y, Z, and P is the point obtained by taking $x = a$, $y = b$, $z = c$, where (x, y, z) is the general point in 3-space.

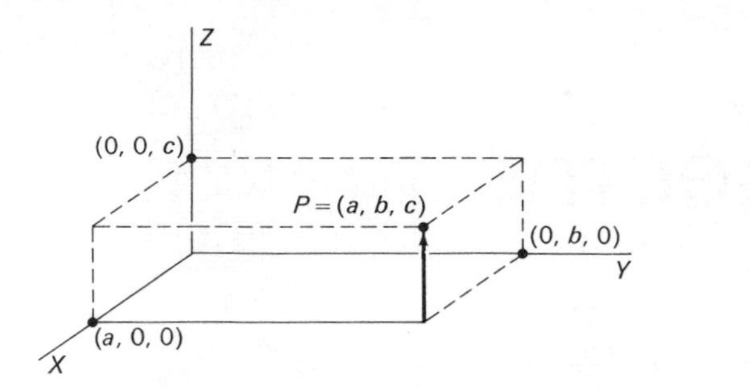

Figure 1–1

The coordinates of a point may be positive or negative; their signs determine which one of the eight octants the point lies in. (See Figure 1–2.) Unlike the quadrants in the plane, the individual octants have no generally agreed upon names, although one might reasonably call the set of all points (x, y, z) with $x \geq 0$, $y \geq 0$, $z \geq 0$ the **positive** octant.

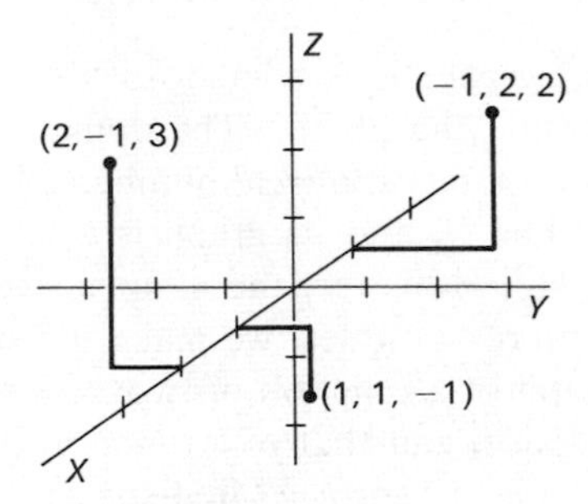

Figure 1–2

It is easy to give a formula for the distance from a point P to the origin $\boldsymbol{O}$. If $P = (x, y, z)$, then

$$\text{dist}(P, \boldsymbol{O}) = |P| = \sqrt{x^2 + y^2 + z^2}\,. \tag{1.1}$$

This is an extension of the Pythagorean theorem, and is easily proved by geometry. (See Exercise 12.) For example, the length of the line segment from the origin to the point $(2, -1, 4)$ is $\sqrt{4 + 1 + 16} = \sqrt{21}$. Generalizing, the Euclidean distance between two points $P_1 = (x_1, y_1, z_1)$ and $P_2 = (x_2, y_2, z_2)$ is

$$\begin{aligned} \text{dist}(P_1, P_2) &= |P_1 - P_2| \\ &= \sqrt{(x_1 - x_2)^2 + (y_1 - y_2)^2 + (z_1 - z_2)^2}\,. \end{aligned} \tag{1.2}$$

This formula is clearly a generalization of the formula for distance in the plane.

Many geometric concepts can be described solely in terms of distance. Thus, a **sphere** (spherical surface) with center P_0 and radius r consists of all points P with $\text{dist}(P, P_0) = r$, and the corresponding spherical ball consists of those points P with $\text{dist}(P, P_0) \leq r$. More precisely, we shall call the latter the **closed** ball with center P_0, radius r; the corresponding **open** ball is defined to be

$$\{\text{all points } P \text{ with dist } (P, P_0) < r\},$$

so that the union of the open ball and the spherical surface is the closed ball. Translating into algebraic formulas by means of (1.2), we have

Equation of sphere with center (a, b, c) and radius r:

$$(x - a)^2 + (y - b)^2 + (z - c)^2 = r^2.$$

Equation of closed ball with center (a, b, c) and radius r:

$$(x - a)^2 + (y - b)^2 + (z - c)^2 \leq r^2.$$

We can use these to give precise meaning to the useful intuitive concept of **neighborhood**. A neighborhood, or vicinity, of a point P_0 ought to include all the points P that are immediately around P_0. For example, the closed (or open) ball with center $(2, 3, -4)$ and radius $\frac{1}{2}$ is a neighborhood of the point $(2, 3, -4)$, and so is any set N that contains this ball as a subset.

Definition 1 A set N is said to be a neighborhood of a point P_0 if N contains as a subset an open ball with center P_0 and some positive radius $r > 0$.

We need to have r strictly positive because an open ball of radius 0 is the empty set. However, r can be as small as we wish, and usually we are primarily interested in what goes on in arbitrarily small neighborhoods of a point.

To illustrate the above definitions, let N be the solid cube of side 3 centered on the origin. (See Figure 1–3.) We assert that it is a neighborhood of the origin because N contains the open ball about $\boldsymbol{O}$ of radius 1. This is easily seen geometrically, since the distance from the origin to the nearest point of the surface of the cube is $\frac{3}{2}$,

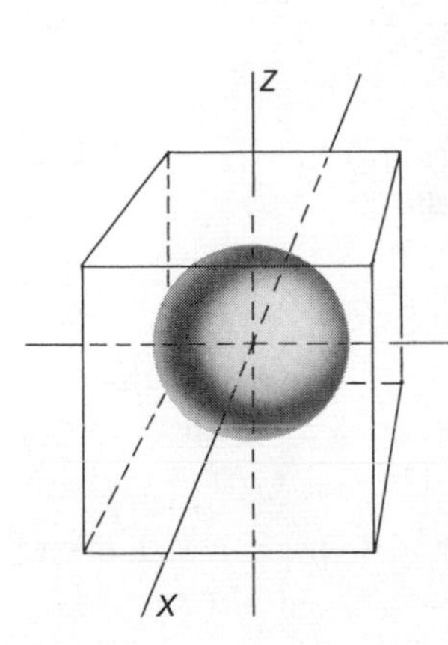

Figure 1–3

whereas every point in the ball is no more than a distance 1 away from the origin. This statement can also be proved analytically—that is, without appealing to a picture. The solid cube can be described by the formula

$$C = \{\text{all } P = (x, y, z) \text{ with } |x| \leq \tfrac{3}{2},\quad |y| \leq \tfrac{3}{2},\quad |z| \leq \tfrac{3}{2}\}.$$

Suppose that $|P| < r$, i.e., $x^2 + y^2 + z^2 < r^2$. Clearly, we must have $x^2 < r^2$, so that $|x| < r$. Likewise, we have $|y| < r$ and $|z| < r$. Accordingly, any point P in the open ball $|P| < r$ will lie in C when $r < \frac{3}{2}$, for then the defining condition describing C is satisfied by P. Hence, C contains all the open balls of radius $r < \frac{3}{2}$, center $\boldsymbol{O}$.

In passing, we point out that this book contains no three-dimensional diagrams, but only flat two-dimensional pictures of three-dimensional diagrams, like Figure 1–3 above. The ability to look at such a picture and convert it into a three-dimensional mental image is apparently a learned skill, and is more difficult for some people than for others. Practice helps, and so does the use of models made of paper, string, or cardboard. You might wish to test your ability at visualization by the following three examples.

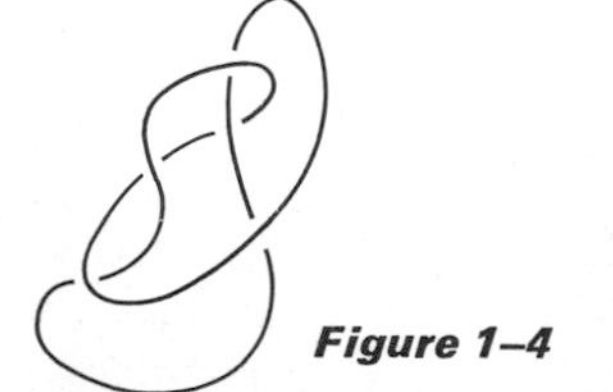

Figure 1–4

(i) Is the loop of cord in Figure 1–4 knotted or merely twisted?
(ii) Can the two loops of cord in Figure 1–5 be separated without cutting either?
(iii) Are the diagrams in Figure 1–6 perspective pictures of real objects?

Figure 1–5

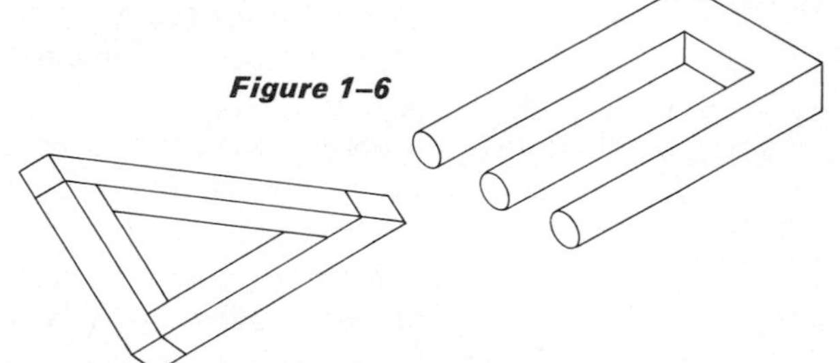

Figure 1–6

Exercises

1 Consider the solid sphere $(x - 1)^2 + (y + 1)^2 + (z - 2)^2 \leqq 9$. Which of the following points are inside the sphere?

(a) the origin $(0, 0, 0)$, (b) $(2, 2, 2)$,
(c) $(3, -1, 1)$, (d) $(-1, 1, 1)$.

2 Is the ball $(x - 1)^2 + (y - 2)^2 + (z - 2)^2 \leqq 1$ inside (a subset of)
(a) $x^2 + y^2 + z^2 \leqq 4$, (b) $x^2 + y^2 + z^2 \leqq 16$?
Can you describe how the three balls are related?

3 What relation do the spherical surfaces S_1 and S_2 have to each other, where S_1, S_2 are defined as follows:

$$S_1 : (x - 1)^2 + (y - 1)^2 + z^2 = 1,$$
$$S_2 : (x - 3)^2 + (y - 2)^2 + (z - 2)^2 = 4?$$

4 How are the two balls

$$S_1 : (x+1)^2 + (y+1)^2 + (z+1)^2 \leqq 1,$$
$$S_2 : (x-1)^2 + (y-1)^2 + (z-1)^2 \leqq 1$$

related? Do they have any points in common?

5 What is the closest distance between the spherical surfaces
(a) $x^2 + y^2 + z^2 = 36$ and $(x+2)^2 + (y-1)^2 + (z+2)^2 = 4$,
(b) $(x-2)^2 + (y+2)^2 + (z-1)^2 = 4$ and $x^2 + y^2 + z^2 = 25$?

6 Consider the two solid spheres S_1 and S_2, where S_1 is given by $x^2 + y^2 + z^2 \leqq 1$ and S_2 by $(x-1)^2 + (y-2)^2 + z^2 \leqq 9$. Is either one inside (a subset of) the other? What can you say about them?

7 There is a spider at A on a box of sides 4, 5, and 6 inches. (See Figure 1–7.) What is the shortest path along the surface of the box to point B? (Both A and B are midpoints of a side of length 4.)

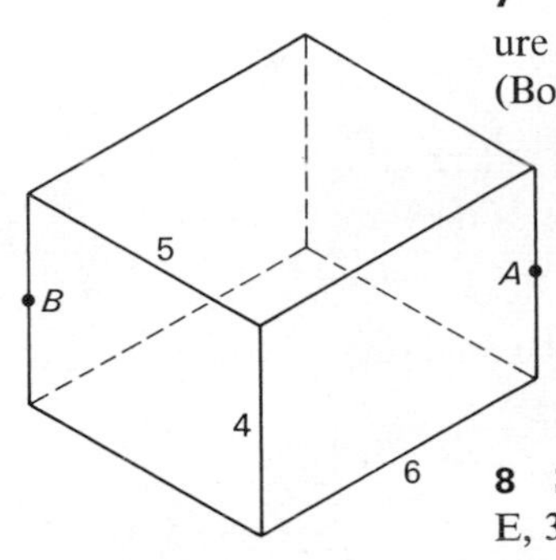

Figure 1–7

8 Starting at the origin, take a trip in 3-space as follows: Proceed 1 mile E, 3 miles S, 5 miles up, 4 miles N, 2 miles W, and 3 miles down. How far are you from the origin?

9 Two rings are connected by a cord as shown in Figure 1–8. Can they be disconnected without cutting the cord?

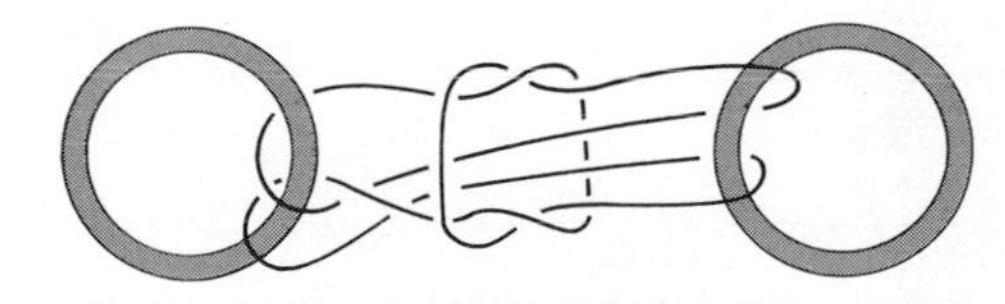

Figure 1–8

10 What common objects cast the shadows, in two perpendicular directions, shown in Figure 1–9?

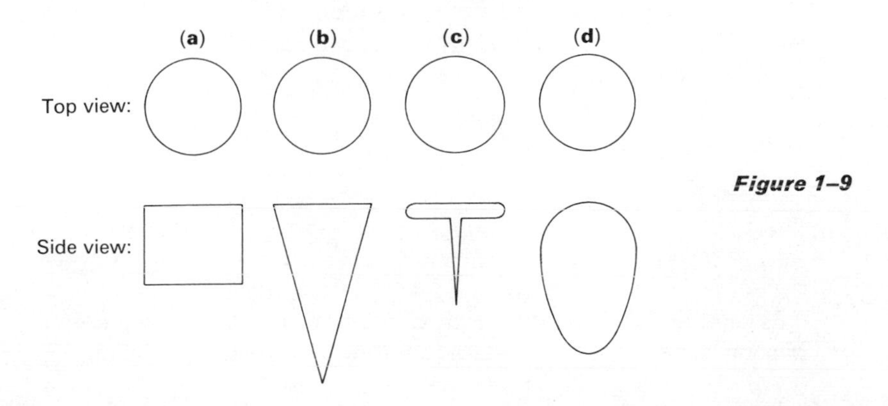

Figure 1–9

11 Write equations for the following:
(a) spherical surface of radius 2 about $(-2, 1, 1)$ as center,
(b) solid ball of radius 3 about $(1, -2, -3)$ as center,
(c) open spherical neighborhood of radius $\frac{1}{2}$ about $(1, 0, -1)$.

12 Prove by elementary geometry that the distance from O to $P = (x, y, z)$ is $\sqrt{x^2 + y^2 + z^2}$, using Figure 1–10.

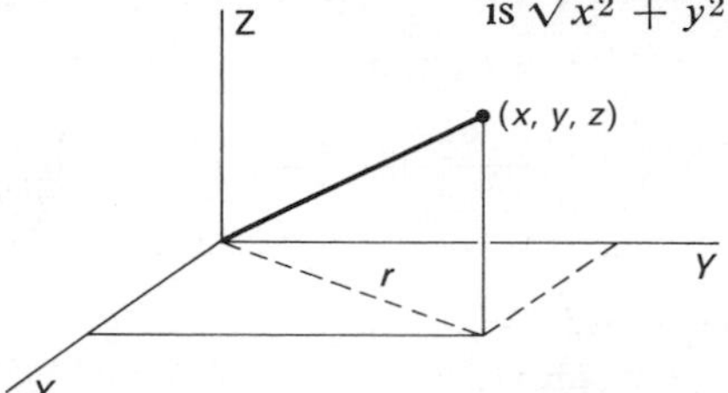

Figure 1–10

13 Given an open ball S about the origin of radius 1, prove that if P_0 belongs to S, then there are points P in S such that $|P| > |P_0|$.

14 Show that every sphere has an equation of the form

$$x^2 + y^2 + z^2 + Ax + By + Cz + D = 0.$$

15 Find the center and radius of each of the following spheres:
(a) $x^2 + y^2 + z^2 + 2x + 6y - 2z = 5$,
(b) $2x^2 + 2y^2 + 2z^2 + 8x - 4y + 4z - 38 = 0$.

16 Find the equation of the sphere which has $(2, -1, 5)$ and $(7, 3, -1)$ as ends of a diameter.

17 Why is the equation $x^2 + y^2 + z^2 - 2x + 4y + 6 = 0$ not the equation of a sphere?

1.2

Vectors

With any point P in space we can associate the directed line segment joining O to P, and call it the **vector** (or **position vector**) belonging to P. We indicate the vector as an arrow, and usually we use the same letter for a point and for its position vector. The addition of vectors is defined geometrically by constructing a parallelogram, as shown in Figure 1–11.

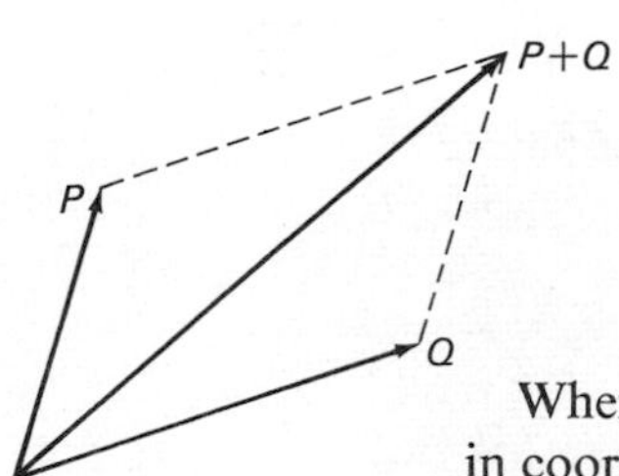

Figure 1–11

When this definition is translated into a formula for adding points in coordinate form, it becomes the following very natural definition:

Definition 2 If $P = (x_1, y_1, z_1)$ and $Q = (x_2, y_2, z_2)$, then

$$P + Q = (x_1 + x_2, y_1 + y_2, z_1 + z_2). \tag{1.3}$$

Similarly, the process for multiplying vectors by positive or negative real numbers leads to a corresponding process for multiplying points by numbers. To multiply a vector by 2, we double its length but leave its direction unchanged; to multiply it by -1, we merely take the position vector of the same length pointing in the opposite direction. Applied to points in coordinate form, this definition becomes

Definition 3 If $P = (x, y, z)$ and α is any real number, then

$$\alpha P = (\alpha x, \alpha y, \alpha z). \tag{1.4}$$

With these definitions, 3-space becomes an example of a very important mathematical system, a **linear space** (or **vector space**). The formal postulates for a real vector space are given below.

Definition 4 A real vector space consists of a set V, whose members may be called points or vectors, and two algebraic operations. With any two vectors u and v is associated a unique vector $u + v$ called their sum, and with any vector u and any real number α is associated a unique vector αu, called the (scalar) product of u by α. In addition, the operations must satisfy the following laws:

(i) For any vectors u and v, $u + v = v + u$.
(ii) For any vectors u, v, and w, $u + (v + w) = (u + v) + w$.
(iii) For any vectors u and v and any number α,

$$\alpha(u + v) = \alpha u + \alpha v.$$

(iv) For any numbers α and β and any vector u,

$$(\alpha + \beta)u = \alpha u + \beta u,$$
$$(\alpha\beta)u = \alpha(\beta u).$$

(v) There is a special vector $\boldsymbol{O}$ such that for any vector u,

$$u + \boldsymbol{O} = \boldsymbol{O} + u = u.$$

(vi) For any vector u, the real numbers 1 and 0 have the property that $0(u) = \boldsymbol{O}$ and $(1)u = u$.

It is easy but tedious to verify that the points in 3-space form a vector space. As we shall see later, so do the continuous functions defined on any region, the solutions of certain important differential equations, and all polynomials of degree at most N. In fact, there are so many examples of vector spaces in analysis that the study of linear algebra and vector spaces is basic to all advanced work in mathematics.

Returning to our examination of 3-space as a vector space, recall formulas (1.1) and (1.2). If $P = (x, y, z)$, then

(1.1) $$\text{dist}\,(\boldsymbol{O}, P) = |P| = \sqrt{x^2 + y^2 + z^2}\,,$$

(1.2) $$\begin{aligned}\text{dist}\,(P_1, P_2) &= |P_2 - P_1| \\ &= \sqrt{(x_2 - x_1)^2 + (y_2 - y_1)^2 + (z_2 - z_1)^2}\,.\end{aligned}$$

Formula (1.1) says that $|P|$ is the length of the vector P. Noting that $P_2 - P_1 = P_2 + (-1)P_1 = (x_2 - x_1,\ y_2 - y_1,\ z_2 - z_1)$, we see that (1.2) says merely that the distance between points P_1 and P_2 is just the length of the vector $P_2 - P_1$. (See Figure 1–12.)

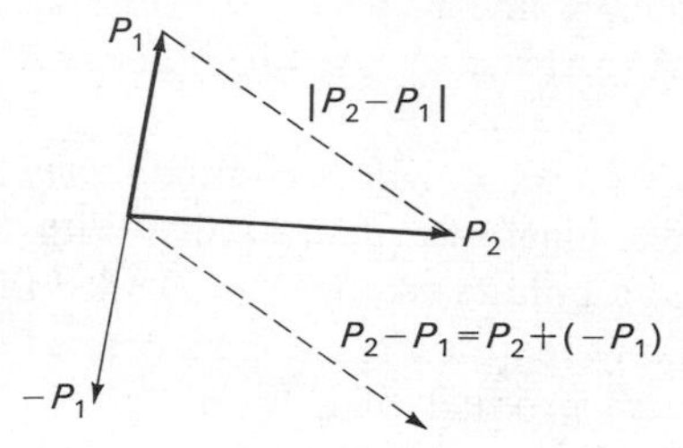

Figure 1–12

In general, a vector space V is said to be a *normed* space if every vector u in V is assigned a length, denoted by $|u|$, and if this length function obeys the following rules:

(i) $|u| \geq 0$ for any u, with equality holding if and only if $u = 0$.
(ii) For any vector u and any real number α,

$$|\alpha u| = |\alpha|\,|u|.$$

(iii) For any vectors u and v,

$$|u + v| \leq |u| + |v|.$$

The last of these rules is called the **triangle inequality**, since as we see from Figure 1–11, it can be interpreted as saying that a side of a triangle can never be longer than the sum of the lengths of the other two sides. The number $|v|$ is often called the **norm** of the point or vector v. In any normed linear space, the distance between two points is defined by

$$\text{dist}\,(u, v) = |u - v| = |v - u|,$$

so that $|v| = |v - \boldsymbol{O}|$ is the distance from $\boldsymbol{O}$ to v.

We can summarize much of the preceding discussion of the nature of ordinary Euclidean space in the following brief statement.

Theorem 1 *Three-dimensional Euclidean space is a normed linear space, with the algebraic operations defined by* (1.3) *and* (1.4) *and the norm defined by* (1.1).

We leave the verification of the properties of the norm as an exercise (Exercise 13).

In plane analytical geometry, the notion of slope played a central role. In space, the situation is more complicated, and the closest analogue to slope is the concept of direction, which can be described best by talking about vectors. Suppose that you are standing at the origin and wish to specify a direction. You could do so by pointing, which amounts to choosing a particular vector, thus indicating the direction along that vector. Clearly, the zero vector $\boldsymbol{O}$ doesn't indicate any particular direction, but all others do. To convert this rather vague idea into something that can be used, let us enclose the origin in a spherical surface of unit radius. Then each direction is specified by pointing toward a particular point on this sphere. (This is essentially the way astronomers locate stars.) If we now associate every such point with its corresponding unit vector, then we have replaced the vague idea of "directions in space" with the set of all unit vectors. The direction of an arbitrary vector $v \neq \boldsymbol{O}$ will be a unit vector u (i.e., $|u| = 1$) which points in the same direction as v. This means that u must be a positive multiple of v, and it is then easily seen that u is uniquely prescribed, and is given by the formula

$$u = \frac{1}{|v|}\,v = \frac{v}{|v|}\,.$$

For example, if $v = (1, 2, -2)$, then $|v| = \sqrt{1 + 4 + 4} = 3$, so that the direction of v is the unit vector $(\frac{1}{3}, \frac{2}{3}, -\frac{2}{3})$.

How do we talk about direction if we are standing at some other point than the origin? More generally, can we use vectors to talk about starting at a point P and taking a trip that goes in a certain direction for a certain distance? The answer to this question will lead us to see that vectors are not used solely as position vectors to locate points.

Restricting the action to the XY-plane for simplicity, we could specify a trip by naming the starting point and the ending point, as in a "trip from (2, 1) to (4, 3)." Does this trip have anything in common with the trip from $(3, -1)$ to $(5, 1)$, or the trip from $(-3, 1)$ to $(-1, 3)$? As is clear from Figure 1–13, each of these is a trip of length $\sqrt{8}$ in a northeast direction; each could be described completely by giving the vector $v = (2, 2)$ and a starting point.

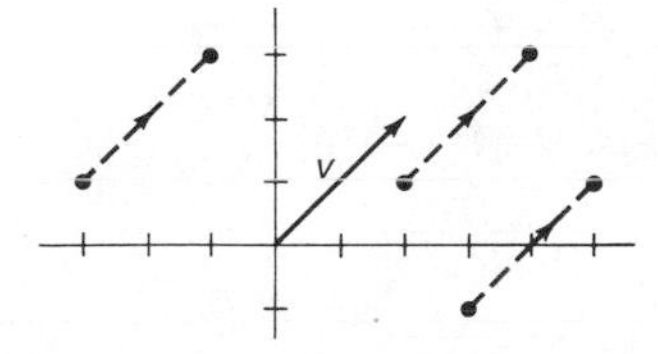

Figure 1–13

As in Figure 1–12, in general a trip that starts at P_1 and ends at P_2 can be described by the vector $v = P_2 - P_1$. Two trips that have the same direction and length but possibly different starting points correspond to the same vector v. Used thus, every vector v serves to define a translation of the plane, which moves each point the distance and in the direction specified by v. Starting at p, we move to $p + v$, and the motion can be indicated by treating v as a "free" vector—placing its initial point at p so that the head of the arrow rests on $p + v$ (see again Figure 1–13). When we use vectors in this way as arrows whose initial point can be placed anywhere, it is customary to say that we are using "free" vectors. In fact the vectors are unchanged, but in this context we are using them in a different way; in the discussion above, we used vectors to represent trips. Later, we shall see how vectors are used (in the free form) to represent the velocity of a moving point and the normal to a surface.

What happens if two trips are combined in the natural way? If we start at p, first take the trip described by v_1 and then take the trip described by v_2 from this new location, what is the description of the combined trip? Clearly, we first arrive at $p + v_1$; then starting from there, we arrive at $(p + v_1) + v_2$. Since this vector is the same as $p + (v_1 + v_2)$, we see that the "sum" of two trips is the trip described by the sum of the two vectors.

What has been discussed for the plane also works in space as well. A trip that starts at $(4, 3, -1)$ and ends at $(6, 1, 2)$ is to be described by the vector $v = (6, 1, 2) - (4, 3, -1) = (2, -2, 3)$, and the length of the trip is $\sqrt{4 + 4 + 9} = \sqrt{17}$. If we take the same trip starting instead at the point $p = (-1, 3, 1)$, we shall arrive at the point

$$p + v = (-1, 3, 1) + (2, -2, 3) = (1, 1, 4).$$

We close this section with two illustrations of the solution of geometric problems using the vector viewpoint.

Theorem 2 *The point $M = \frac{1}{2}P + \frac{1}{2}Q$ is the midpoint of the segment joining P and Q.*

Proof We give two arguments. First, using vectors we observe that the midpoint of the segment PQ can be obtained by adding the vector P and half of the vector from P to Q (Figure 1–14). Since the vector describing the trip from P to Q is $Q - P$, we have

$$M = P + \tfrac{1}{2}(Q - P) = P + \tfrac{1}{2}Q - \tfrac{1}{2}P = \tfrac{1}{2}P + \tfrac{1}{2}Q.$$

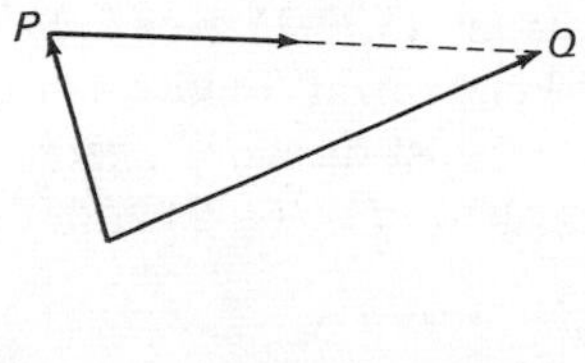

P Q

Figure 1–14

The second proof depends on properties of distance. If

$$M = \tfrac{1}{2}P + \tfrac{1}{2}Q,$$

then $|M - P| = |\tfrac{1}{2}P + \tfrac{1}{2}Q - P| = |\tfrac{1}{2}Q - \tfrac{1}{2}P| = \tfrac{1}{2}|Q - P|.$

Also, $|M - Q| = |\tfrac{1}{2}P + \tfrac{1}{2}Q - Q| = |\tfrac{1}{2}P - \tfrac{1}{2}Q| = \tfrac{1}{2}|P - Q|.$

Since $|P - Q| = |Q - P| = \text{dist}(P, Q)$, we see that M is a point which is exactly halfway between P and Q, and is therefore the midpoint of PQ.

As another example of the vector-space viewpoint, we have the following. (See Figure 1–15.)

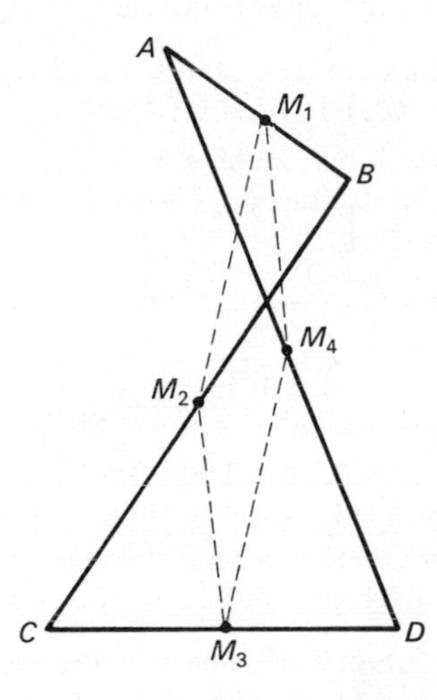

Figure 1–15

Theorem 3 *Let A, B, C, D be four points in space forming a quadrilateral in that order. Then the midpoints of its sides are the vertices of a parallelogram.*

Proof Let M_1, M_2, M_3, M_4 be the midpoints of sides AB, BC, CD, DA respectively. We can show that they form a parallelogram in several ways: for example, by proving that the line segment M_1M_2 is parallel and equal in length to M_3M_4. This is true if the vectors from M_1 to M_2 and from M_4 to M_3 are equal. Using the midpoint formula for M_i, $i = 1, \cdots 4$, we have

$$M_2 - M_1 = \frac{B + C}{2} - \frac{A + B}{2} = \frac{C - A}{2},$$

and

$$M_3 - M_4 = \frac{C + D}{2} - \frac{D + A}{2} = \frac{C - A}{2}.$$

Exercises

1 Given $A = (2, -1, 1)$ and $B = (-4, 3, 1)$, find
(a) $(A + B)/2$, (b) $(A - B)/2$, (c) $|3A + 2B|$.

2 Given that $A = (1, -4, 3)$ and $A + 2B = (2, 4, -7)$, what is $A - 2B$?

3 Given the pair of equations $A - B = (1, -1, 1)$ and $A + 2B = (7, -4, -2)$, find A and B.

4 Find the unit vector in the direction of each of the following vectors:
(a) $(2, -1, 2)$, (b) $(6, -6, 3)$, (c) $(1, -1, 1)$, (d) $(0, -.5, 1.2)$.

5 Find the unit vector in the direction of each of the following vectors:
(a) $(-4, 0, 3)$, (b) $(-1, 2, 1)$, (c) $(2, -\sqrt{3}, \sqrt{2})$,
(d) $(1, -1, -.5)$, (e) $(.4, -1.8, -1.2)$.

6 Show that the vectors $P = (.05, -.4, .2)$ and $Q = (-\frac{2}{3}, \frac{16}{3}, -\frac{8}{3})$ point in exactly opposite directions.

7 (a) Given any two points A and B, let $C = (2A + B)/3$. Show that $\text{dist}\,(B, C) = 2\,\text{dist}\,(A, C)$.
(b) Take a real number λ, $0 \le \lambda \le 1$, and let $C = \lambda A + (1 - \lambda)B$. Show that $\text{dist}\,(A, C) = (1 - \lambda)\,\text{dist}\,(A, B)$ and that $\text{dist}\,(C, B) = \lambda\,\text{dist}\,(A, B)$.

8 Let vectors $v_1 = (1, 2)$ and $v_2 = (2, -1)$ describe two trips T_1, T_2 in the plane.
(a) Make a sketch in the plane to illustrate the following: Starting from the point $A = (3, 0)$, take trip T_1, arriving at a point A', and from there take trip T_2, arriving at A''. Then compare this with the result of starting at A and taking trip T_2 and then T_1.
(b) Formulate a general theorem about combining a number of trips in different orders. Can you prove it?

9 A translation T of 3-space by the vector v is the function (transformation) which sends a point p into the point $p' = p + v$, called the image of p under T. (Note that p' is just the point reached from p by the trip described by v.) A transformation is said to be distance preserving if $\text{dist}\,(p, q) = \text{dist}\,(p', q')$, where p' and q' are the images under T of the points p and q.
(a) With $v = (-3, 1, 3)$, find the images under T of the points $A = (2, 0, 3)$, $B = (3, -4, 5)$.
(b) Verify that $\text{dist}\,(A, B) = \text{dist}\,(A', B')$.
(c) Show that translation by $v = (-3, 1, 3)$ is distance preserving.
(d) Is every translation of 3-space distance preserving?

10 (a) Let $A = (-2, 1)$ and $B = (2, -3)$. Is there a vector v such that translation by v sends A to a point on the X-axis and at the same time sends B to a point on the Y-axis?
(b) Is there a translation of 3-space that sends $(3, -1, 2)$ to a point on the X-axis and $(-1, 2, 0)$ to a point on the Y-axis?

11 If $P_1 = (2, -1, 3, 1)$ and $P_2 = (4, 4, 1, 5)$, what is the unit vector in 4-space in the direction from P_1 toward P_2?

12 Justify part of Theorem 1 by illustrating with diagrams in the plane the first four properties listed in Definition 4.

13 Justify part of Theorem 1 by verifying that the norm defined in equation (1.1) satisfies the three properties required of a general norm in the paragraph that precedes the statement of the theorem. (Check the first two by algebra and the third by geometry.)

14 Use the third property of a norm to show that in general

$$\text{dist}\,(A, C) \leq \text{dist}\,(A, B) + \text{dist}\,(B, C).$$

15 Prove Theorem 3 in a different way by using the fact that a quadrilateral is a parallelogram if and only if its diagonals bisect each other.

16 Given the points $A = (2, 1, -1)$, $B = (-1, 3, 0)$, $C = (1, -2, 3)$, can you find a point D such that all four points form the vertices of a parallelogram? Is D unique?

17 Prove that four points A, B, C, D are vertices of some parallelogram if and only if one of the three following relations hold: $A + C = B + D$, $A + B = C + D$, $A + D = B + C$.

1.3

Inner Product

If $p_1 = (x_1, y_1, z_1)$ and $p_2 = (x_2, y_2, z_2)$, then the **inner product** (also called the **dot** product or **scalar** product) of these vectors is defined to be

$$p_1 \cdot p_2 = x_1x_2 + y_1y_2 + z_1z_2. \tag{1.5}$$

Thus, the inner product of $(2, -1, 3)$ and $(1, 4, -1)$ is -5. The inner product has an immediate geometric interpretation in terms of the angle θ between the directions of the two vectors. We first note that the inner product obeys several algebraic rules:

$$\begin{aligned} p \cdot (u + v) &= p \cdot u + p \cdot v, \\ p \cdot (\alpha q) &= (\alpha p) \cdot q = \alpha(p \cdot q), \\ p \cdot q &= q \cdot p. \end{aligned} \tag{1.6}$$

We also note that

$$|p|^2 = (x_1)^2 + (y_1)^2 + (z_1)^2 = p \cdot p.$$

Theorem 4 *If θ is the angle between the directions of the vectors p and q, then*

$$p \cdot q = |p|\,|q| \cos\theta. \tag{1.7}$$

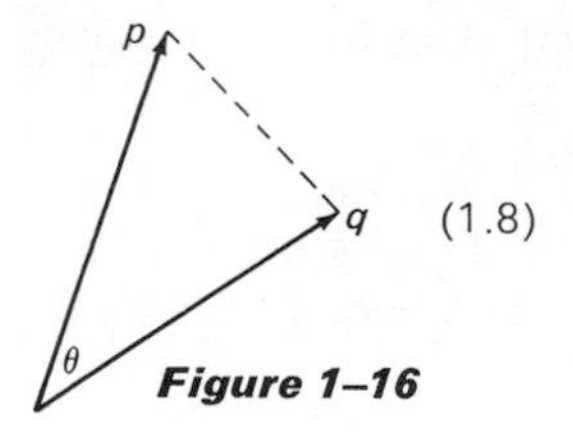

Figure 1–16

Proof This comes at once from the law of cosines of elementary trigonometry. Looking at Figure 1–16, we see that

$$\{\text{dist}(p, q)\}^2 = |p|^2 + |q|^2 - 2|p|\,|q| \cos \theta. \tag{1.8}$$

Using the fact that distance is expressible in terms of the inner product and then using the identities in (1.6), we can write

$$\begin{aligned} \{\text{dist}(p, q)\}^2 &= |p - q|^2 = (p - q)\cdot(p - q) \\ &= p\cdot p - p\cdot q - q\cdot p + q\cdot q \\ &= |p|^2 - 2(p\cdot q) + |q|^2. \end{aligned}$$

Using (1.8), we have

$$|p|^2 - 2(p\cdot q) + |q|^2 = |p|^2 + |q|^2 - 2|p|\,|q| \cos \theta,$$

from which (1.7) comes at once.

There are several useful results that follow from (1.7). For example, if we solve for $\cos \theta$, we have

$$\cos \theta = \frac{p\cdot q}{|p|\,|q|}.$$

Note that this relation makes it possible to find the angle between two given vectors neither of which is $\boldsymbol{O}$. For example, if $p = (-2, 6, 9)$ and $q = (1, -2, 2)$, then $p\cdot q = 4$, $|p| = 11$, $|q| = 3$, and $\theta = \cos^{-1}(\frac{4}{33})$.

Again, if we choose q as a unit vector u, then (1.7) yields the relation

$$p\cdot u = |p| \cos \theta, \tag{1.9}$$

which can be interpreted as the geometric statement that $p\cdot u$ is the **projection** of the vector p in the direction u. (See Figure 1–17.)

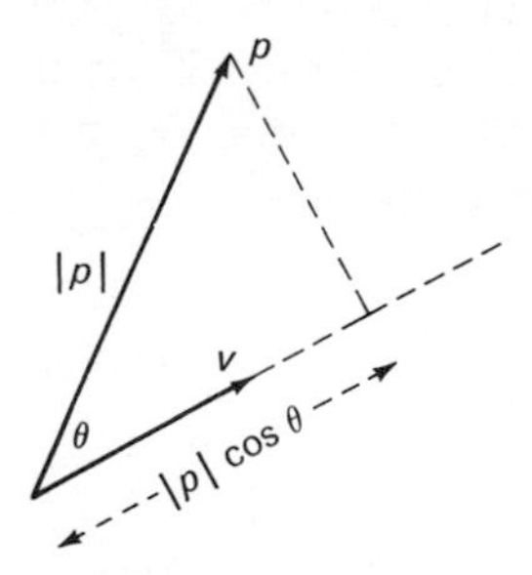

Figure 1–17

If two vectors p and q are perpendicular, then their angle θ is 90°, so that $\cos \theta = 0$. This gives us at once:

Corollary 1 *Two nonzero vectors p and q are orthogonal (perpendicular) if and only if $p\cdot q = 0$.*

An example is given by the pair $(1, 3, -2)$ and $(4, 2, 5)$; these vectors and the vector $(-19, 13, 10)$ form a mutually orthogonal triple. Exercise 4 shows that there cannot exist four such mutually orthogonal vectors in 3-space.

There is a simple inequality which can be proved for vectors in 3-space merely by observing that $-1 \leq \cos\theta \leq 1$ for any choice of θ.

Corollary 2 *If p and q are vectors in space, then*

$$-|p|\,|q| \leq p \cdot q \leq |p|\,|q|.$$

All of the geometrical ideas so far discussed can be generalized so that they apply to 4-space or indeed to n-space for any n. A point in n-space is identified with an ordered set of n numbers, as

$$p = (a_1, a_2, \ldots, a_n), \quad q = (b_1, b_2, \ldots, b_n).$$

The sum of p and q is

$$p + q = (a_1 + b_1, a_2 + b_2, \ldots, a_n + b_n),$$

and for any real number λ,

$$\lambda p = (\lambda a_1, \lambda a_2, \lambda a_3, \ldots, \lambda a_n).$$

With these formulas, n-space becomes another example of a vector space.

We define an **inner product** in n-space:

$$p \cdot q = a_1b_1 + a_2b_2 + \cdots + a_nb_n. \tag{1.10}$$

For example, $(1, 3, -1, 2, 2) \cdot (0, 1, 4, -1, 2) = 1$. Guided by our experience in 3-space, we define distance by

$$\text{dist}\,(p, q) = |p - q|, \tag{1.11}$$

where, for any point p, we define the length of p as

$$|p| = \sqrt{p \cdot p} = \sqrt{a_1^2 + a_2^2 + \cdots + a_n^2}\,. \tag{1.12}$$

Thus in 5-space the distance between (1, 3, 4, 0, 1) and (2, 0, 5, 2, 1) is

$$\sqrt{(1-2)^2 + (3-0)^2 + (4-5)^2 + (0-2)^2 + (1-1)^2} = \sqrt{15}\,.$$

Although we cannot draw pictures to display n-space events for $n \geq 4$, as we do with $n = 3$, to some extent we can still depend upon geometric and algebraic analogies to help us. For example, in 4-space the sphere with center (1, 1, 1, 1) and radius 3 will have the equation

$$(x-1)^2 + (y-1)^2 + (z-1)^2 + (w-1)^2 = 9,$$

since this says precisely that the point $P = (x, y, z, w)$ is a distance 3 from the point $C = (1, 1, 1, 1)$, i.e., $|P - C| = 3$. Likewise, the ball consisting of all points P with $|P - C| < .01$ is a neighborhood of the point C. The notation we have chosen for such things as distance and inner product has been selected so that many formulas have the same appearance whether we are dealing with 2-space or 5-space. Indeed, if you go back to 1-space, where a point p is just a real number x, then the general formula for $|p|$ becomes $|p| = \sqrt{x^2} = |x|$, the absolute value of x.

We now ask if we can continue to operate by analogy and adopt as definitions the two statements

(i) In n-space, the angle θ between two vectors p and q is determined by $\cos\theta = p \cdot q/|p|\,|q|$,
(ii) Two vectors p and q are orthogonal if and only if $p \cdot q = 0$.

How do we know that we won't run into trouble somewhere? For example, could it ever happen that two points p and q would give values of $|p|$, $|q|$, and $p \cdot q$ so that we would be looking for an angle θ for which $\cos\theta = 1.5$? The assurance that this will never happen, that therefore these definitions make sense and permit us to speak of the angle between vectors in n-space comes from a special theorem known variously as Schwarz's Inequality, Cauchy's Inequality, and Bunyakovski's Inequality. The theorem states that the inner product operation in n-space has the same property given in Corollary 2 of Theorem 4 for 3-space.

Theorem 5 *Using the definitions in* (1.12) *and* (1.10) *for* $|p|$, $|q|$, *and* $p \cdot q$, *it is true that for all points* p *and* q *(neither* $\mathbf{O}$*)*,

$$-1 \leq \frac{p \cdot q}{|p|\,|q|} \leq 1.$$

Since this ratio lies between -1 and 1, we can find a unique angle θ between 0° and 180° such that $\cos\theta = (p \cdot q)|p|^{-1}|q|^{-1}$, which we call the angle between p and q.

If the conclusion of the theorem above is restated in terms of the coordinates of p and q, (1.10), and (1.12), then it can be written: *For any real numbers* $a_1, a_2, \ldots, a_n$ *and* $b_1, b_2, \ldots, b_n$,

$$(a_1b_1 + a_2b_2 + \cdots + a_nb_n)^2 \leq (a_1^2 + a_2^2 + \cdots + a_n^2)(b_1^2 + \cdots + b_n^2).$$

The theorem can be proved in this form as a direct exercise in algebra, and it can also be proved in other ways. We choose to leave it unproved here, since it will not be used often and can be left for later courses.

Exercises

1 Find the angle θ between the vectors P and Q and sketch P, Q.

(a) $P = (1, 1, 0)$, $Q = (1, 2, -2)$;
(b) $P = (1, 1, 2)$, $Q = (1, -2, -1)$;
(c) $P = (1, 2, 3)$, $Q = (-1, 2, -1)$;
(d) $P = (1, 0, 1)$, $Q = (1, 1, 2)$.

2 Find the cosine of the angle between the vectors P and Q, P and $P + Q$, Q and $P + Q$, when

(a) $P = (-2, 4, 4)$, $Q = (3, 0, 4)$;
(b) $P = (1, -4, 8)$, $Q = (1, 8, -4)$.

3 Show that if two different vectors P and Q have equal length, then each makes the same angle with $P + Q$.

4 Prove that there cannot be 4 mutually perpendicular nonzero vectors in 3-space.

5 Let B be the closed ball in 4-space consisting of points (x, y, z, w) such that $(x - 1)^2 + (y - 2)^2 + (z + 3)^2 + w^2 \leq 14$. Decide whether each of the following points is inside, outside, or on the surface of B:

(a) $O = (0, 0, 0, 0)$, (b) $(1, 1, 1, 1)$, (c) $(1, 1, -1, 1)$.

6 In 4-space, consider the four points $A = (1, 0, 2, 3)$, $B = (-1, 2, 0, -1)$, $C = (2, 3, -1, 0)$, $D = (1, 1, -1, -1)$. Verify that

$$\text{dist}\left(\frac{A + B}{2}, \frac{B + C}{2}\right) = \text{dist}\left(\frac{C + D}{2}, \frac{A + D}{2}\right),$$

$$\text{dist}\left(\frac{B + C}{2}, \frac{C + D}{2}\right) = \text{dist}\left(\frac{A + D}{2}, \frac{A + B}{2}\right).$$

What does this mean about the midpoints of the sides of the quadrilateral with vertices A, B, C, D?

7 What is the center and radius of the "hypersphere"

$$x^2 + y^2 + z^2 + w^2 - 4x + 2y - 4w + 5 = 0?$$

8 If $V = (a, b, c)$ is a nonzero vector, the angles θ_x, θ_y, θ_z between V and the positive coordinate axes are called the direction angles for V (see Figure 1–18). The triple $\cos\theta_x$, $\cos\theta_y$, $\cos\theta_z$ are called the direction cosines for V. Find the direction cosines for each of the following vectors:

(a) $(1, 2, -1)$, (b) $(2, 0, 4)$, (c) $(-3, 2, 1)$.

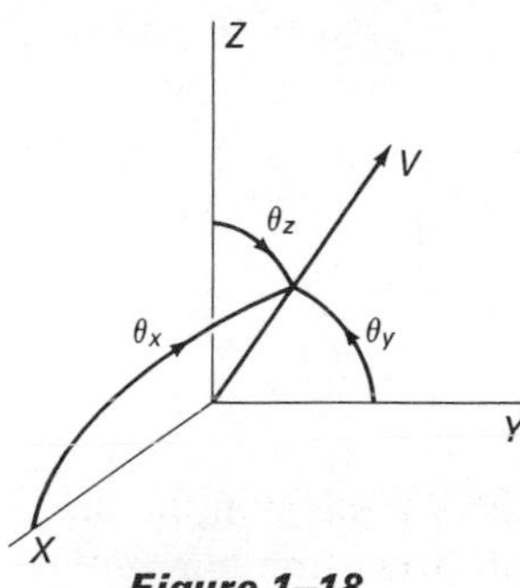

Figure 1–18

9 (a) Show that for any V,

$$(\cos\theta_x)^2 + (\cos\theta_y)^2 + (\cos\theta_z)^2 = 1.$$

(b) Show that

$$(\cos\theta_x, \cos\theta_y, \cos\theta_z) = \frac{V}{|V|}.$$

10 Show that the points $(2, 1, -3)$, $(3, 4, -1)$, $(1, 2, -4)$, $(2, 5, -2)$ form a rectangle in space.

11 Find a rectangle in space that has one vertex on the Z-axis and has $(1, 4, 1)$ and $(0, -2, -1)$ as two other vertices.

1.4

Graphing in 3-Space

In the study of functions of two variables, it is useful to be able to visualize the graphs of equations such as $z = x^2 - y^3$ or $x^2 - z^2 + 2xy = 0$. One expects such graphs to be surfaces in space, and the task of drawing them is harder than that of drawing curves in the plane. For example, if one is drawing a flat perspective picture of a graph in 3-space, it is of very little help to plot a collection of individual points. (See Exercise 9.) In most cases, the best approach is to examine a number of cross sections obtained by choosing specific values for one of the variables, and use these space curves as guides to the shape of the graph. We shall discuss a number of examples, some easy and some hard.

(i) Consider the equation $2x + 4y + 3z = 12$.

Setting $x = 0$, we find that the intersection of the graph and the YZ-plane is a line, $4y + 3z = 12$. Similarly, the plane $x = 1$, parallel to the YZ-plane, intersects the graph in a line with the equation $4y + 3z = 10$. Continuing with $x = 2, 3, \ldots, 6$, we obtain the information displayed in Figure 1–19. If we try setting $y = 0$, we find that the graph meets the XZ-plane in the line $2x + 3z = 12$; setting $z = 0$, we see that the graph meets the XY-plane in the line $2x + 4y = 12$. These are the dotted lines in Figure 1–19. It is clear that the graph is a plane through the points $(6, 0, 0)$, $(0, 3, 0)$, and $(0, 0, 4)$.

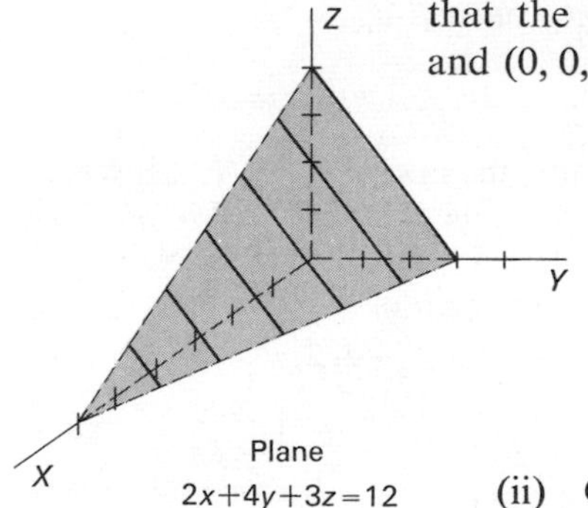

Figure 1–19

(ii) Consider the equation $z^2 = 2y - 3x$.

Setting $z = 0$, we see that the graph meets the XY-plane in the line $y = \frac{3}{2}x$. Setting $z = 2$, we see that the graph meets the horizontal plane $z = 2$ in the line $y = \frac{3}{2}x + 2$. In general, each plane $z = C$ meets the graph in a line whose equation is $y = \frac{3}{2}x + \frac{1}{2}C^2$. If we take $x = 0$, we find that the endpoints of the line segments trace out the parabola $z^2 = 2y$. We display this information in Figure 1–20.

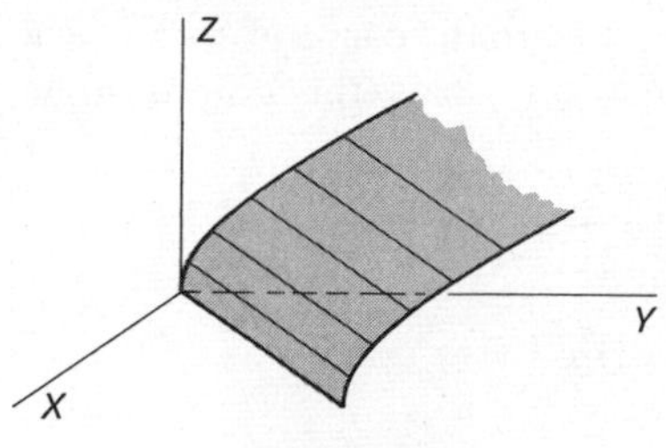

Figure 1–20

Parabolic cylinder
$z^2 = 2y - 3x$

(iii) Consider $z = 4x^2 + y^2$.

For each choice of $z = C$, we obtain the equation of an ellipse, $4x^2 + y^2 = C$. Setting $y = 0$, we have the parabola $z = 4x^2$; setting $x = 0$, we have the parabola $z = y^2$. Thus, the surface meets the XZ- and YZ-planes in parabolas, and every horizontal section is an ellipse. Note that there can be no points on the graph with negative z, since $4x^2 + y^2 \geq 0$ for all (x, y). (See Figure 1–21.)

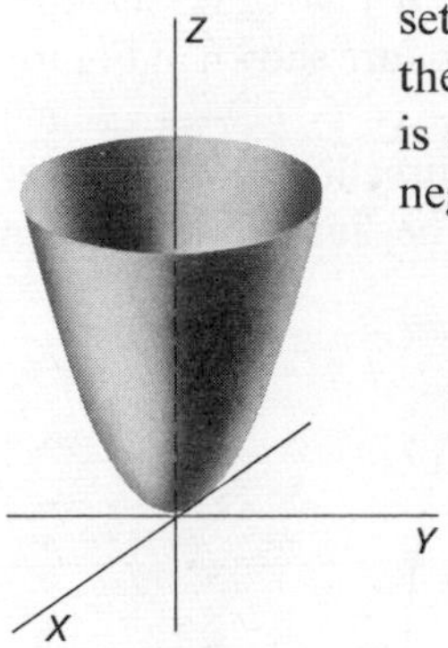

Figure 1–21

Elliptic paraboloid
$z = 4x^2 + y^2$

(iv) Consider the equation $x^2 + y^2 - z^2 = 0$.

If $z = C$, then $x^2 + y^2 = C^2$, the equation of a circle of radius C. If $x = 0$, then $y^2 = z^2$, so that $z = \pm y$. The graph is therefore a full cone, that is, a cone of two nappes (Figure 1–22). The equation of the top half alone is $z = \sqrt{x^2 + y^2}$.

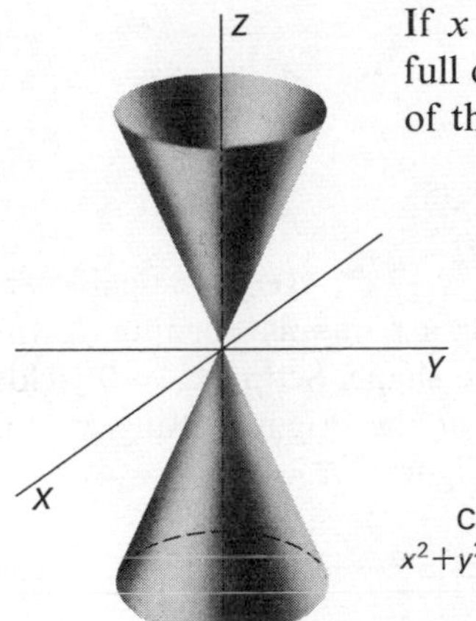

Figure 1–22

Cone
$x^2 + y^2 - z^2 = 0$

(v) Consider the equation $x^2 - y^2 + z^2 + 4 = 0$.

Rewriting this as $x^2 + z^2 = y^2 - 4$, we see that any choice of $y = C$ will yield an equation of the form $x^2 + z^2 = C$, which has either a circle, a point, or the empty set as its graph. The last occurs for all y with $-2 < y < 2$. Setting $x = 0$, we see that the graph we are seeking meets the YZ-plane in the curve $y^2 - z^2 = 4$, which is a hyperbola. This information yields the graph shown in Figure 1–23.

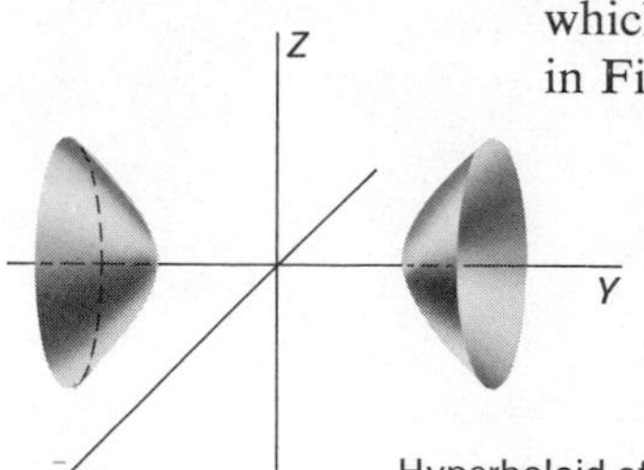

Figure 1–23

Hyperboloid of two sheets

(vi) Consider the equation $x^2 - y^2 - z + 4 = 0$.

If $z = 0$, we obtain a hyperbola with equation $y^2 - x^2 = 4$. If $x = 0$, we have a parabola $z = 4 - y^2$. If $y = 0$, we have another parabola $z = x^2 + 4$. These cross sections are shown in Figure 1–24. If we set $z = C$, then we obtain $x^2 - y^2 = C - 4$, which is a hyperbola. Whether it opens out in the direction of the Y-axis or of the X-axis depends on the size of C. The final graph is shown in Figure 1–24.

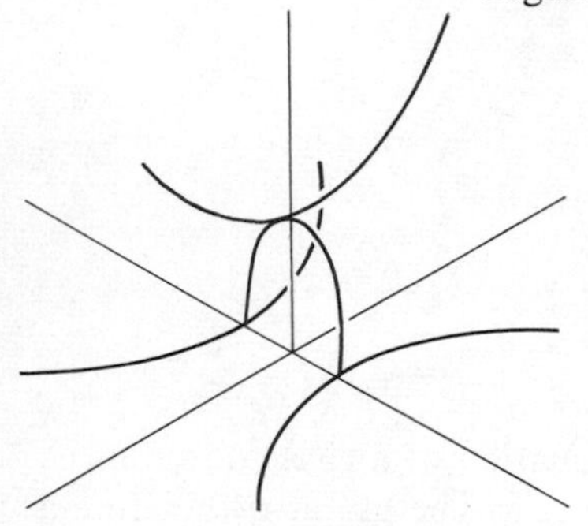

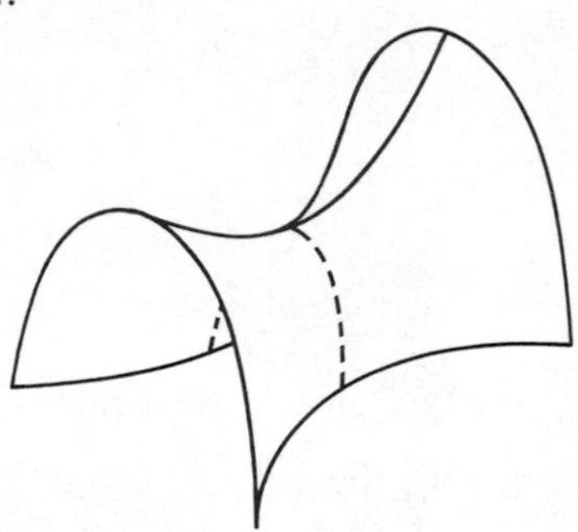

Figure 1–24

Hyperbolic paraboloid

(vii) Consider $z = y^3 - x^2$.

Since $y = C$ yields a parabola $z = C^3 - x^2$, every vertical (perpendicular to the Y-axis) section yields a parabola opening downwards. These parabolas are of the same shape. Setting $z = 0$ yields the curve $y^3 = x^2$, which has a cusp at the origin. Setting $x = 0$ yields the cubic $z = y^3$. The result is Figure 1–25.

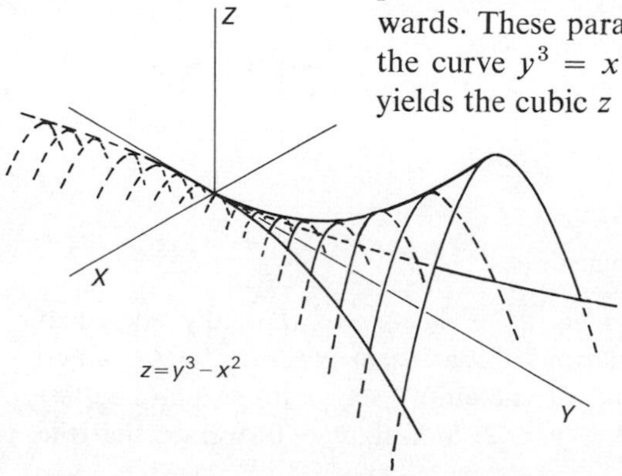

Figure 1–25

(viii) Consider $4z = xy^2$.

For any choice of C, $y = C$ yields $z = \frac{1}{4}C^2x$, a straight line. The graph is therefore a curved surface made up of straight lines; such a surface is called a **ruled** surface. In Figure 1–26, these lines are shown for a number of values of C. For any choice of $x = C$, we have a parabola $z = (C/4)y^2$. A portion of the final graph is shown in Figure 1–26. Examples (i), (ii),(iv), and (vi) are also ruled surfaces.

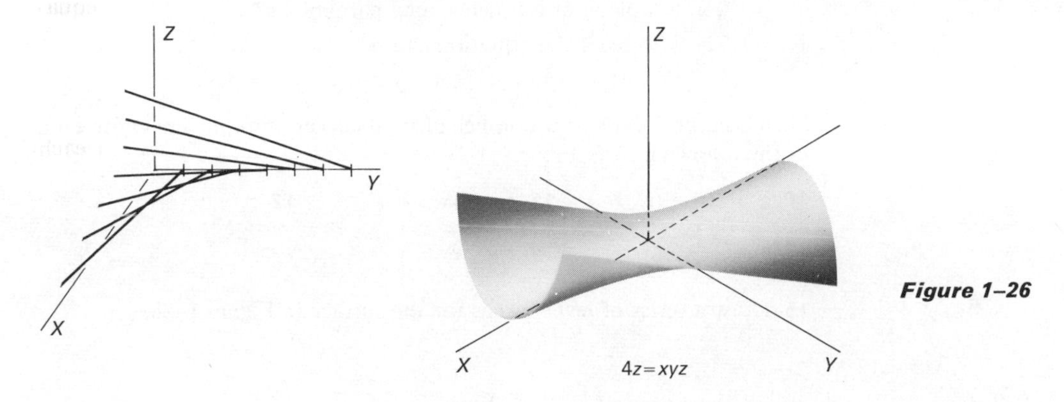

Figure 1–26

An alternate way to visualize a surface is by means of its contour or level lines, which are the cross sections obtained by taking $z = C$ for a sequence of choices of C. With experience, we obtain an immediate picture of the surface from the contour line display. For example, in Figure 1–27 the left side represents the surface shown at the right. The contour line display can also be used to build a three-dimensional replica of the surface.

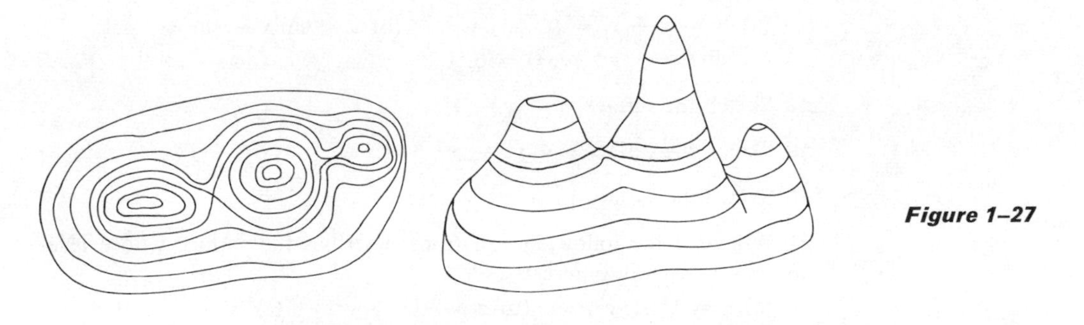

Figure 1–27

Exercises

Sketch the graph in 3-space for each of the following equations:

1 $x^2 - y^2 = 1$.

2 $z = \sin x$.

3 $z^2 = x^2 - 2xy + y^2$.

4 $\dfrac{x^2}{16} + \dfrac{y^2}{9} + z^2 = 1$.

5 $y^2 + z^2 - 4z = 0.$ **6** $\dfrac{x^2}{9} + \dfrac{y^2}{4} - z^2 = 1.$

7 $\frac{1}{4}z^2 - \frac{4}{9}x^2 - y^2 = 1.$ **8** $z = e^{-(x^2+y^2)}.$

9 (a) Plot the following points: $(2, -6, 10)$, $(2, 2, 2)$, $(3, 1, \frac{5}{2})$, $(4, -2, 5)$, $(3, -5, \frac{17}{2})$, $(-1, -7, \frac{25}{2})$, which are given as satisfying a certain equation.

(b) Does your sketch look as though the surface defined by the equation is a plane or a bowl-shaped paraboloid?

(c) Check to see if the equation can be

$$4z = x^2 + y^2 \quad \text{or} \quad \tfrac{1}{2}x + y + z = 5.$$

Sketch in the XY-plane a number of level curves (contour lines) for each of the following surfaces:

10 $z = 2x + y.$ **11** $z = 4x^2 - y.$ **12** $z = x^2 + y^2.$

13 $z = \dfrac{x - y}{x + y}.$ **14** $z = xy.$

15 Draw a series of level curves for the surface in Figure 1–28.

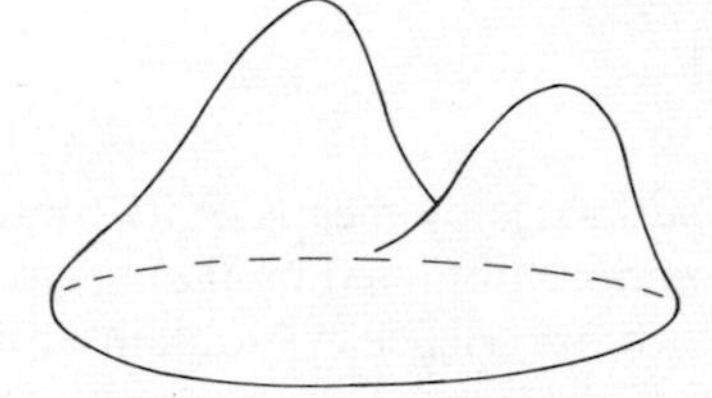

Figure 1–28

***16** Graph several contours for the equation $|x| + y^2 = z$.[1]

[1] *An asterisk indicates an exercise of more than average difficulty.*

***17** Give a verbal description of the following three surfaces (no pictures necessary):

(a) $|x - y + z| = 1,$ (b) $z = \sin x - \sin y,$

(c) $\sin(x^2 + y^2 + z^2) = 0.$

18 Sketch the surface $x = |y - 1|.$

19 Sketch the surface $z = y^2 - x^2 + 1.$

***20** Sketch the surface $y^2 + z^2 = \sin x.$

21 Which of the following equations describes the surface whose level curves are given in Figure 1–29?

(a) $z = x^2 + y^2,$ (b) $z = 9x^2 + y^2,$ (c) $z = x^2 + 9y^2.$

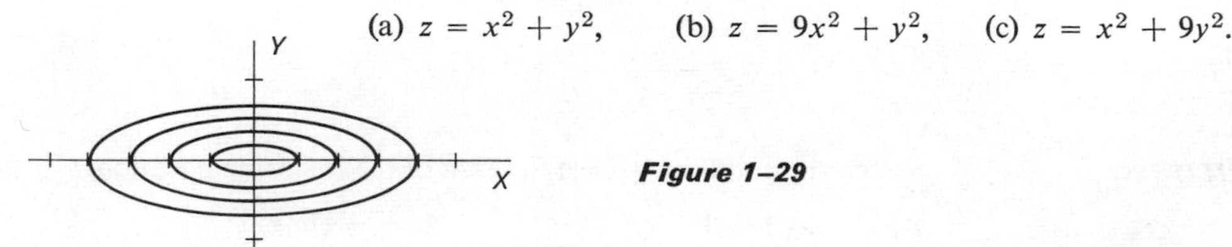

Figure 1–29

1.5

Planes In this section, we shall use the topic of planes in space to illustrate the interplay between geometric and algebraic ideas in analytic geometry.

A plane can be specified by a point p_0 on it and a direction u thought of as the direction of a **normal** (perpendicular vector) to the plane. (See Figure 1–30.) The plane then consists of all points p such that the vector from p_0 to p is orthogonal to u. Then $u \cdot (p - p_0) = 0$ and

$$u \cdot p = u \cdot p_0. \tag{1.13}$$

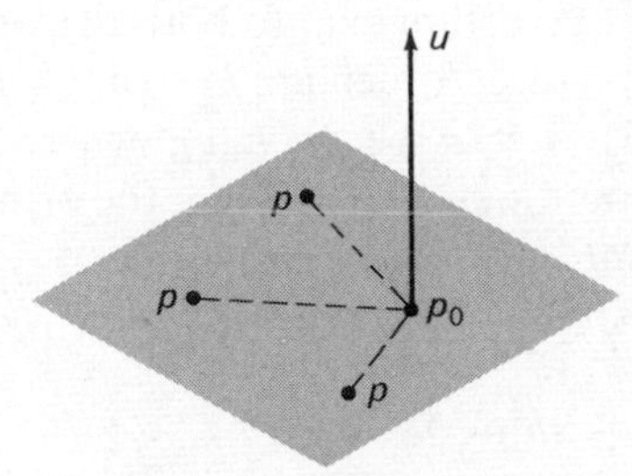

Figure 1–30

Setting $u = (a, b, c)$, $p = (x, y, z)$, and $p_0 = (x_0, y_0, z_0)$, we have the standard equation

$$a(x - x_0) + b(y - y_0) + c(z - z_0) = 0,$$

or

$$ax + by + cz = d, \tag{1.14}$$

where $d = u \cdot p_0$.

Conversely, any equation of the form (1.14) with $(a, b, c) \neq (0, 0, 0)$ will have a plane as its graph. If $v = (a, b, c)$, then (1.14) is equivalent to the relation $v \cdot p = d$. If p_0 is any particular point that satisfies this relation, then we have $v \cdot p_0 = d$, so that we can rewrite equation (1.14) as

$$v \cdot p = v \cdot p_0,$$

or

$$v \cdot (p - p_0) = 0.$$

All that remains is to show that there does exist such a point p_0. Set $p_0 = \lambda v$; then $v \cdot p_0 = \lambda(v \cdot v) = \lambda|v|^2$, and to get $v \cdot p_0 = d$, we take $\lambda = d/|v|^2$.

Geometrically, a plane can also be specified by giving three noncollinear points on it. How can we go about finding the equation of such a plane? We show two ways, one motivated by geometry and one that is purely algebraic in viewpoint. To be concrete, let the given points be $(1, -1, 4)$, $(3, 1, 0)$, $(0, 2, 1)$, and denote them as p_0, P, Q respectively. (See Figure 1–31.)

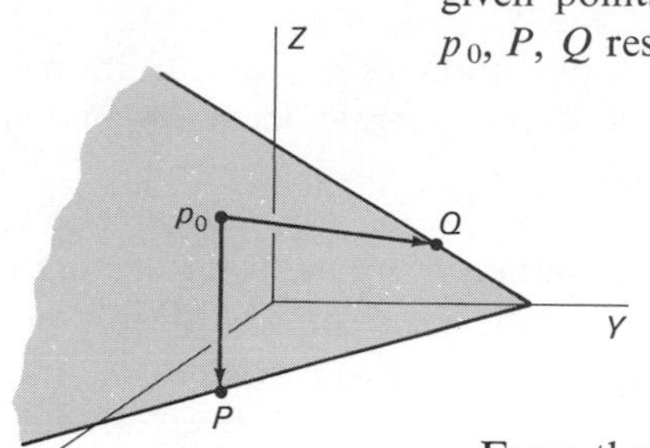

Figure 1–31

From the geometric viewpoint, our task is to find a normal vector u; having the normal and the point p_0, we can then write an equation for the plane. Clearly, u must be orthogonal to both the vector from p_0 to P and the vector from p_0 to Q, which are $P - p_0 = (2, 2, -4)$ and $Q - p_0 = (-1, 3, -3)$. Let u be (a, b, c), where our only requirement is that not all coordinates be 0. Then, the orthogonality conditions become the pair of linear equations

$$\begin{aligned} 2a + 2b - 4c &= 0, \\ -a + 3b - 3c &= 0. \end{aligned}$$

Adding twice the second to the first, we have $8b - 10c = 0$. If we don't mind fractions as solutions, we can set $c = 1$ and obtain $b = \frac{5}{4}$, and then obtain $a = \frac{3}{4}$ from the original equation. Thus, a possible normal vector u is $(\frac{3}{4}, \frac{5}{4}, 1)$. Any nonzero multiple of u will also be a normal, so we can use instead $u = (3, 5, 4)$, and the resulting equation for the plane is $3x + 5y + 4z = u \cdot p_0 = 14$. We note that there are standard procedures for finding a vector orthogonal to two given vectors; one of these, using determinants, is given in Exercise 7, and applied to the current problem yields

$$u = \left(\begin{vmatrix} 2 & -4 \\ 3 & -3 \end{vmatrix}, \begin{vmatrix} -4 & 2 \\ -3 & -1 \end{vmatrix}, \begin{vmatrix} 2 & 2 \\ -1 & 3 \end{vmatrix} \right) = (6, 10, 8).$$

In the algebraic viewpoint, the problem is rephrased. Since all planes have equations like (1.14), we seek coefficients a, b, c, d so that the three given points satisfy the equation $ax + by + cz = d$. This leads at once to the system of three linear equations in four unknowns

$$\begin{aligned} a - b + 4c &= d, \\ 3a + b &= d, \\ 2b + c &= d. \end{aligned}$$

Standard techniques again yield a solution. (The underlying equivalence of the two approaches can be seen by observing that if the variable d is eliminated by subtracting the first equation from the others, we obtain the same pair of linear equations that arose in the first solution.)

Depending upon analogy, we can discuss "planes" in 4-space similarly. Given a nonzero vector u and a point p_0, we define the **hyperplane** through p_0 normal to u to be the set of all points p such that $u \cdot (p - p_0) = 0$. If $u = (a, b, c, d)$ and $p = (x, y, z, w)$, then the same argument as above shows that the equation of a hyperplane is

$$ax + by + cz + dw = e. \tag{1.15}$$

How many points in 4-space determine a hyperplane? The obvious guess is four; its correctness is most easily seen by using results from linear algebra about the solution of systems of homogeneous equations. In the same way, it is also easy to generalize all of the above to hyperplanes in n-space, and to show that there is exactly one hyperplane containing a collection of n distinct points, provided that the condition which corresponds to "noncollinearity" is satisfied. In 4-space, this condition can be understood as meaning that there is no two-dimensional plane that passes through all four of the given points.

What happens in 3-space if the three given points *are* collinear? It is geometrically evident that there will not be a unique orthogonal vector u, and that there will be a whole **pencil** of planes containing the three given points (Figure 1–32). In Exercise 11, you are asked to look at this situation algebraically.

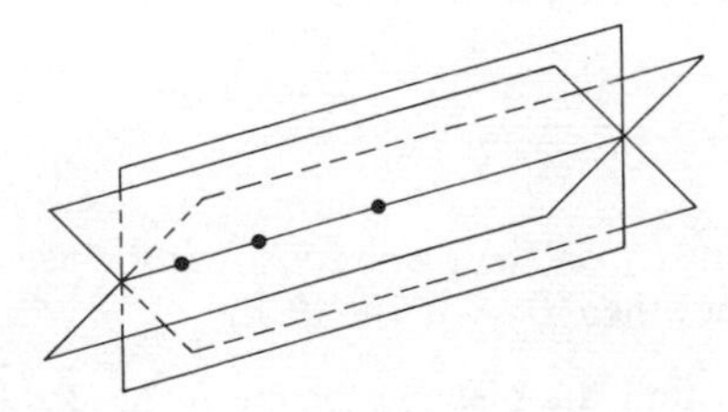

Figure 1–32

Exercises

For the following planes, find the intersection with the coordinate axes and the normal direction, and sketch:

1 $2x + 3y = 12.$

2 $3y + 4z = 12.$

3 $7x + 4y + 4z = 28.$

4 $3x - 2y - 6z = 6.$

5 $x + 2y - 2z + 2 = 0.$

6 Write an equation for the plane through $(1, 2, 3)$ parallel to the plane $2x - 3y + 4z = 6.$

7 Given vectors $V = (V_1, V_2, V_3)$ and $W = (W_1, W_2, W_3)$, show that a vector $U = (U_1, U_2, U_3)$ orthogonal to both V and W is given by

$$(U_1, U_2, U_3) = \left(\begin{vmatrix} V_2 & V_3 \\ W_2 & W_3 \end{vmatrix}, \begin{vmatrix} V_3 & V_1 \\ W_3 & W_1 \end{vmatrix}, \begin{vmatrix} V_1 & V_2 \\ W_1 & W_2 \end{vmatrix} \right).$$

(*Hint:* A 3×3 determinant is zero if any 2 rows are identical.)

8 Given vectors $V = (2, 4, 3)$ and $W = (-2, 3, -1)$, find U orthogonal to both V and W using the result of Exercise 7.

9 Using Exercise 7, find a vector U perpendicular to the plane determined by the 3 points $A = (1, 2, 1)$, $B = (-1, 2, -2)$, and $C = (-3, -1, 1)$.

10 Find coefficients a, b, c, d for the plane $ax + by + cz + d = 0$ which goes through the points A, B, C of Exercise 9. Use the algebraic method of undetermined coefficients illustrated in this section.

11 Given points $A = (1, 3, -2)$, $B = (-2, -3, 7)$, and $C = (0, 1, 1)$:

(a) Use the algebraic method of undetermined coefficients to find an equation of a plane through A, B, C. How do you interpret the result? (*Hint:* Consider Figure 1–32.)

(b) Prove that these points A, B, and C are collinear by showing that the vectors $A - B$ and $A - C$ have the same direction.

12 Using the algebraic method of undetermined coefficients, find an equation of the sphere $x^2 + y^2 + z^2 + Ax + By + Cz + D = 0$ (recall Exercise 14, Section 1.1) which goes through the points $(0, 0, 1)$, $(-1, 0, 4)$, $(3, -1, 1)$, $(3, -4, 2)$.

13 Given a plane $ax + by + cz + d = 0$ and a point $P_0 = (x_0, y_0, z_0)$, show that the distance from P_0 to the plane is given by

$$D = \frac{|ax_0 + by_0 + cz_0 + d|}{\sqrt{a^2 + b^2 + c^2}}.$$

(*Hint:* Look for a point P_1 on the plane such that the vector $P_1 - P_0$ is orthogonal to the plane; then $D = |P_1 - P_0|$.)

14 Use Exercise 13 to find the distance of the plane $3x + 4y - 5 = 0$ from the points (a) $(2, 1, 4)$, (b) $(0, 0, 0)$, (c) $(-3, 1, 6)$. Draw the plane and the points.

15 Use Exercise 13 to find the distance to the plane $x - 2y + 2z - 7 = 0$ from the points (a) $(4, -3, 3)$, (b) $(0, 0, 0)$, (c) $(2, 1, -1)$. Can you find, using the development in Exercise 13, the foot of the perpendicular from $(4, -3, 3)$ to the plane?

16 Write equations for the planes through $A = (12, 3, 1)$ and through $B = (4, -1, 2)$ each normal to $u = (1, 3, -2)$. Show algebraically that these planes do not intersect.

17 Find an equation for the hyperplane in 4-space going through the points $(3, 1, 1, 2)$, $(1, -1, -1, 0)$, $(0, -2, -1, 1)$, and $(1, 0, 1, 0)$, using the algebraic method of undetermined coefficients.

18 Can you guess a formula for the distance from a point

$$P_0 = (x_0, y_0, z_0, w_0)$$

to the hyperplane $ax + by + cz + dw + e = 0$?

1.6

Other Coordinate Systems

From a generalized point of view, any system that uses numbers (measurements) to locate uniquely any point in a plane might be called a coordinate system for the plane. Such a system will be useful only if there is present a certain quality of continuity, so that nearby points have nearly the same coordinate labels. (See Exercise 12.) The usual Cartesian system is one example, and so is the familiar polar coordinate system in which a point is located by an angle and radial distance (Figure 1–33). When, as in this case, we have two coordinate systems applying to the same plane, then we can write equations to convert from one system to the other, so that if we know how a point is labeled in one system, we can find its label in the other. The translation equations or coordinate transformation going from polar coordinates to Cartesian coordinates is

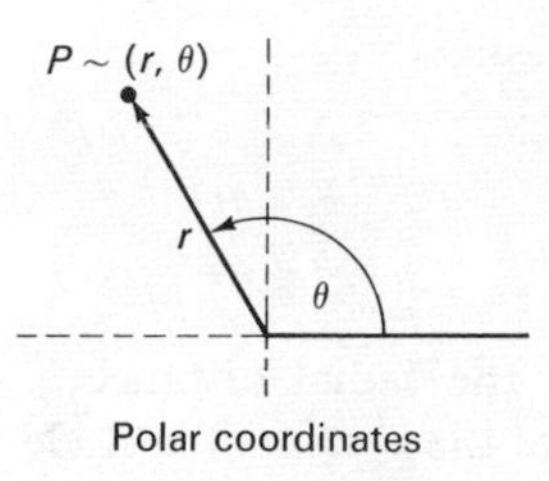

Polar coordinates

Figure 1–33

$$\begin{cases} x = r\cos\theta, \\ y = r\sin\theta. \end{cases}$$

These two systems are not the only coordinate systems that occur in practice. For example, if we choose two base points A and B a known distance apart, then the location of any point P in the plane (with certain exceptions) can be specified by giving an ordered pair of angles, the bearing of P from A and B respectively. Such a method would be used in a direction finder to locate an illegal transmitter

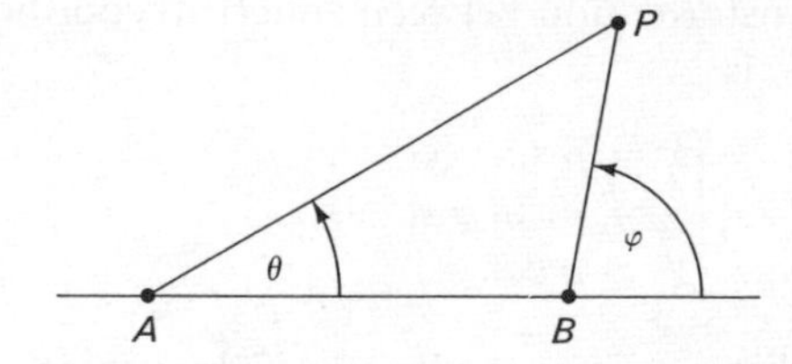

Figure 1–34

(Figure 1–34). Again, with the same base points we might try to locate a point P by giving its distance from A and from B; note however that the system thus constructed would be ambiguous, since two different points correspond to the same ordered pair of distances (Exercise 9).

In space, aside from the Cartesian system, the most commonly used systems are **cylindrical coordinate** and spherical coordinate systems. The first combines polar coordinates in the base plane with a vertical axis and is clearly illustrated in Figure 1–35.

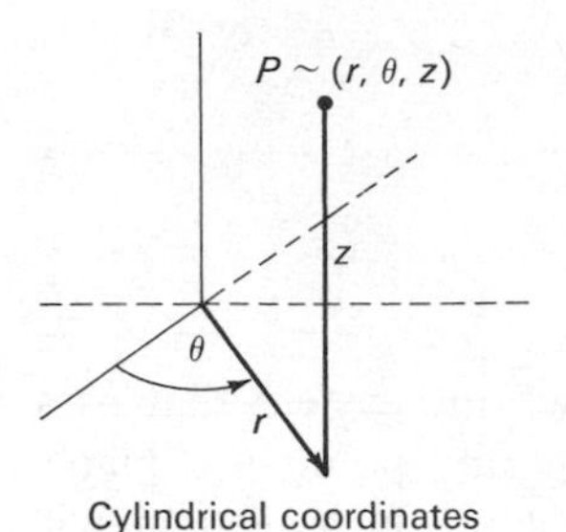

Cylindrical coordinates

Figure 1–35

The coordinate transformation from cylindrical coordinates to Cartesian coordinates is

$$\begin{cases} x = r\cos\theta, \\ y = r\sin\theta, \\ z = z. \end{cases} \tag{1.16}$$

Spherical coordinates use two angles and the radial distance ρ, as shown in Figure 1–36. It is a modification of the system of latitude and longitude. Note that $0 \le \rho$, $0 \le \theta \le 2\pi$, and $0 \le \varphi \le \pi$. As with polar coordinates, some points have many different spherical coordinate labels.

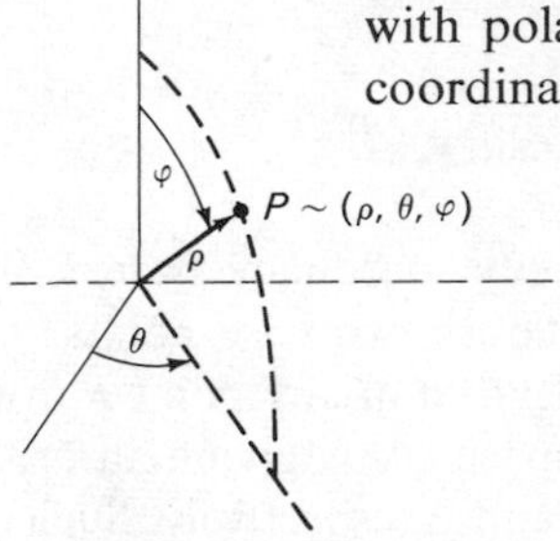

Spherical coordinates

Figure 1–36

The coordinate transformation between spherical coordinates and Cartesian coordinates is

$$\begin{cases} x = \rho\sin\varphi\cos\theta, \\ y = \rho\sin\varphi\sin\theta, \\ z = \rho\cos\varphi. \end{cases} \tag{1.17}$$

To check this, note that $\rho \sin \varphi$ is the length of the projection on the horizontal plane of the segment from $\boldsymbol{O}$ to P.

The reason one might choose to use some coordinate system other than the usual Cartesian one is usually implicit in the problem itself. For example, there may be symmetries which make it easier to describe certain aspects of the problem in terms of spherical coordinates. Thus, a cone can be described by the equation $\varphi = \varphi_0$, and a sphere merely by $\rho = \rho_0$. On the other hand, geometric

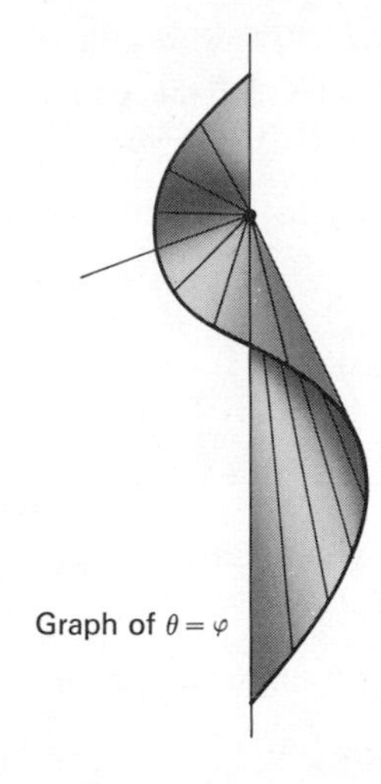

Figure 1–37

objects that have simple equations in Cartesian coordinates often turn out to have complicated equations in spherical coordinates. Thus,

$$\rho = \frac{5}{(2\cos\theta + 3\sin\theta)\sin\varphi + 4\cos\varphi}$$

is the equation in spherical coordinates for the plane $2x + 3y + 4z = 5$. The equation $\theta = \varphi$ describes a complex surface, a portion of which is shown in Figure 1–37. In cylindrical coordinates, the equation $z = C\theta$ describes the screw-like surface shown in Figure 1–38.

Sometimes a special coordinate system is chosen to fit a particular situation. For example, the natural system for locating points on the surface of a torus (doughnut) is one which uses two angles, as shown in Figure 1–39.

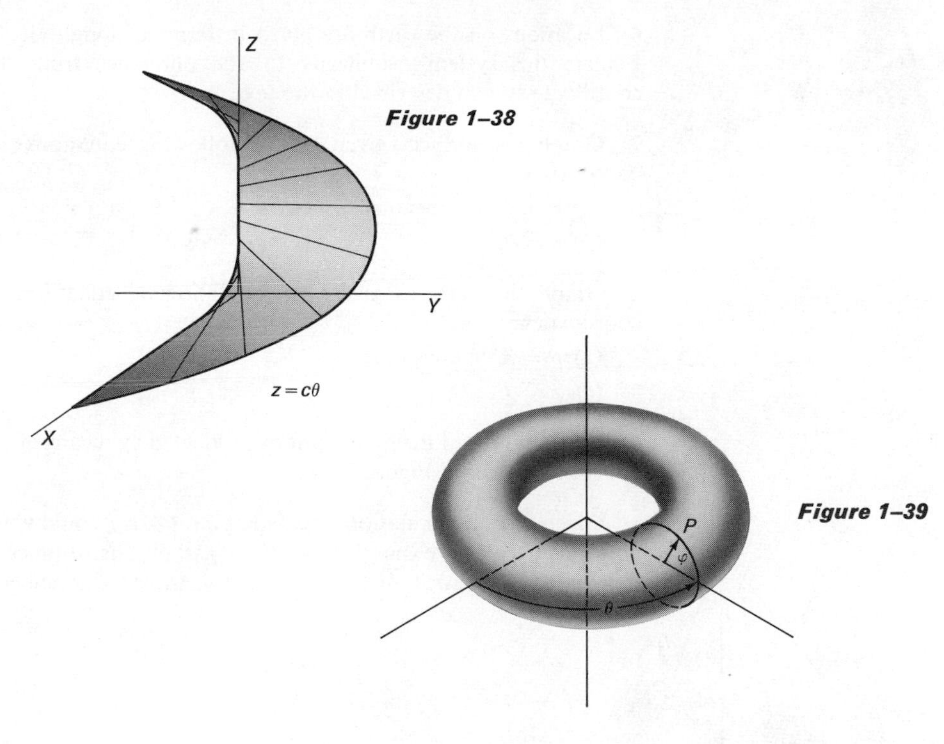

Figure 1–38

Figure 1–39

Exercises

1 Plot the following points given in cylindrical coordinates and find their corresponding Cartesian coordinates:

$$A = \left(4, \frac{\pi}{3}, -3\right), \quad B = \left(2, \frac{5\pi}{4}, 2\right).$$

2 For cylindrical coordinates, show that $x^2 + y^2 = r^2$, $z = z$, and $\theta = \arctan(y/x)$. Find a set of cylindrical coordinates for the points whose Cartesian coordinates are $A = (2, -2, 1)$, $B = (1, \sqrt{3}, -2)$.

3 Plot the following points given in spherical coordinates and find their corresponding rectangular coordinates:

$$A = \left(4, \frac{\pi}{3}, \frac{\pi}{6}\right), \quad B = \left(2, \frac{5\pi}{3}, \frac{3\pi}{4}\right).$$

4 (a) For spherical coordinates, show that $x^2 + y^2 + z^2 = \rho^2$, $\theta = \arctan(y/x)$, and $\varphi = \arccos(z/\sqrt{x^2 + y^2 + z^2})$.

(b) Find spherical coordinates for the points whose Cartesian coordinates are $A = (2, -2, -2)$, $B = (-\sqrt{3}, 1, 2)$.

5 (a) Devise a coordinate system to locate a point P on an upper half cone (Figure 1–40).

(b) What if you had a full cone instead?

Figure 1–40

6 Locations on the earth are given in terms of longitude and latitude. Explain this system's similarity to and difference from the spherical coordinate system described in the text.

7 Graph the surfaces given by the following equations in cylindrical coordinates:

(a) $r = C$ (a constant), (b) $\theta = C$, (c) $r = z$,

(d) $r = \theta$, (e) $r = \cos\theta$.

8 Graph the surfaces given by the following equations in spherical coordinates:

(a) $\rho = C$ (a constant), (b) $\theta = C$,

(c) $\varphi = C$, (d) $\rho = \varphi$.

9 Which points of the plane cannot be located by means of the two-angle system described in Figure 1–34?

10 Given three radar stations located at P_1, P_2, P_3, could you locate each point in space above the plane of P_1, P_2, P_3 by its distance d_i from P_i? How? (See Figure 1–41.) Would every triple of numbers determine a point?

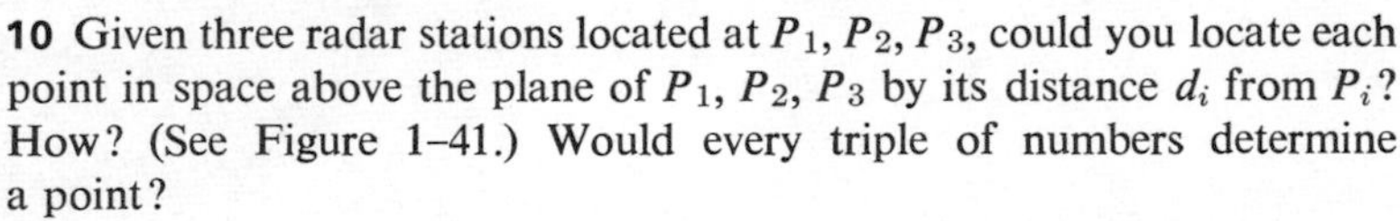

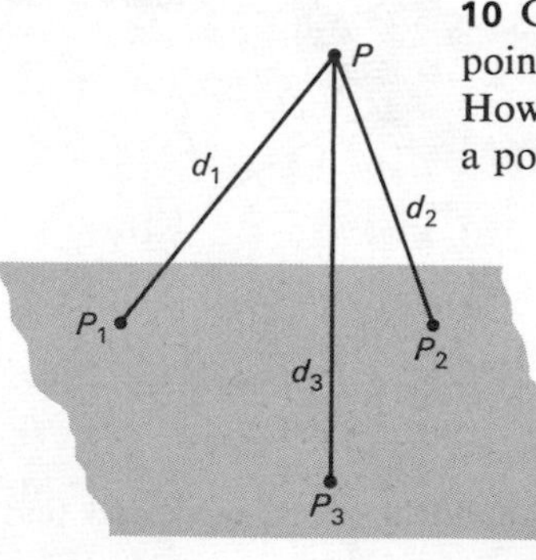

Figure 1–41

11 Another system of coordinates for 3-space might be that illustrated in Figure 1–42, using distances d_1 and d_2, and angle θ. Could all points P in space be so located? Would a location be uniquely determined?

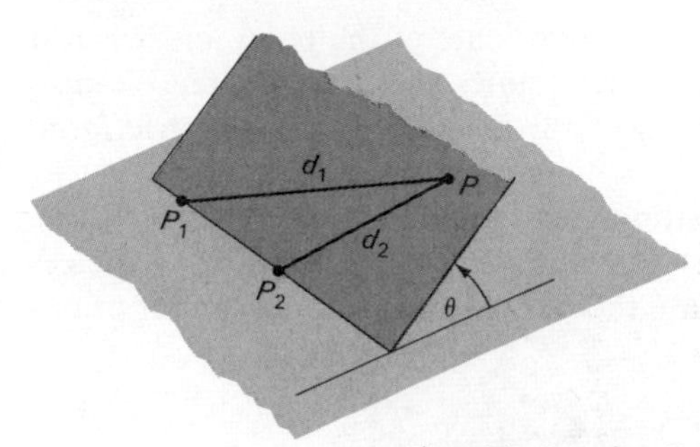

Figure 1–42

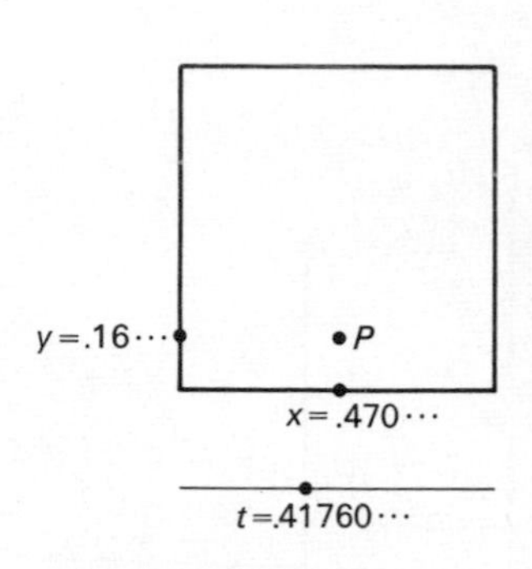

Figure 1–43

***12** Each point P in a unit square of the plane can be located by giving only one real number t between 0 and 1 as follows: Take $P = (x, y)$ (Figure 1–43) with $0 \leq x \leq 1$, $0 \leq y \leq 1$. Write x and y in decimal form: $x = .x_1x_2x_3x_4 \ldots$, $y = .y_1y_2y_3y_4 \ldots$ (where any terminating string of 9's is replaced by 0's, that is, .1837999999 . . . is replaced by .1838000000 . . .). Then assign to P the single number t (a mono-coordinate), where $t = .x_1y_1x_2y_2x_3y_3x_4y_4 \ldots$. Show that this assignment of coordinates for the square violates the "continuity" requirement for a workable coordinate system; that is, arbitrarily close points of the square can have widely different mono-coordinates t.

13 If P has cylindrical coordinates (r, θ, z) and spherical coordinates (ρ, θ, φ), show that $r = \rho \sin \varphi$ and $z^2 = \rho^2 - r^2$.

14 Show that the surface with equation $z^2 = x^2 - 2xy + y^2$ has the equation $z^2 = r^2(1 - \sin 2\theta)$ in cylindrical coordinates.

15 A surface has the equation $\rho = 2 \sin \varphi \sin \theta$ in spherical coordinates. What is its equation in Cartesian coordinates? Describe the surface.

16 Show that the Cartesian equation $x^2 - y^2 = 1$ becomes in cylindrical coordinates the equation $r^2 \cos 2\theta = 1$.

17 Show that the Cartesian equation $z^2 - x^2 - y^2 = 1$ becomes in spherical coordinates the equation $\rho^2 \cos 2\varphi = 1$.

1.7

Curves in Space

The simplest way to describe a curve in space is to regard the curve as the track of a moving point and to write equations that tell where the moving point is at each moment of time, as

$$\begin{cases} x = f(t), \\ y = g(t), \qquad a \leq t \leq b. \\ z = h(t), \end{cases} \tag{1.18}$$

In order to simplify the notation, it is sometimes convenient to write merely $x = x(t)$, $y = y(t)$, etc., and hope that the context will make it clear when "x" refers to the coordinate of the point and when it refers to the function which determines that coordinate. We also say that an equation such as (1.18) is a curve in **parametric form,** with t as the parameter.

Given such a set of equations, we can sketch the curve by listing the values of $P = (x, y, z)$ corresponding to a sequence of choices of the parameter and joining the plotted points in order. For example, given the equations

$$x = 4 - t^2,$$
$$y = 2t - 1, \quad (1.19)$$
$$z = t^2 - 1,$$

we have

$t =$	$x =$	$y =$	$z =$
-2	0	-5	3
-1	3	-3	0
0	4	-1	-1
1	3	1	0
2	0	3	3

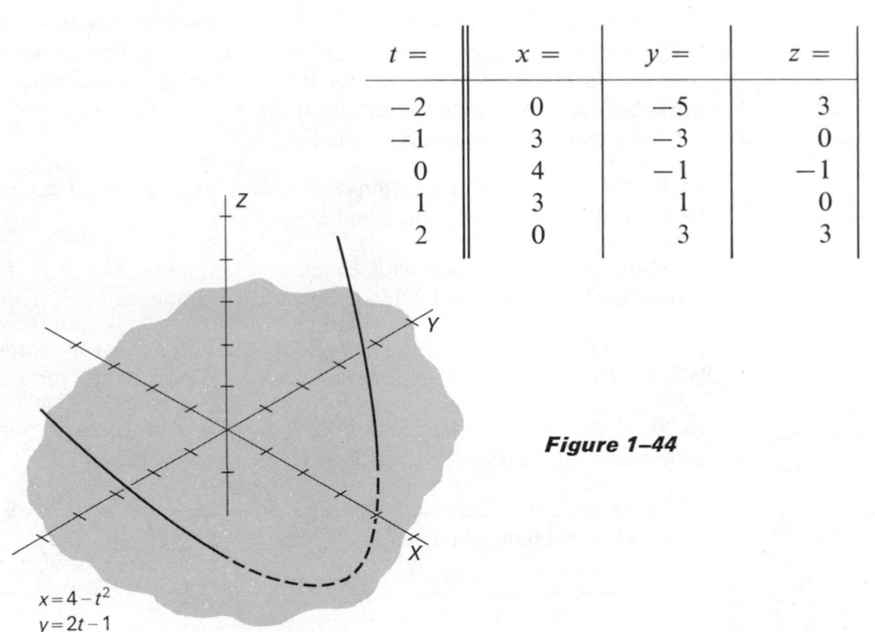

Figure 1–44

and the corresponding sketch is Figure 1–44. Experience will show that it is usually better to join the points as they are plotted, rather than to wait until all have been plotted, since the order in which they are reached is crucial. Different parametric equations can yield the same geometric curve, for each may describe a different way for a point to move along the same track. If in (1.19) we replace t by $2t$, we have $x = 4 - 4t^2$, $y = 4t - 1$, $z = 4t^2 - 1$, which describes the path of a point on the same parabolic path but going twice as fast. Again, the same geometric curve is given by the parametric equations $x = 3 - 4s^2 + 4s$, $y = 4s - 3$, $z = 4s^2 - 4s$, which were obtained from (1.19) by replacing t by $2s - 1$.

Straight lines are the simplest curves, and may be given by the standard parametric form

$$\begin{cases} x = at + x_0, \\ y = bt + y_0, \\ z = ct + z_0, \end{cases} \qquad -\infty < t < \infty, \tag{1.20}$$

which are the parametric equations of the line through $p_0 = (x_0, y_0, z_0)$ in the direction of the vector $u = (a, b, c)$. Indeed, (1.20) can be rewritten as $p = tu + p_0$ and pictured as in Figure 1–45.

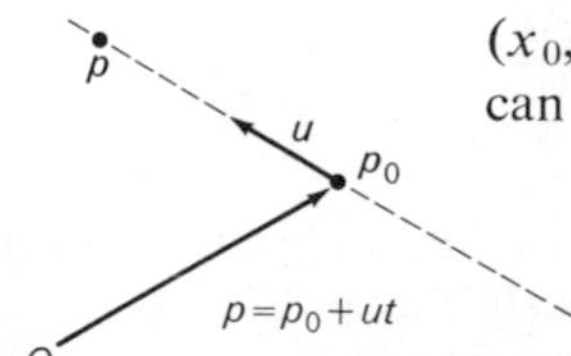

Figure 1–45

Let us try several typical problems in which one uses equations of curves.

(i) Where does the line through $(1, -1, 1)$ and $(-3, 5, 3)$ intersect the surface $x^2 + y^2 - 2z = 1$?

To find the direction of the line, we subtract the vectors corresponding to the points, getting $u = (-3, 5, 3) - (1, -1, 1) = (-4, 6, 2)$. For simplicity, we use instead $\frac{1}{2}(-4, 6, 2)$ (this vector has the same direction), and thus we have as equations for the line

$$\begin{aligned} x &= -2t + 1, \\ y &= 3t - 1, \\ z &= t + 1. \end{aligned} \tag{1.21}$$

We want to determine the value of t for which the point $p = (x, y, z)$ lies on the surface with equation $x^2 + y^2 - 2z = 1$, so we substitute into this equation from (1.21), obtaining the quadratic equation

$$13t^2 - 12t = 1.$$

Solving, we obtain $t = 1$ and $t = -1/13$. Using these values in (1.21), we find two points of intersection: $(-1, 2, 2)$ corresponding to $t = 1$, and $(15/13, -16/13, 12/13)$ corresponding to $t = -1/13$.

(ii) Does the curve given by $x = t$, $y = 2t^2$, $z = t - 1$ intersect the line through the points $(-1, 1, 1)$ and $(1, 2, 2)$?

As above, we find equations for the line to be $x = 2t - 1$, $y = t + 1$, $z = t + 1$. We want to know if there is a point that lies on both the line and the curve; is it correct to equate the formulas for x, y, and z, as $t = 2t - 1$, $2t^2 = t + 1$, $t - 1 = t + 1$ and try to solve for t? A moment's reflection will convince you that this is the same as asking if the two moving particles collide, not merely if their tracks cross. In the latter case, we do not require that the moving points get to the same point at the same value of the parameter t, but we are satisfied if they reach the same point at different values of the parameter. We therefore replace t by s in the equations of the line, and then ask if there is a solution to the equations

$$\begin{aligned} t &= 2s - 1, \\ 2t^2 &= s + 1, \\ t - 1 &= s + 1. \end{aligned}$$

From the first and the last equations, we find only one possibility, $t = 5$ and $s = 3$. Since these do not satisfy the middle equation, we conclude that the curve and the line do not meet.

In practice, curves often arise as the intersection of two surfaces. Such a characterization of a curve is not unique, since many pairs of surfaces may intersect in the same curve. Given a pair of surfaces, it is often possible to find an explicit parametric equation for their curve of intersection.

(iii) Find a parametric equation for the curve of intersection of the parabolic cylinder $z^2 = y - x$ and the plane $x + y + z = 6$. (See Figure 1–46.)

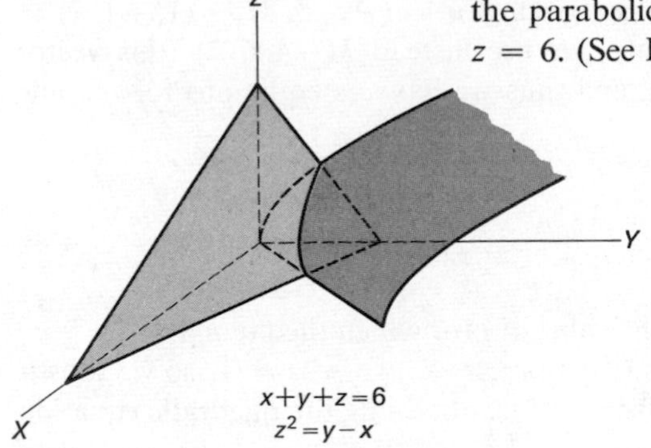

Figure 1–46

Our objective is to arrive at equations like (1.18) such that both of the given equations hold as identities when x, y, z are replaced by $f(t)$, $g(t)$, $h(t)$. Let us set $z = t$ arbitrarily, and then seek x and y so that

$$\begin{aligned} y - x &= t^2, \\ x + y &= 6 - t. \end{aligned}$$

These are linear and easily solved, yielding $x = 3 - \frac{1}{2}t - \frac{1}{2}t^2$ and $y = 3 - \frac{1}{2}t + \frac{1}{2}t^2$. Thus, the desired parametric equations are

$$\begin{cases} x = 3 - \frac{1}{2}t - \frac{1}{2}t^2, \\ y = 3 - \frac{1}{2}t + \frac{1}{2}t^2, \\ z = t. \end{cases}$$

The same algebraic method can clearly be followed to find an equation for the line of intersection of two planes. However, in this case it is instructive to observe that another, geometrically motivated method can be used. Suppose that the planes are $2x + 3y + 4z = 12$ and $3x - y + z = 2$. Since the line ℓ of intersection must lie in each plane, its direction must be orthogonal to the normal of each plane. We therefore seek a vector u that is orthogonal to $(2, 3, 4)$ and to $(3, -1, 1)$. The method explained earlier (see Exercise 7, Section 1.5) yields $u = (7, 10, -11)$. We next find a point on the line ℓ by setting $z = 1$ and solving for a pair x and y that satisfy the equations of both planes, obtaining $(1, 2, 1)$. Finally, the equation of the line is

$$\begin{aligned}(x, y, z) &= (7t + 1, 10t + 2, -11t + 1),\\ &= (7, 10, -11)t + (1, 2, 1).\end{aligned}$$

Exercises

1 Show that the two given lines intersect:

$$L_1: \begin{cases} x = 2 - 4t,\\ y = 3 + 4t,\\ z = -1 + 2t; \end{cases} \qquad L_2: \begin{cases} x = 1 + s,\\ y = -1 + 4s,\\ z = 2 - 3s. \end{cases}$$

2 Find an equation for a line through the origin normal to the plane

$$2x - y + 3z + 6 = 0.$$

3 Find the point of intersection of the curve

$$\mathcal{C}: \begin{cases} x = t^2 + 2t,\\ y = t + 3,\\ z = 3 - t^2, \end{cases} \qquad \text{and the plane } P: x - y + z + 3 = 0.$$

4 Where does the curve $\mathcal{C}$: $x = t^2 + 1$, $y = t^2 - 2$, $z = 2t$ intersect the sphere S: $(x - 1)^2 + y^2 + z^2 = 4$?

5 Find a parametric equation for the line through $(2, -1, 3)$ and $(1, -4, 4)$.

6 Find an equation for the line through $(4, -1, 2)$ which is parallel to the line described by $x = 1 - t$, $y = 3t - 5$, $z = 0$.

7 Following the method of the text, find an equation for the line of intersection of the planes given by

$$x + y + 2z - 9 = 0 \quad \text{and} \quad 2x - 3y - z + 2 = 0.$$

8 Show that the lines described in Exercises 5–7 all pass through the point $(3, 2, 2)$.

9 Graph the curve $x = t^2$, $y = 2 - t^2$, $z = t$.

10 Sketch the curve

$$x = \sin t,$$
$$y = \cos t,$$
$$z = t/\pi. \quad (\textit{Hint: } \sin^2 t + \cos^2 t = 1.)$$

11 Find a parametric equation for the curve $4x^2 + y^2 = 1$ in the XY-plane.

***12** Without resorting to the plotting of points, describe the curve

$$x = t^2 - 2t + 1,$$
$$y = 2t^2 - 4t + 1,$$
$$z = 2 + 2t - t^2. \quad (\textit{Hint: } s = (t-1)^2.)$$

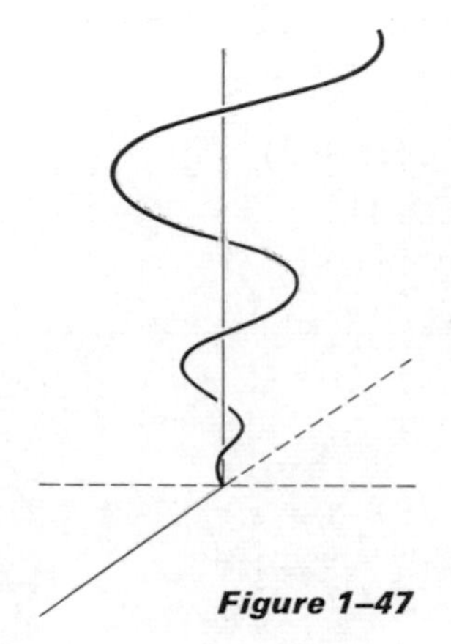

Figure 1–47

13 Sketch the curve whose equation in cylindrical coordinates is $r = 1$, $\theta = t$, $z = t/2$.

14 Sketch the curve whose equation in cylindrical coordinates is $r = t$, $\theta = \pi/4$, $z = t^2$.

15 Decide which of the following is the equation in spherical coordinates of the curve shown in Figure 1–47:

(a) $\rho = t$, $\theta = t + 1$, $\varphi = \pi/2$; (b) $\rho = t$, $\theta = t^2$, $\varphi = t + 1$; (c) $\rho = t$, $\theta = 2t$, $\varphi = \pi/6$.

16 Show that the line through $(4, -7, 5)$ and $(-1, 3, 0)$ intersects the line through $(-1, 1, 8)$ and $(4, -4, -7)$ at right angles, and find the point at which the lines meet.

1.8 Topology

Euclidean geometry is concerned primarily with such things as distance, angle, similarity, congruence; its deepest theorems are concerned with relations such as concurrence and collinearity. However, there are many other properties of sets in the plane and in space which did not become topics of mathematical investigation until the 19th and 20th centuries. The study of these topics was first called analysis situs and then topology; it is now a subject that is central to most analysis and has diffused into every other branch of mathematics. On the elementary level, it has simplified the discussion of calculus by supplying the concepts and terminology that we take up in this section.

Sequences of numbers and the notion of convergence and divergence are basic to much of analysis. One says that $\{a_n\}$ converges to L if the distance from a_n to L can be made as small as one wishes by taking n sufficiently large; specifically, given any number $\epsilon > 0$, one may find a number N such that $|a_n - L| < \epsilon$ for all subscripts

$n \geq N$. The extension to sequences of points is simple and immediate. A sequence of points $\{p_n\}$ is an assignment of a point p_k to each positive integer k; this is usually done by an explicit rule, such as

$$p_k = \left(2 + \frac{1}{k}, \frac{k+4}{k+7}, \frac{(-1)^k}{k}\right), \tag{1.22}$$

which defines a sequence of points in 3-space. A sequence $\{p_n\}$ is convergent to the point P when the distance from p_n to P can be made as small as you wish, merely by taking n sufficiently large. The formal definition reads exactly the same as that for sequences of numbers.

Definition 5 The sequence $\{p_n\}$ converges to P if and only if given $\epsilon > 0$ there is a number N such that $|p_n - P| < \epsilon$ for all subscripts $n \geq N$. In this case, we write

$$\lim_{n \to \infty} p_n = P.$$

The following result shows that convergence of sequences of points reduces to convergence of sequences of numbers (see Exercise 4.)

Theorem 6 *Let $p_n = (a_n, b_n, c_n)$. Then the sequence $\{p_n\}$ converges with limit P if and only if $P = (a, b, c)$ and*

$$\lim a_n = a,$$
$$\lim b_n = b,$$
$$\lim c_n = c.$$

Because of this theorem, any tests for convergence of sequences of numbers can be used to test the convergence of sequences of points. For example, the sequence (1.22) above is convergent and has limit $(2, 1, 0)$.

In your study of functions of one variable, open and closed intervals on the line were very important; you may even have encountered a half-open interval $(a, b]$ consisting of all numbers x with $a < x \leq b$. In the plane, and more generally in n-space, the sets we work with are much more complicated than intervals; however, we still deal with open sets, closed sets, and sometimes sets that are neither open nor closed.

Definition 6 A set D is called an open set if it is a neighborhood for each of its points—i.e., if each point p_0 in D is surrounded by an open ball that is itself a subset of D.

Definition 7 A set E is called a closed set if it contains the limit of every converging sequence of points taken from E.

Some sets satisfy neither requirement, and thus are neither open nor closed; they have no special name. A special relationship between open sets and closed sets is sometimes useful. Suppose that we are dealing with sets that lie in the plane. Then the **complement** of a set S is the set consisting of all the points of the plane that are not in S. It may be shown that a set in the plane is closed if and only if its complement is open. The same is true for subsets of n-space as well. Proofs of these statements are left to a more advanced course.

As examples, each of the following describes an *open* set in the plane.

(i) The open right half plane, consisting of all (x, y) with $x > 0$.
(ii) The interior of an ellipse, consisting of all (x, y) with $x^2 + 4y^2 < 1$.
(iii) The shaded region in Figure 1–48, consisting of all (x, y) with $x^2 > y^2$.

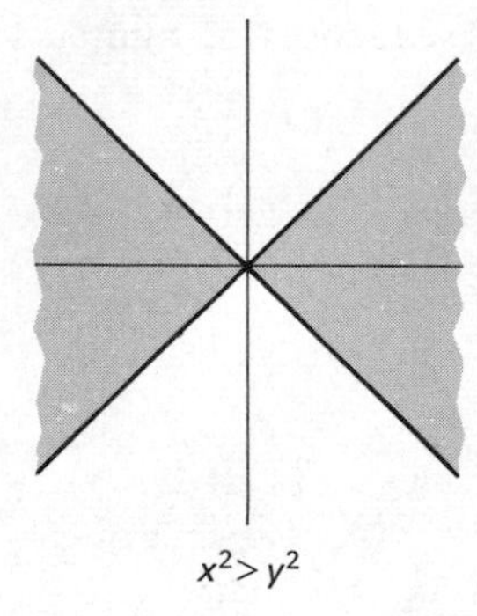

Figure 1–48

Sets described by polynomial inequalities using $<$ but not $\leq$, as in the examples above, always yield open sets.

The complements of each of these sets with respect to the plane will be *closed* sets in the plane:

(i)′ The closed left half plane, consisting of all (x, y) with $x \leq 0$.
(ii)′ The exterior of an ellipse, including the curve itself, described by $x^2 + 4y^2 \geq 1$.
(iii)′ The unshaded region in Figure 1–48, including the two lines themselves, described by the equation $x^2 \leq y^2$.

Sets given by polynomial inequalities using $\leq$ always yield closed sets. The reason for this and the corresponding fact stated above about open sets will be found in the next chapter, when we discuss continuous functions and their properties.

With any set, we associate another set called its boundary, or frontier. Intuitively, the boundary is the edge that separates a set from its complement, and in simple cases it consists of one or more curves and perhaps a scattering of individual points. Thus, the boundary of the sets described above in (i) and (i)′ is the vertical line $x = 0$; the boundary of the sets in (ii) and (ii)′ is the ellipse $x^2 + 4y^2 = 1$. The technical definition is the following:

Definition 8 The boundary of a set S is the collection of points p_0 having the property that every neighborhood about p_0 contains at least one point in S and at least one point in the complement of S.

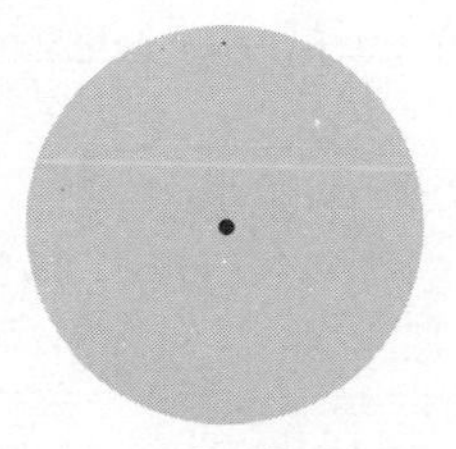

$0<x^2+y^2<1$

Figure 1–49

The effect of this definition is that any point on the boundary of a set S can be approached along a sequence of points in S and also along a sequence of points in the complement of S. There is one situation that needs more explanation. Suppose that S is an open unit disc with its center removed, namely the set of all (x, y) with $0 < x^2 + y^2 < 1$. (See Figure 1–49.) The complement of this set will be the origin and the closed set consisting of all (x, y) with $x^2 + y^2 \geq 1$. What is the boundary of S? Clearly, it must contain the circle $x^2 + y^2 = 1$; what is perhaps not so clear is that the origin itself is also a boundary point for S. To see that this is required by the definition, take any neighborhood of the origin. It certainly will contain points of S; it will also contain a point in the complement of S, namely the origin itself! Finally, we remark that a set is open if it contains *none* of its boundary and is closed if it contains *all* of its boundary. It is instructive to see what these terms mean in 1-space, the line. For example, the boundary of an interval is the set consisting of both endpoints, and an interval is closed when it contains both endpoints.

One possible additional property of a set S is boundedness—which has nothing to do with the linguistically similar term boundary. A set S is **bounded** if it has finite size, meaning that there is a finite upper bound to the distance between points in S. This implies that there is some ball of radius R which contains all of the set S; analytically, for some radius R, it is true that $|p| < R$ for all $p \in S$. Of the examples given above, only (ii) is a bounded set; the others are all unbounded and contain points that are arbitrarily far from the origin.

There are many other properties of sets which are important mathematically and have special names. It is quite difficult to give formal definitions of some of them, even though they have simple intuitive meanings. For example, the open set that is shaded in Figure 1–48 is made up of two separated pieces, whereas the open set in example (ii) is in just one piece. The technical term that is used here is **connected**, and we would say that (ii) describes a connected set, but (iii) does not; we would also add that the set in (iii) has two **components**, the term used to refer to connected pieces of a set. We shall not give a formal definition for the term "connected," but suggest that you think of a connected set as one in which you can join any two points by a curve, without going outside the set.

Two connected open sets in the plane may also have quite different properties. In each of the sets pictured in Figure 1–50, there are paths that join the points P and Q. However, in the first all such paths are similar: You could push any path from P to Q into any other path from P to Q without leaving the set itself. (Think of the paths as pieces of flexible string.) For the set in the second diagram, however, you cannot push either of the paths shown so that it lies on top of the other without letting it cross over the hole or releasing it at P or at Q.

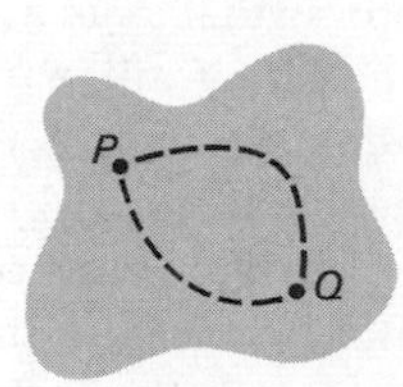

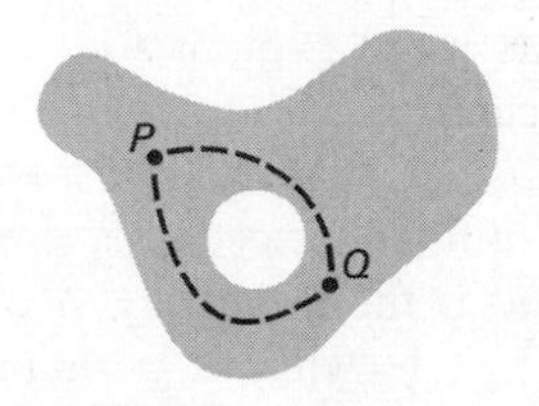

Figure 1–50

To indicate this difference, we call the first set simply connected and the second set multiply connected. The set shown below in Figure 1–51 is also multiply connected, but is basically different from the set on the right in Figure 1–50. Can you tell what the difference is in terms of paths?

Figure 1–51

This discussion has dealt solely with sets in the plane, in order to keep the illustrations easy to visualize. All of the ideas apply equally to sets in space. Some practice in thus applying them is given in the exercises.

Exercises

1 Given the sequence of points $\{p_k\}$ with

$$p_n = \left(\frac{n-1}{n+2}, \frac{(-1)^n}{n}, \frac{1}{2^n}\right),$$

what is $P = \lim_{k\to\infty} p_k$?

2 Show that the open sphere $x^2 + y^2 + z^2 < r^2$ is an open set in the sense of Definition 6. (See also Exercise 13, Section 1.1.)

3 Show that the complement of the closed sphere $x^2 + y^2 + z^2 \leqq r^2$ is an open set in the sense of Definition 6.

4 Prove Theorem 6: The sequence $\{P_n\} = \{(a_n, b_n, c_n)\}$ converges to the limit $P = (a, b, c)$ if and only if $\lim_{n\to\infty} a_n = a$, $\lim_{n\to\infty} b_n = b$, and $\lim_{n\to\infty} c_n = c$. (*Hint:* $|a_n - a| \leq |P_n - P|$ and $|P_n - P| \leqq |a_n - a| + |b_n - b| + |c_n - c|$.)

For the following sets in space, decide whether the sets are open, closed, or neither; connected or disconnected; simply connected or multiply connected; bounded or unbounded. Describe the boundary. No proofs are required.

5 The set of (x, y, z) where $1 < x^2 + y^2 + z^2 < 2$.

6 The set of (x, y, z) where $z^2 > x^2 + y^2$.

7 The set of (x, y, z) where $y^2 - x^2 - z^2 = 1$.

8 The set of points (x, y, z) with $x^2 + y^2 + z^2 \leqq 1$ which are *not* on the plane $x + y + z = 1$.

9 Given the curve described by the equations $x = t$, $y = t^2$, $z = t^3$, find a sequence of points on the curve which converges to $\boldsymbol{O}$.

limits and continuity

chapter 2

2.1

Functions Although the main concern in this chapter is numerical-valued functions of two or more real variables defined on a region in the plane or in space, we shall start by discussing again the abstract concept of a general function. Given two sets A and B whose members might be anything at all, a function F "on A to B" (i.e., having A as its domain and taking values in B) is a mapping or correspondence which assigns to each member of A some member of B. If $a \in A$, then the member of B that is assigned by F to a is denoted by $F(a)$.

In general, one might make these assignments in an arbitrary way. For example, one might divide the set A into two disjoint parts A_1 and A_2, select two distinct members of B, b_1 and b_2, and then define F on A to B by

$$F(a) = \begin{cases} b_1 & \text{if} \quad a \in A_1, \\ b_2 & \text{if} \quad a \in A_2. \end{cases}$$

Usually, however, such functions have little interest; the functions of greatest interest are those that have other special properties, such as continuity, linearity, convexity, etc.

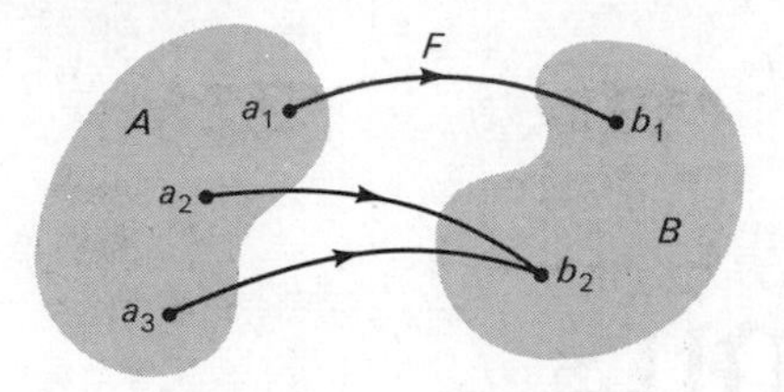

Figure 2–1

We shall sometimes write $F\colon A \to B$ to indicate that F has domain A and that it maps members of A into members of B. This can also be indicated by a picture as in Figure 2–1. The **graph** of a function F is the set of all the ordered pairs (x, y) where x belongs to the domain of F and where y is $F(x)$, the image of x under the mapping F. If dom (F), the domain of F, is A and the values of F lie in B, then each point on the graph of F is a pair (a, b), where $a \in A$ and $b \in B$. If $A \times B$ is the set of all pairs (a, b), $a \in A$, $b \in B$, then the graph of F is a special subset of $A \times B$, consisting of the pairs $(a, F(a))$. If we use the diagram in Figure 2–2 to display this, we can see that the graph of F has the property that every vertical line cuts it once, but none cuts it twice.

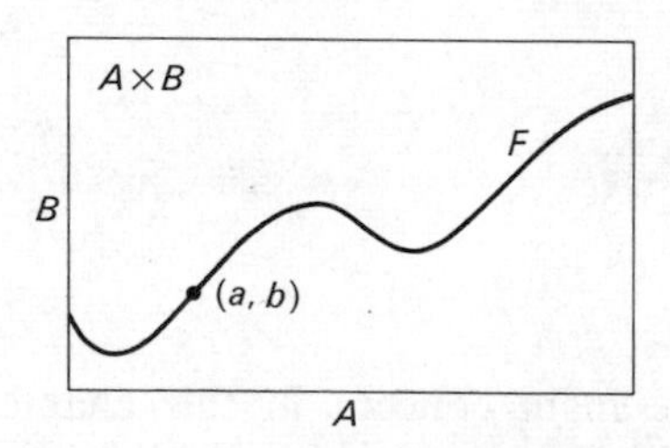

Figure 2–2

In the study of functions of *one* variable, functions whose domain was an interval of numbers, the graph played a central role. Differentiation was discussed in terms of the slope of a tangent to the graph, and integration could be related to the area between the graph and the horizontal axis. Some analogous relationships are true for functions of two variables. For functions of three or more variables, the usefulness of graphs is very minute, and we must resort to other ways of picturing their global behavior.

Let us begin with a concrete illustration. Consider the function F whose domain is the set of all points $p = (x, y)$ in the plane and which is defined by the equation

$$F(p) = F(x, y) = 2x^2 + y^2.$$

Write z for the value $F(p)$. The graph of F will consist of all ordered pairs $(p, F(p)) = (p, z)$ for all possible choices of p. Since $p = (x, y)$, we can interpret this to mean that the graph of F consists of all points (x, y, z), where $z = F(x, y) = 2x^2 + y^2$. This, of course, is a bowl-shaped surface in 3-space.(Figure 2–3).

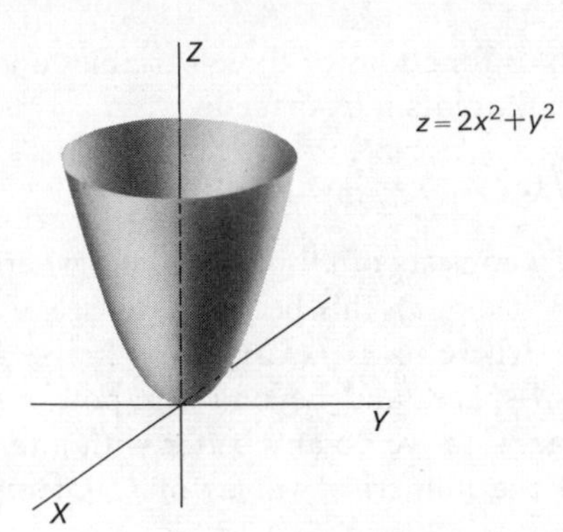

Figure 2–3

We can obtain a great deal of information about F by looking at its graph. However, much of the same information can be obtained by forgetting about the graph of F and taking an entirely different approach.

Suppose that we interpret the number z which F assigns to a point p in the plane as the temperature at the point p. A temperature chart is given in Figure 2–4 for the function in this example. What can we tell about the nature of F by looking at this chart? First, we see (and confirm by looking at the equation for F) that the temperature increases steadily as we move away from the origin in any direction and that the minimum temperature is at the origin.

Temperature chart
$F(p) = 2x^2+y^2$

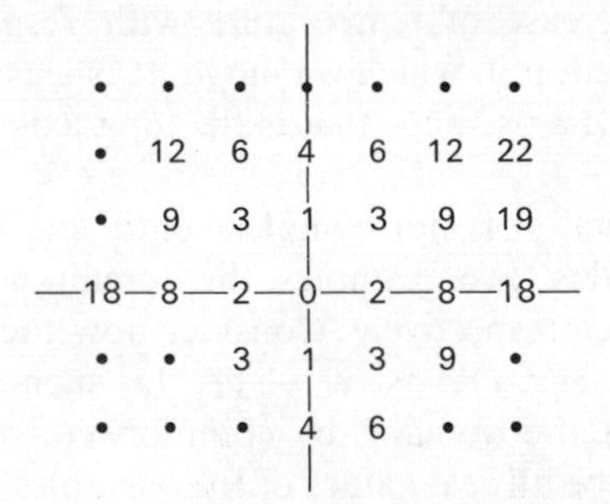

Figure 2–4

Examining the chart more carefully and drawing a few equithermal lines, we notice that the rate at which the temperature changes depends on the direction in which we move, and that the greatest rate of increase seems to occur when we move away from the origin along the X-axis (Figure 2–5).

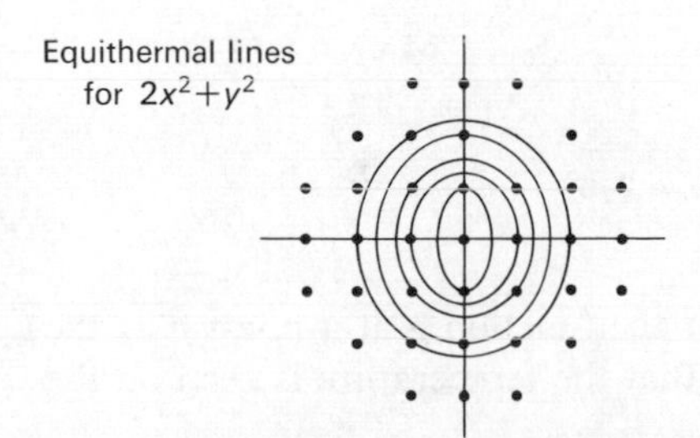

Figure 2–5

Suppose we now turn to functions of three variables and consider the example defined at all points p in space by

$$f(p) = f(x, y, z) = 4x^2 + 4y^2 + z^2.$$

As before, the graph of f consists of all pairs (p, w), where we have set $w = f(p)$. Since $p = (x, y, z)$, this becomes the set of all points (x, y, z, w) in 4-space, where $w = f(x, y, z) = 4x^2 + 4y^2 + z^2$. This is a "surface" in 4-space, and beyond our power to picture except as a mental image. Can we do any better with a temperature chart? Let us interpret the numerical values of f as temperatures, and write the equation

$$T = f(p) = f(x, y, z) = 4x^2 + 4y^2 + z^2,$$

giving the temperature at each point p of space. At $(1, 1, 0)$ the temperature is 8, at $(0, 1, 1)$ it is 5, and so on. If we stand at the origin and move in any direction, the temperature will increase. In general, the farther we are from the origin, the hotter it is. If we choose a temperature T_0 and look at the corresponding equithermal surface consisting of all the points at which the temperature is T_0, we see that the surface is an ellipsoid $4x^2 + 4y^2 + z^2 = T_0$ and that as T_0 increases, the ellipsoid expands. Since we have here ellipsoids, not spheres, the rate of increase of temperature with respect to distance depends on the direction in which we move. It is least if we move in the direction of the Z-axis, since that is the long axis of the ellipsoids.

One's knowledge of a function is not complete until one knows what the domain is. For the last two examples, the domain was the entire plane and all of 3-space respectively. Consider now the function given by the formula $g(x, y) = \sqrt{x^2 - y^2}$. In such cases, where nothing is said about the domain, by common convention the domain is understood to be all real values of the variables which make mathematical sense and give real values for g. In this case, we allow any choice of x and y for which $x^2 - y^2 \geq 0$. This set, shown in Figure 2–6, is normally understood to be the domain of g.

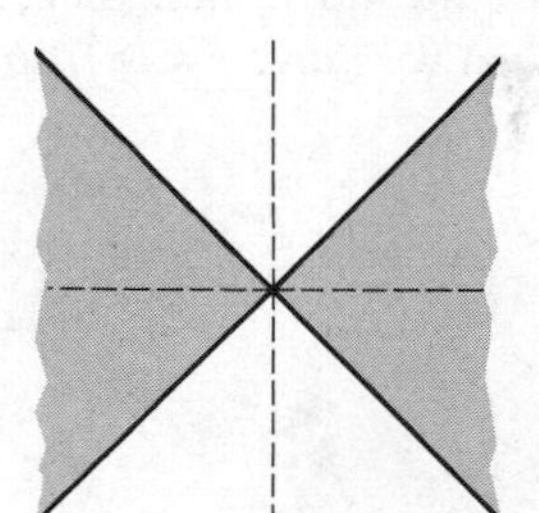

Domain of $g(x, y) = \sqrt{x^2 - y^2}$

Figure 2–6

If we again interpret the value of the function g at a point p as the temperature of p, we would say that the temperature is zero on the edges of the shaded triangular regions, increases as you move from an edge toward the center line of a region, and further increases as you move away from the origin along either center line (i.e., along the X-axis.)

There are also special cases in which the domain of a function is restricted to a set smaller than the set for which the formula describing the function makes sense. A simple illustration for a function of one variable is given by the formula for the volume of the box made by cutting a square of side x from each corner of a 3 ft. by 5 ft. piece of tin; here $V = x(3 - 2x)(5 - 2x)$ is defined for all x, $-\infty < x < \infty$, but the physical limitations restrict the domain of the function to $0 \leq x \leq \frac{3}{2}$. Again, a function is often constructed by piecing together several different formulas, each used for part of the total domain. The gravitational attraction on a unit mass by the earth is given by

$$F = \begin{cases} \dfrac{k}{r^2} & \text{when } r \geq R \text{ (i.e., off the earth),} \\ \dfrac{k}{R^3}\, r & \text{when } 0 \leq r < R \text{ (inside the earth).} \end{cases}$$

The number of variables in a function is sometimes a matter of choice or viewpoint. For example, we might write

$$f(x, y) = ax^2 + bxy + cy^2,$$

or we might write

$$g(x, y, a, b, c) = ax^2 + bxy + cy^2.$$

The first equation indicates that we intend to regard a, b, c as fixed throughout the discussion and that we thus intend to treat the problem as one involving a function of two variables. The second equation indicates that all five variables may be subject to change and that we may be interested in the results of a change in any one of them.

It should be easy to see how one might encounter a function of a large number of variables. For example, a general polynomial of degree N in one variable has $N + 1$ coefficients, and thus might be regarded as a function of $N + 2$ variables.

Finally, we remark that in later sections it will be useful to talk about the *graph* of a function of many variables, even though this is something we cannot picture in diagrams and sketches. We shall depend upon your ability to use geometric intuition, to generalize concepts, and to create mental images that are the analogues of ordinary three-dimensional diagrams.

Exercises

1 (a) Draw a temperature chart for $f(x, y) = 2x - 3y$.

(b) Describe in words the behavior of the temperature as $|p| \to \infty$ in various ways.

2 Repeat Exercise 1 using instead $f(x, y) = y^2 - 2x$.

3 Sketch the equithermal curves for the functions of Exercises 1 and 2.

4 Give a verbal description of the equithermal surfaces for $f(x, y, z) = x^2 + y^2 - z^2$. Describe the behavior of the temperature as $|p| \to \infty$ in various ways.

5 Draw pictures to show the "natural" domains for the functions given below.

(a) $F(x, y) = \sqrt{x + y}$,

(b) $F(x, y) = \dfrac{1}{x^2 - y^2}$,

(c) $F(x, y, z) = \dfrac{1}{x^2 + y^2 - z^2}$,

(d) $F(x, y, z) = \sqrt{(x - z)(2x - y)}$,

(e) $F(x, y) = \sqrt{\sin(x - y)}$.

***6** Sketch the graph of the function f where

$$f(x) = \begin{cases} x^2 & \text{when } x \text{ is rational,} \\ x & \text{when } x \text{ is irrational.} \end{cases}$$

Does the picture look like a function? Is it a well-defined function? Can you describe the picture more accurately than you can draw it?

7 Put $f(u, v) = u^2v - 3uv$, $u = x - y$, $v = x + y$ to define $F(x, y) = f(x - y, x + y)$. Show that

$$F(x, y) = x^3 - x^2y - xy^2 + y^3 - 3x^2 + 3y^2.$$

8 Let $F(x, y, z) = x^2y + 2yz - 3xz^2$. Observe that if $y = 1$, $z = 2$, then a function of one variable is obtained:

$$f(x) = F(x, 1, 2) = x^2 - 12x + 4.$$

(a) Is there a choice of numerical values for y and z so that F becomes
(i) $f(x) = 2x^2 - 3x - 4$,
(ii) $f(x) = 3x^2 + 4x + 2$?

(b) Evaluate $F(-1, 2, x)$.

***9** In a function $F(x, y, z, w)$ of 4 variables we can regard w as time and thus think of F as yielding a temperature distribution dependent on time. Check the following statements about

$$F(x, y, z, w) = x^2 + y^2 + z^2 + w \quad \text{with } w = t.$$

(a) At any point, the temperature increases with time.

(b) At any time, the temperature increases as $|p| \to \infty$.

(c) If you are moving toward the origin, you *might* find yourself at the same temperature all the time, provided you did not move at a constant speed.

***10** A function g whose domain is $A \times A$, where A is any set, obeys the identity

$$g(x, y) + g(y, z) = g(x, z).$$

Can you discover several examples for g? Can you find all such functions g?

2.2

Nonnumerical Functions

All of the functions used as illustrations in the last section were numerical valued; that is, they were defined on some set A in n-space and took values in 1-space, the real numbers. This is certainly not a necessary property of functions. Indeed, the treatment of curves in the last chapter (Section 1.7) was in fact a discussion of functions whose domain was a set on 1-space but whose values lay in 3-space. The function defined by

$$F(t) = (f(t), g(t), h(t)) \tag{2.1}$$

is exactly the same as the set of parametric equations

$$\begin{cases} x = f(t), \\ y = g(t), \\ z = h(t), \end{cases}$$

describing a curve in space. Since each point (x, y, z) can be regarded as a position vector, we may call such a function F given by (2.1) a **vector-valued** function. Thinking of the parameter t as time, we see that F is the proper mathematical model for a moving particle, and thus we might write $P = F(t)$ to express this.

By analogy, there ought to be useful functions of several variables whose values are not numbers but points in space—vector-valued functions of several variables. Such functions are easy to motivate by physical considerations.

Let us consider how we might use mathematics to describe a moving stream of fluid such as a river. We can think of the fluid as composed of an infinite collection of individual particles of water each moving according to its own pattern. Suppose first that the stream has reached a "steady state," so that when we stare at one location in the stream, we see that every particle that passes through that spot is moving at the same speed and in the same direction. If we decide to indicate this velocity by a vector giving the direction and magnitude of the motion, then we thereby associate a velocity vector with each location within the stream. This assignment of vectors to locations defines a function F such that at each point $p = (x, y, z)$ within the stream the value $F(p)$ is v, the velocity vector there. The formula for such a function is of the form $F(p) = F(x, y, z) = v = (v_x, v_y, v_z)$, or

$$F: \begin{cases} v_x = f(x, y, z), \\ v_y = g(x, y, z), \\ v_z = h(x, y, z), \end{cases} \tag{2.2}$$

where v_x, v_y, v_z are the components of the velocity vector v in the axis directions. (See Figure 2–7.)

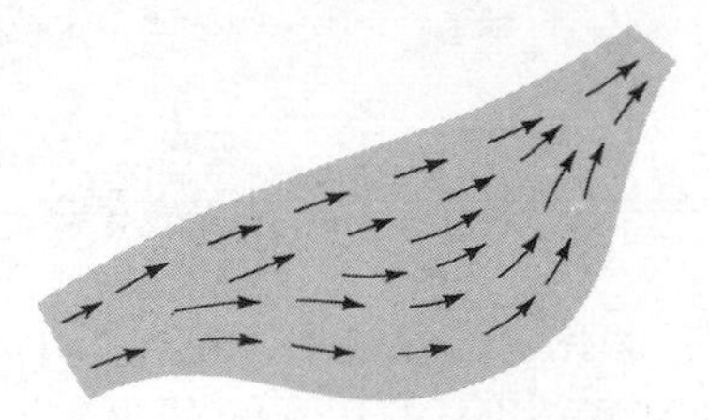

Figure 2–7

In fact, a moving stream is not apt to be in a steady state; the velocity of the fluid passing a particular point of the river will change with time. The only change this makes in the mathematical model is that the value of v will depend both upon the point $p = (x, y, z)$ at which we measure it and upon the time t. We shall therefore write $v = F(p, t) = F(x, y, z, t)$, or

$$F: \begin{cases} v_x = f(x, y, z, t), \\ v_y = g(x, y, z, t), \\ v_z = h(x, y, z, t), \end{cases} \tag{2.3}$$

so that we are dealing with a vector-valued function of four variables. A satisfactory device for displaying such a function would be a moving picture each frame of which shows a picture of the velocity field at a particular moment of time, as in Figure 2–7.

Repulsive force field

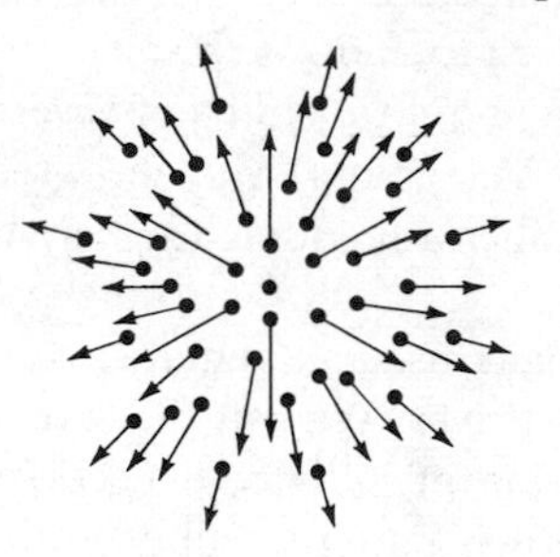

Figure 2–8

Vector-valued functions arise in many other ways in the physical sciences. For example, Figure 2–8 shows the type of function that might be associated with a repulsive force field. Any point responding to this field would be pushed away from the central source: The closer the point is to the center, the greater is the force of repulsion.

It is also easy to give examples of functions whose values are points in n-space, with $n > 3$. Suppose we wanted to give a single formula to describe mathematically the position and motion of all the planets in the solar system. At each moment t, we must give the position P_i and the velocity V_i of planet i, for $i = 1, 2, \ldots, 9$. P_i has three coordinates and V_i has three components, so that we must record six numbers for each planet, and therefore a total of 54 numbers. We can therefore create a single function f from the real numbers to 54-space by setting

$$f(t) = (P_1, V_1, P_2, V_2, \ldots, P_n, V_n) = (a_1, a_2, \ldots, a_{54}). \tag{2.4}$$

In advanced physics, the same thing is done to describe the individual positions and motions of a collection of N particles by one function

$$f(t) = (a_1, a_2, \ldots, a_{6N}). \tag{2.5}$$

This example provides an interesting illustration of the power of mathematics to simplify a complex situation; if we forget the source that led to (2.5), we see that (2.5) can be described as a parametric curve in $6N$-space. Thus, the combined motions of a collection of 100,000 individual molecules of a gas each moving independently can be reduced to the motion of a single point along a curve in space of dimension 600,000.

Exercises

1 Sketch the vector field defined by $F(x, y) = (y, x)$.

2 Sketch the vector field defined by $F(x, y) = (1 - y, 1 + x)$.

3 (a) Sketch the curve described parametrically by

$$f(t) = (2 \cos t, 3 \sin t).$$

(b) What is the graph of the function in (a)?

4 Sketch the curve described parametrically by

$$F(t) = (3 \sin 2t, 4 \cos 2t, t/2\pi) \quad \text{for } 0 \le t \le 2\pi.$$

5 Given $F(x, y) = (2x - y, x + y)$, calculate $F(1, 1)$, $F(2, 1)$, $F(1, 2)$, $F(F(2, 1))$, $F(F(1, 1))$, and $F(F(x, y))$.

6 Given $F(x, y) = (x^2 - xy, xy + y^2)$ and $G(x, y) = (x + y, x - y)$, calculate $F(2, 1)$, $G(2, 1)$, $F(G(2, 1))$, $G(F(2, 1))$, $F(G(a, b))$, and $G(F(a, b))$.

7 An attractive force field has equations

$$F_x = \frac{-2x}{x^2 + y^2}, \quad F_y = \frac{-2y}{x^2 + y^2}.$$

Sketch the vectors associated with the points $(\pm 1, \pm 1)$, $(\pm 2, 0)$, $(0, \pm 2)$, $(\pm 3, \pm 1)$.

8 Water at a faucet is turned off at $t = 0$. The flow in a trough that was carrying off the water has velocity vectors described by $v_x = x/(1 + t)$, $v_y = 2(1 - y)/(1 + t)$. Draw the vectors for the trough $0 \leqq x \leqq 4$, $0 \leqq y \leqq 2$ at the points $(0, \frac{1}{2})$ $(1, \frac{3}{2})$, $(2, 1)$, $(3, 2)$, $(4, 1)$ when $t = 0$, $t = 1$, and $t = 2$.

2.3 Continuity

For the remainder of this chapter, we shall be discussing special properties of functions, and it will not matter whether they are functions of one, three, or 100 variables. We therefore simplify our notation, using p for any point in the domain of a function f and $f(p)$ for the value of f at p, usually a number. Sometimes in applications the domain of a function f will be a set in the plane, and we can also write $p = (x, y)$; in other cases, the domain may be a set in 3-space, and p could be replaced by (x, y, z). Since the basic concepts are always the same, it is important to learn to think about them as properties true for functions of a point p, not stressing the coordinates of p.

Intuitively, a function f is continuous if small changes in the input always result in small changes in the values. Returning to the temperature interpretation, we might say that $T = f(p)$ describes a continuous temperature distribution in the set $D = \text{dom}(f)$ if nearby points have almost the same temperatures. For example, with $p = (x, y, z)$, suppose $f(p) = x^2y + 2yz$. Then, we observe that $f(1, 1, 2) = 5$, $f(1.1, 0.9, 2.1) = 4.87$ and $f(1.01, 0.99, 2.01) = 4.9897$.

How can we make this intuitive idea explicit and precise? Guided partly by earlier discussions of continuity, we are led to the following formal definition:

Definition 1 A real-valued function f is continuous at a point p_0 in its domain D if given any $\epsilon > 0$ there is a neighborhood $\mathfrak{N}$ of p_0 such that $|f(p) - f(p_0)| < \epsilon$ for every point p of D that lies in $\mathfrak{N}$; f is continuous on D if f is continuous at each point p_0 in D.

In most cases, the size of the neighborhood $\mathfrak{N}$ will depend upon the size of the number ϵ; the smaller ϵ is, the smaller $\mathfrak{N}$ will have to be. If we use spherical neighborhoods,

$$\mathfrak{N} = \{\text{all } p \text{ with } |p - p_0| < \delta\},$$

then the definition above takes on an equally common form: f is continuous at $p_0 \in D$ if given any number $\epsilon > 0$ there is a number δ such that $|f(p) - f(p_0)| < \epsilon$ for every point $p \in D$ such that $|p - p_0| < \delta$.

To illustrate the meaning of this definition, we first prove the following elementary result.

Theorem 1 *The function defined on the plane by $f(x, y) = y$ is continuous everywhere.*

Proof If $p = (x, y)$ and $p_0 = (x_0, y_0)$, then $|f(p) - f(p_0)| = |y - y_0|$. Given a number ϵ, we want to be able to specify a number δ so that $|y - y_0| < \epsilon$ for any choice of the point p within a distance δ of p_0. Since $|y - y_0| \leqq |p - p_0|$ (see Figure 2–9), it is clear that we can take $\delta = \epsilon$.

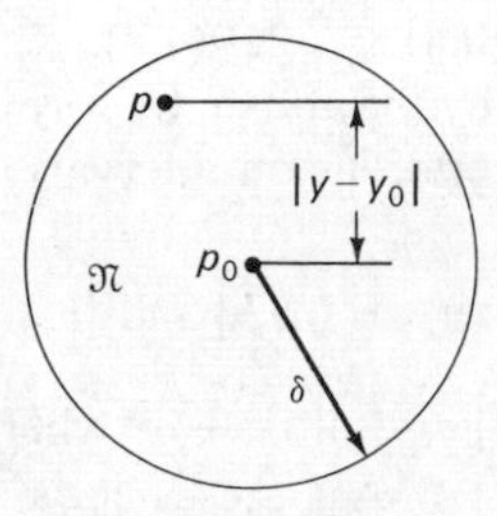

Figure 2–9

A similar argument could be given to show that the function $g(x, y) = x$ is continuous everywhere in the plane, and a more complicated proof would show that $h(x, y) = x^2y + 3x - 4y^2$ is also continuous on the plane. We don't want to have to check the continuity of each specific function we want to use, so we need some general theorems that take care of infinitely many functions, all at once—e.g., a theorem that states that *all* polynomials in x and y are continuous on the plane.

Before we look at such theorems, we digress to take up a special topic called *interval arithmetic*; although the basic idea is very old, it has become of interest only since the development of the high-speed electronic computer. We shall use it here to simplify some proofs of theorems about continuity.

Ordinary arithmetic deals with addition, multiplication, and division of single real numbers. Interval arithmetic carries out the same operations on intervals of numbers. The motivation here is that in practice we seldom know any numerical value exactly. If we want to find the area of a rectangle and measure the sides to be 2.3 and 4.7, we may in fact know only that the width lies between 2.25 and 2.35 and the length, between 4.65 and 4.75. Using the standard notation for intervals, we should write

$$\text{width} = [2.25, 2.35],$$
$$\text{length} = [4.65, 4.75].$$

Since

$$\text{area} = (\text{width}) \times (\text{length}),$$

what we want is to calculate the product of these two intervals. This we do by calculating the products of each possible width by each possible length measurement to see that the resulting numbers (which represent all possible values of the area) fill an interval $[a, b]$ with $a = (2.25)(4.65)$ and $b = (2.35)(4.75)$. Thus, the result of our calculation is

$$\text{area} = [2.25, 2.35] \times [4.65, 4.75] = [10.4625, 11.1625].$$

The answer gives us the complete range of possible values of the area, assuming of course that the information about the width and length is accurate; note that, as expected, the area interval contains the number $10.81 = (2.3)(4.7)$.

Motivated by the above, we formulate the following definitions for arithmetic operations on intervals of numbers.

Definition 2 Given two intervals $[a, b]$ and $[c, d]$,

$$[a, b] + [c, d] = \{\text{all } x + y \text{ where } a \le x \le b \text{ and } c \le y \le d\},$$
$$[a, b] \times [c, d] = \{\text{all } xy \text{ where } a \le x \le b \text{ and } c \le y \le d\}.$$

As usual, division has some limitations, and we must set up a special requirement.

Definition 3 Given an interval $[a, b]$ and an interval $[c, d]$ which does not contain the number 0, then

$$[a, b] \div [c, d] = \frac{[a, b]}{[c, d]} = \left\{\text{all } \frac{x}{y} \text{ with } x \in [a, b] \text{ and } y \in [c, d]\right\}.$$

The requirement ensures that y can never be 0.

It is easy to learn to do interval arithmetic. The basic rules are given in the following theorem, whose proof we leave as an exercise (Exercise 3).

Theorem 2 *Let $I = [a, b]$ and $J = [c, d]$. Say that an interval is positive if all its members are positive, negative if all are negative, and posineg if some are negative and some positive. Then we have the following general calculations:*

(i) *In all cases, $I + J = [a + c, b + d]$.*
(ii) *If I is positive, then*

$$I \times J = \begin{cases} [ac, bd] & \text{if } J \text{ is positive,} \\ [bc, bd] & \text{if } J \text{ is posineg,} \\ [bc, ad] & \text{if } J \text{ is negative.} \end{cases}$$

(iii) *If I and J are both posineg, then*

$$I \times J = [u, v], \text{ where } u = \text{minimum of } ad \text{ and } bc,$$
$$v = \text{maximum of } ac \text{ and } bd.$$

(iv) *Assuming that J is either positive or negative,*

$$I \div J = I \times (1/J), \qquad \text{where}$$

$$1/J = \left[\frac{1}{d}, \frac{1}{c}\right].$$

To illustrate, here are some sample calculations.

$$\begin{aligned}
[3, 4] + [-1, 6] &= [2, 10], \\
[3, 4] \times [2, 5] &= [6, 20], \\
[3, 4] \times [-1, 6] &= [-4, 24], \\
[3, 4] \times [-2, -1] &= [-8, -3], \\
[-3, 4] \times [-2, 3] &= [-9, 12], \\
[-3, 4] \div [2, 3] &= [-3, 4] \times [\tfrac{1}{3}, \tfrac{1}{2}] = [-\tfrac{3}{2}, 2].
\end{aligned}$$

In the last calculation, we have also used the fact that $I \times J = J \times I$. It is not important for this course that you acquire skill in doing such calculations, although they form a useful tool in estimating the results of routine calculations and in working with approximations. You do not in fact have to know the rules in Theorem 2, since—as you may have noticed—the product of two intervals always turns out to be merely the largest interval having its endpoints among the numbers ac, ad, bc, bd.

We shall use the idea involved in interval arithmetic to prove the important theorem about sums, products, and quotients of continuous functions.

Theorem 3 *Let f and g be continuous functions defined on the domain D. Then*

(i) *$f + g$ and fg are continuous on D,*
(ii) *f/g is continuous at each point p_0 in D where $g(p_0) \neq 0$.*

To show what this general theorem does for us, we shall first obtain the following.

Corollary *Any polynomial function in x and y (e.g., $f(x, y) = 7x^2y^3 + 3xy^2 - 8x^2$) is continuous everywhere, and every rational function (e.g., $h(x, y) = (x^2y - 3xy^4 + 1)/(x^3 - 7xy^2)$) is continuous at all points of the plane except those at which the bottom polynomial takes the value 0.*

Proof This result is obtained by applying the theorem to the functions $g(x, y) = x$ and $f(x, y) = y$, already known to be continuous on the plane. By (i), $x^2 = (x)(x)$ is continuous, as is any function built from x and y by addition and multiplication; since a constant function—one defined by $h(x, y) = C$ for all points (x, y)—is also continuous, we see that any polynomial function

$$P(x, y) = a_{00} + a_{10}x + a_{01}y + a_{20}x^2 + a_{11}xy + a_{02}y^2 + \cdots + a_{nm}x^n y^m$$

is continuous everywhere in the plane. Finally, by (ii) any rational function $R(x, y) = P(x, y)/Q(x, y)$, being a quotient of polynomials, is continuous at all the points of the plane except those points $p = (x_0, y_0)$ where $Q(x_0, y_0) = 0$.

Proof (of Theorem 3) Let us first see intuitively why it is true. For $f + g$ to be continuous, it is required that $f(p) + g(p)$ be nearly the same value as $f(p_0) + g(p_0)$ whenever the point p is sufficiently near p_0. Since f is continuous at p_0, the values $f(p)$ all lie in some small interval I containing the number $f(p_0)$ when p is near p_0; likewise, $g(p)$ lies in a small interval J for p near p_0. The sum $f(p) + g(p)$ will certainly be a number in the interval $I + J$. But the sum of two small intervals is again small, so that $f(p) + g(p)$ will be near $f(p_0) + g(p_0)$.

A similar nonrigorous argument works for the product fg. We need to show that $f(p)g(p)$ is nearly equal to $f(p_0)g(p_0)$ when p is near to the point p_0. As above, we see that in fact $f(p)g(p)$ will lie in the product of the intervals I and J, where I, J are again small intervals about the numbers $f(p_0)$ and $g(p_0)$. But, $I \times J$ will be a small interval containing $f(p_0)g(p_0)$.

The argument for the quotient f/g is harder, and in fact requires a special fact about continuous functions that will not be proved until Section 2.5 (Theorem 8). This is the so-called inertia property of a continuous function; if g is continuous at a point p_0 and $g(p_0) \neq 0$, then there will be a neighborhood about p_0 on which the sign of $g(p)$ is the same as that of $g(p_0)$, and in fact a number $c > 0$ such that $|g(p)| > c$ for all choices of p near p_0. Returning now to the continuity of f/g, we note that f/g can be written as $(f)(1/g)$, so that the result we want will follow if we can show that the function $1/g$ is continuous at every point p_0 in D where $g(p_0) \neq 0$. As before, we can assume that the values $g(p)$ lie in a very small interval J containing $g(p_0)$ for all points p sufficiently near p_0. The inertia property assures us that J is not a posineg interval, and we know that the numbers $1/g(p)$ will lie in the interval $1/J$ which contains $1/g(p_0)$. If we had known initially that $1/J$ was a small interval, then we would now have completed the intuitive justification of the continuity of $1/g$ and hence that of f/g.

What prevents these arguments from being rigorous mathematical proofs? We find the answer by rereading the formal definition for continuity (Definition 1) at the start of this section and observing that the arguments above have left out all the quantitative parts. How small a neighborhood must we take about p_0? How does its size depend upon how widely the number $f(p)g(p)$ is allowed to vary from the number $f(p_0)g(p_0)$? How does the size of an interval J affect the size of the interval $1/J$? All of these objections are removed by the following completed proof of Theorem 3.

Because f and g are continuous at p_0, if we are given $\epsilon_0 > 0$, a neighborhood of p_0 can be found such that $|f(p) - f(p_0)| < \epsilon_0$ for all p in D that lies in this neighborhood; likewise, a second neighborhood of p_0 can be found which does the same for g. Let us denote by $\mathfrak{N}_0$ the smaller of these. Then for any p in D that lies in $\mathfrak{N}_0$, we have both

$$|f(p) - f(p_0)| < \epsilon_0 \quad \text{and} \quad |g(p) - g(p_0)| < \epsilon_0.$$

This implies that the intervals I and J in which the numbers $f(p)$ and $g(p)$ lie have lengths at most $2\epsilon_0$. As the intuitive analysis of Theorem 3 showed, we shall have proved (i) if we can construct the neighborhood $\mathfrak{N}_0$ so that in one case $I + J$ has length at most 2ϵ and in the other $I \times J$ has length at most 2ϵ.

Set $f(p_0) = a$ and $g(p_0) = b$, so that we may write

$$I = [a - \epsilon_0, a + \epsilon_0] \quad \text{and} \quad J = [b - \epsilon_0, b + \epsilon_0].$$

Clearly, $I + J = [a + b - 2\epsilon_0, a + b + 2\epsilon_0]$, and its length is $4\epsilon_0$. Thus, to prove that $f + g$ is continuous, we select $\epsilon_0 = (\frac{1}{2})\epsilon$, and then take $\mathfrak{N}$ to be the corresponding neighborhood $\mathfrak{N}_0$.

The computation for the product is slightly harder. However, it is easily seen that the interval $I \times J$ is contained in the interval

$$[ab - (|a| + |b| + 1)\epsilon_0, ab + (|a| + |b| + 1)\epsilon_0]$$

when ϵ_0 is small (see Exercise 8), so that the length of $I \times J$ is less than $2(|a| + |b| + 1)\epsilon_0$. To make this less than 2ϵ we select ϵ_0 smaller than $1/(|a| + |b| + 1)\epsilon$ and using $\mathfrak{N}$ again as the corresponding $\mathfrak{N}_0$, the continuity of fg at p_0 follows.

For the proof of continuity of the reciprocal $1/g$, we note that we want to make the length of $1/J$ small by controlling J. We must do more than just make J itself small, for if $J = [.001, .002]$, then $1/J = [500, 1000]$. Again we use the inertia property. If $g(p_0) \neq 0$ (we shall assume $g(p_0) > 0$), then g will yield only positive values near p_0, and in fact there will be a number $c > 0$ such that $g(p) \geq c$ for all p sufficiently near p_0. Proceeding as before, given any $\epsilon_0 > 0$ we can choose $\mathfrak{N}_0$ so that $g(p) \in J$ for all $p \in \mathfrak{N}_0$, and then make $\mathfrak{N}_0$ smaller if necessary so that J contains only numbers larger than c.

Since $J = [b - \epsilon_0, b + \epsilon_0]$, $1/J = [1/(b + \epsilon_0), 1/(b - \epsilon_0)]$. The length of $1/J$ is therefore

$$\frac{1}{b - \epsilon_0} - \frac{1}{b + \epsilon_0} = \frac{2\epsilon_0}{(b - \epsilon_0)(b + \epsilon_0)} < \frac{2\epsilon_0}{c^2},$$

where in the last step we used the fact that $b + \epsilon_0 > b - \epsilon_0 \geq c > 0$. To be sure that $1/J$ has length less than 2ϵ, we choose ϵ_0 less than $c^2\epsilon$. Thus we have found a constructive way to select a neighborhood $\mathfrak{N} = \mathfrak{N}_0$ so that $1/J$ has length less than a preassigned number, and we have therefore shown that $1/g$ is continuous at p_0.

The proof of Theorem 3 has been organized so that it applies to functions of any number of variables and so that it includes the proof that the sum (or difference) of continuous vector-valued functions is also continuous; for vector-valued functions the rest of the theorem has no meaning, since we do not multiply or divide general vector-valued functions.

Although Theorem 3 is strong enough to tell us that the function

$$f(x, y, z) = \frac{3x^2yz - 4y^3z^2 + 7xyz}{x^2 + y^2 + z^2}$$

is continuous everywhere in space except at the origin, it is not strong enough to tell us anything about the continuity of $h(x, y) = \sin(x + y)$. To take care of such functions, we need a theorem about the continuity of functions that arise as the composition of other functions, here the sine function and the polynomial $P(x, y) = x + y$. There is such a theorem, and it states in general that composition preserves continuity; any continuous function of continuous functions is itself continuous. However, it would be time-consuming to formulate and prove this result, and we hope that sufficient illustration will be provided by the proof of the following special case:

Theorem 4 *Let g be a function of three variables, defined and continuous on a domain D. Let f be a function of one variable, defined and continuous on an interval K, and suppose that all the values of g for $p \in D$ lie in K. Then, the function h defined on D by $h(x, y, z) = f(g(x, y, z))$ is continuous on D.*

Proof Intuitively, the argument is clear. Put $u = g(x, y, z) = g(p)$ and $v = f(u)$, so that $h(p) = v$. If $p_0 \in D$ and $g(p_0) = u_0$, then $u = g(p)$ will be near u_0 when p is near p_0. Likewise, $f(u)$ will be near $v_0 = f(u_0)$ when u is near u_0. Thus, combining these, $f(u)$ (which is the same as $h(p)$) will be near v_0 (which is $h(p_0)$) when p is near p_0. Hence, h is continuous at p_0. To make this proof valid, we have to make it quantitative, by attaching specific measures to all the words "near."

Given any ϵ, we invoke the continuity of f to select a number δ such that $|f(u) - v_0| < \epsilon$ whenever $|u - u_0| < \delta$. Since g is continuous, we can select a neighborhood $\mathfrak{N}$ about p_0 so that $|g(p) - u_0| < \delta$ whenever $p \in D$ and p lies in $\mathfrak{N}$. Combining these results we have $|h(p) - v_0| < \epsilon$ whenever $p \in D$ and p lies in $\mathfrak{N}$, for we have $h(p) = f(g(p)) = f(u)$ and $v_0 = f(u_0) = f(g(p_0))$. (Please note that this proof does not depend specifically upon g being a function of three variables; the result is true for g a function of n variables.)

As a second illustration, we give without proof another theorem about composition.

Theorem 5 *Let f be a function of three variables, defined and continuous on a domain K. Let A, B, C each be a function of four variables defined and continuous on a domain D. Suppose also that the point with coordinates*

$$(A(x, y, z, w), B(x, y, z, w), C(x, y, z, w))$$

lies in K for each choice of $p = (x, y, z, w)$ in D. Then, the function defined on D by

$$h(x, y, z, w) = f(A(x, y, z, w), B(x, y, z, w), C(x, y, z, w))$$

is continuous everywhere in D.

Exercises

1 Given $F(x, y) = xy - 2x + y$, calculate $F(x, y) - F(x_0, y_0)$ with $(x_0, y_0) = (1, 2)$ and (x, y) equal to each of the four points $(x, y) = (1 \pm 0.1, 2 \pm 0.1)$.

2 Calculate
(a) $[2, 3.5] \times [-1, 3]$, (b) $[-1, 3] \times [-2, 4]$,
(c) $[1.5, 3] \div [2, 3]$, (d) $[-3, 1] \div [-2, -1]$,
(e) $([3, 6] + [-1, 2]) \times [2, 3]$,
(f) $([3, 6] \times [2, 3]) + ([-1, 2] \times [2, 3])$.
Note that the distributive law fails for interval multiplication.

3 Prove any two parts of Theorem 2.

4 Formulate a rule for calculating $I \times J$ when
(a) both $I = [a, b]$ and $J = [c, d]$ are negative intervals,
(b) when $I = [a, b]$ is negative and $J = [c, d]$ is posineg.

5 Define $I - J$ to be $\{\text{all } x - y \text{ where } x \in I = [a, b] \text{ and } y \in J = [c, d]\}$. Is $I - J = I + ([-1, -1] \times J)$?

6 Prove that the function $f(x, y) = x$ is continuous everywhere.

7 Describe the regions where the functions defined by the following equations are continuous:

(a) $f(x, y) = \dfrac{2x^2 + y^2 - 7}{x^2 - y^2}$, (b) $f(x, y) = \dfrac{3x^2 - 2y^2 + 7}{2x^2 + 3y^2 + 4}$,

(c) $f(x, y) = \sqrt{x^2 - y}$, (d) $f(x, y, z) = \sqrt[3]{z^2(x^4 - y^4)}$,

(e) $f(x, y, z) = \dfrac{(x^2 - y)(y^2 - z)}{(z^2 - x)}$.

8 If $0 < \epsilon < 1$, $I = [a - \epsilon, a + \epsilon]$, $J = [b - \epsilon, b + \epsilon]$, show that every number z in $I \times J$ obeys

$$|z - ab| \leq (|a| + |b| + 1)\epsilon.$$

9 Give an intuitive argument to prove Theorem 5.

10 Without proving it, state a theorem like Theorem 5 about the continuity of

(a) $G(t) = g(A(t), B(t), C(t))$,

*(b) $H(x, y, z) = f(U(x), V(y), W(z))$.

***11** (a) Let

$$f(x) = \begin{cases} 1 & \text{when } x \text{ is rational,} \\ 0 & \phantom{\text{when }} x \text{ is irrational.} \end{cases}$$

Why is f continuous nowhere?

(b) Where is the function g defined below continuous?

$$g(x, y) = \begin{cases} 1 & \text{when } x - y \text{ is rational,} \\ 0 & \phantom{\text{when }} x - y \text{ is irrational.} \end{cases}$$

***12** Can you have two functions f and g both continuous nowhere such that $h(x) = f(g(x))$ is continuous everywhere?

2.4

Limits of Functions

In most books on elementary calculus, it has been traditional to talk about limits of functions before one gets to continuity, and in fact to define continuity in terms of limits. This is natural, since the focus of attention is apt to be on the derivative, which can be introduced easily by means of limits. From the viewpoint of modern analysis, however, continuity is more basic and in some ways it is both an easier and a more intuitive concept than limits. We have chosen to treat limits after continuity.[1] There are also some special reasons for preferring this order when one turns to functions of several variables.

[1] *This approach for functions of one variable has also been adopted in Willcox et al.,* Introduction to Calculus 1. *(Boston: Houghton Mifflin Co., 1971).*

For example, a standard example of limits for one variable might be

$$\lim_{x \to 1} \frac{x^3 - 1}{x^2 - 1}.$$

To evaluate this (without using something like L'Hospital's Rule), we can remove the common factor $x - 1$ from top and bottom of the fraction and turn this example into

$$\lim_{x\to 1} \frac{x^2 + x + 1}{x + 1} = \frac{3}{2},$$

which we have evaluated by substitution.

Contrast this problem, however, with either of the following problems dealing with functions of two variables:

(i) Find, if possible,

$$\lim_{(x,y)\to(0,0)} \frac{x^2 y}{x^2 + y^2}.$$

(ii) Find, if possible,

$$\lim_{(x,y)\to(1,0)} \frac{x + y^2 - 1}{(x - 1)^2 + y^2}.$$

In neither (i) nor (ii) can you cancel any useful factors to simplify the problem.

There is also another fundamental difference between one- and two-variable limits. There are now an infinite number of directions in which a moving point p can approach a point p_0, including approaches along curves and even along spirals toward p_0. As we shall see, one fact remains true: For continuous functions, limits can always be evaluated merely by substitution—

$$\lim_{p\to p_0} f(p) = f(p_0).$$

As a starting point, let us look at the limit of functions along curves. To be explicit, suppose f is a function that is defined on a neighborhood of the origin; how it may be defined at the origin will turn out to be irrelevant. Suppose also that $\mathcal{C}$ is a curve whose starting point is the origin. We can take $\mathcal{C}$ to be defined by equations $x = X(t)$, $y = Y(t)$, and assume that $0 \leq t \leq 1$, with $t = 0$ yielding $(x, y) = (0, 0)$. Then, we make the following evident definition.

Definition 4

$$\lim_{\substack{p\to \mathbf{0} \\ \text{along } \mathcal{C}}} f(p) = L = \lim_{t\to 0} f(X(t), Y(t)).$$

Since the right-hand expression asks for the limit of a function of one variable t, we have reduced the original problem to one that we know how to solve. As a first illustration, take $f(x, y) = x^2y^2/(x^2 + y^2)$, and examine the limits along lines through the origin. If such a line $\mathcal{L}$ goes through (a, b), we can take its equation to be $x = at$, $y = bt$. Then, applying the definition, we have

[2]*The notation "$t \downarrow 0$" means that t is approaching 0 from the right—that is, from values larger than 0. The notation "$t \to 0+$" is also used.*

$$\lim_{\substack{p\to O\\ \text{on } \mathcal{L}}} f(p) = \lim_{t\downarrow 0} f(at, bt) = \lim_{t\downarrow 0} \frac{a^2b^2t^4}{(a^2+b^2)t^2}$$

$$= \lim_{t\downarrow 0} \frac{a^2b^2}{a^2+b^2}\, t^2 = 0.^2$$

Since the answer does not depend on a or b, we see that the limiting value exists and is the same for all lines.

Next, replace this function with $f(x, y) = x/(x^2 + y^2)$, and again consider limits along lines. On the line $\mathcal{L}$, we have

$$\lim_{\substack{p\to O\\ \text{on } \mathcal{L}}} f(p) = \lim_{t\downarrow 0} f(at, bt) = \lim_{t\downarrow 0} \frac{abt^2}{(a^2+b^2)t^2}$$

$$= \frac{ab}{a^2+b^2}\,.$$

This time, a limit exists along any given line, but the value of the limit depends on the direction of the line.

A more complicated type of behavior can occur. We now take $f(x, y) = x^2y/(x^4 + y^2)$. Then, along lines as before, we have

$$\lim_{\substack{p\to O\\ \text{on } \mathcal{L}}} f(p) = \lim_{t\downarrow 0} \frac{a^2bt^3}{a^4t^4+b^2t^2} = \lim_{t\downarrow 0} \frac{a^2bt}{a^4t^2+b^2}$$

$$= \frac{0}{0+b^2} = 0$$

when $b \neq 0$. On the other hand, when $b = 0$, so that $\mathcal{L}$ is a horizontal line $x = at$, $y = 0$, we have $f(x, y) = f(at, 0) = 0$, and $\lim_{p\to O} f(p) = 0$. Thus, again we can say that the limit exists and is zero along every line through the origin. However, this function has the remarkable property that although it tends to 0 along every line, it does not have limit 0 when we approach the origin along the right sort of curve!

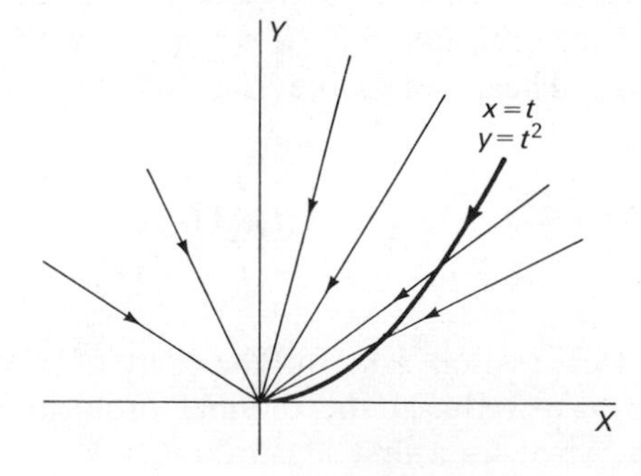

Figure 2–10

To see this, take the parabolic curve $x = t$, $y = t^2$. (See Figure 2–10.) This approaches the origin tangent to the horizontal axis, along which we have

$$\lim_{\substack{p \to \mathbf{O} \\ \text{on } \mathcal{L}}} f(p) = \lim_{t \downarrow 0} f(t, t^2) = \lim_{t \downarrow 0} \frac{t^2(t^2)}{t^4 + (t^2)^2}$$

$$= \lim_{t \downarrow 0} \frac{t^4}{2t^4} = \frac{1}{2}.$$

(In fact, by taking different parabolas, you can get any limiting value L, $-\frac{1}{2} \leq L \leq \frac{1}{2}$!)

The existence of apparently uncomplicated functions which behave in such paradoxical ways is one reason why the general theory of limits and limiting values in several variables is more difficult than the theory of limits for functions of one variable. However, if we restrict ourselves to continuous functions, everything remains simple.

Theorem 6 *Let f be defined on a neighborhood of P and continuous at P. If $\{p_n\}$ is any sequence of points converging to P or if $\mathcal{C}$ is any curve ending at P, then*

$$\lim_{\substack{p \to P \\ \text{on } \mathcal{C}}} f(p) = \lim_{n \to \infty} f(p_n) = f(P).$$

Thus, for a function f continuous at a point P and defined on a neighborhood of p, we may say that the limit of $f(x)$ as $x \to P$ is L (written $\lim_{x \to P} f(x) = L$) without ambiguity, since the limit is independent of the curve or sequence of points used to calculate it.

The theorem says simply that for continuous functions, limits can be calculated merely by substitution. For example, the function $f(x, y) = x^2y/(x^2 + y^2 + 1)$ is continuous at the origin, so $\lim_{p \to \mathbf{O}} f(p) = f(0, 0) = 0/1 = 0$.

To prove this theorem, we must recall both the definition of continuity and the meaning of convergence of a sequence of numbers, and of a sequence of points. Given $\epsilon > 0$, we can choose a neighborhood $\mathfrak{N}$ about P such that $|f(p) - f(P)| < \epsilon$ whenever p lies in $\mathfrak{N}$. Since the sequence $\{p_n\}$ is assumed to converge to P, there must exist a number N such that p_n lies in $\mathfrak{N}$ for all $n \geq N$. Thus we find that $|f(p_n) - f(P)| < \epsilon$ for all $n \geq N$; since this inequality holds for any choice of ϵ, $\lim f(p_n) = f(P)$. The case of a curve $\mathcal{C}$ is similar, and the proof will be omitted.

What is equally useful, at least in developing the theory of functions, is the converse of this result, which we state but do not prove.

Theorem 7 *If a function f is defined on a neighborhood of P, and if $f(P) = \lim_{n \to \infty} f(p_n)$ for every sequence of points $\{p_n\}$ converging to P, then f is continuous at P.*

Finally, it is helpful to have a term to describe a type of discontinuity that is encountered frequently:

Definition 5 A function f that is defined and continuous at all points of an open set D except the point p_0 is said to have a **removable discontinuity** at p_0 if it is possible to define (or redefine) the value of f at p_0 so that f becomes continuous everywhere in D, including p_0. If it is not possible to find such a value for $f(p_0)$, then f is said to have an **essential discontinuity** at p_0.

Putting together the results of Theorems 6 and 7, we arrive at a simple criterion: A function f has a removable discontinuity at a point p_0 if and only if $\{f(p_n)\}$ converges to the same number for every sequence $\{p_n\}$ that converges to p_0. Thus, to prove that a function cannot be defined in any way at a point p_0 so that it becomes continuous there, all one has to do is produce two sequences of points p_n' and p_n'' which converge to p_0 such that

$$\lim f(p_n') \neq \lim f(p_n'').$$

Applying this test to the examples given at the start of this section, we see that $f(x, y) = xy/(x^2 + y^2)$ cannot be defined at the origin $(0, 0)$ to make it continuous there, although f is continuous at all other points in the plane. Expressed more intuitively, there is no "natural" definition for this function at $(0, 0)$. This is not the case for the function $f(x, y) = x^2y^2/(x^2 + y^2)$. This function is not defined at $(0, 0)$, but if we set $f(0, 0) = 0$, f can then be shown to be continuous everywhere; we have "removed" the discontinuity at the origin. To prove this, we need only show that $\lim_{n\to\infty} f(p_n) = 0$ for every sequence $\{p_n\}$ that converges to $(0, 0)$. The easiest way to do this is to look at the form that the function f takes when we use polar coordinates for the point p, with $x = r\cos\theta$, $y = r\sin\theta$. Accordingly, we have

$$\begin{aligned} f(p) &= \frac{(r\cos\theta)^2(r\sin\theta)^2}{(r\cos\theta)^2 + (r\sin\theta)^2} \\ &= \frac{(\cos\theta)^2(\sin\theta)^2r^4}{r^2} \\ &= (\cos\theta)^2(\sin\theta)^2r^2, \end{aligned}$$

so that for any point p, $|f(p)| \leq r^2 = |p|^2$. If $\{p_n\}$ is any sequence converging to $\boldsymbol{O}$, then the distance from p_n to $\boldsymbol{O}$ must become small as n increases, so that $\lim_{n\to\infty} |p_n| = 0$. But $|f(p_n)| \leq |p_n|^2$, so we see that $\lim_{n\to\infty} f(p_n) = 0$.

Perhaps we can throw more light on this difficult topic of continuity and limits of functions by looking at the graphs of the above functions. First, let us recall how simple the situation is when one is dealing with functions of one variable. Consider the function $g(x) = (x^2 - 1)/(x - 1)$ which is defined and continuous on the whole line except at the point $x = 1$. Its graph is shown in Fig-

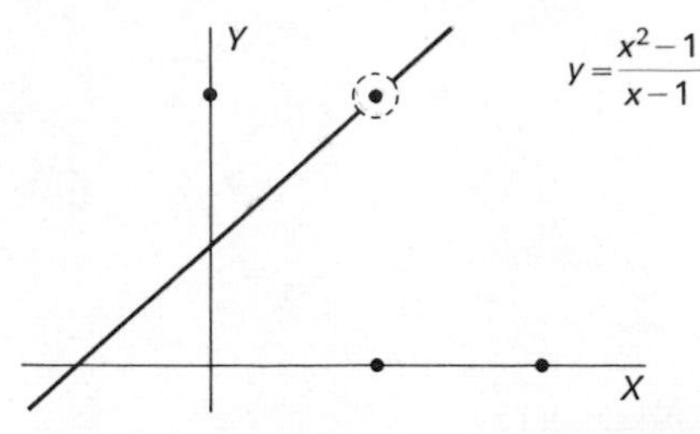

Figure 2–11

ure 2–11 and is simply a line $y = x + 1$ with one point deleted. (Clearly, by factoring, $g(x) = x + 1$ when $x \neq 1$.) The point $x = 1$ is a removable discontinuity for g, and if we set $g(1) = 2$, then g becomes continuous everywhere. Speaking geometrically, the graph of g can be changed into the graph of a continuous function by filling in the point (1, 2) which is "missing" from its graph, and no other assignment of a value for g at $x = 1$ will do this. Thus, $g(1) = 2$ is the unique "natural" way to define g at the missing point.

The same remarks apply exactly to the function we studied above, $f(x, y) = x^2y^2/(x^2 + y^2)$, except that the graph is much harder to construct and to visualize, since there is no convenient factor $x - 1$ to remove. Following the procedures suggested in Chapter 1, we obtain the surface sketched in Figure 2–12.

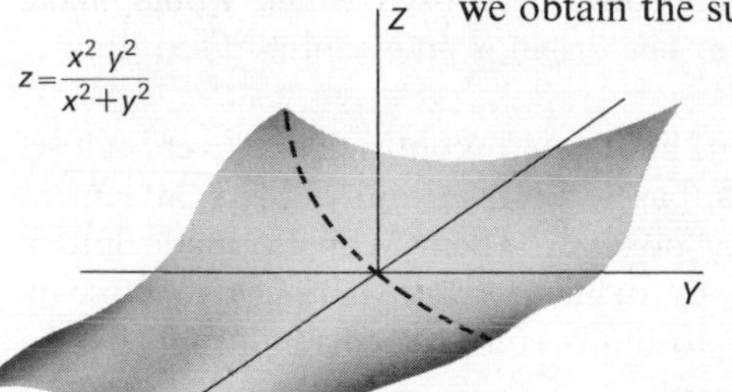

Figure 2–12

As suggested in the picture the surface is generally rather flat near the origin, resembling a paper square whose corners have been lifted up. Since the function f is not defined at the origin, there is no point on the surface where the Z-axis appears. The removal of a discontinuity here merely means that we make the definition $f(0, 0) = 0$, thereby putting the point (0, 0, 0) into the graph. The resulting graph is that of a continuous function.

As a contrast, let us now look at the second function given above,

$$f(x, y) = \frac{xy}{x^2 + y^2} \cdot$$

The graph $z = f(x, y)$ of this function is much harder to sketch, especially near the origin. Observe that if $x = y = a$, then $f(x, y) = f(a, a) = \frac{1}{2}$, so that all points on the graph that lie above the line $y = x$ in the XY-plane must have Z-coordinate $\frac{1}{2}$. Similarly, all those below the line $y = -x$ must have Z-coordinate $-\frac{1}{2}$. Using this information and looking at some of the cross sections $x =$ const and $y =$ const, we obtain the picture in Figure 2–13.

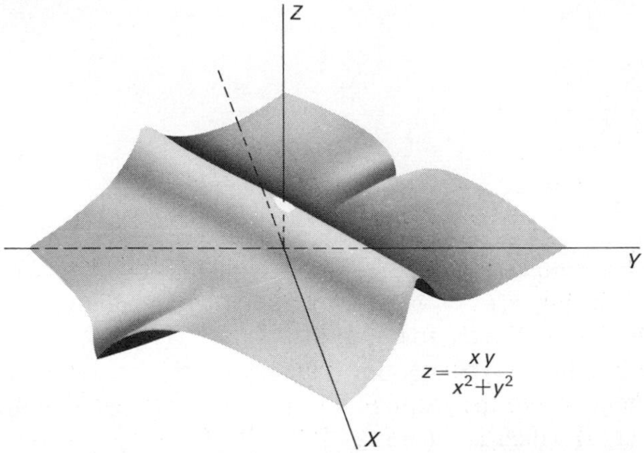

$z = \frac{xy}{x^2 + y^2}$

Figure 2–13

Please note that there is a ridge of height $\frac{1}{2}$ crossing from right to left and a trough of the same depth going from left to right. They reflect the fact that the limiting value of f, as you approach the origin along a line, depends upon the direction of approach you are taking. Again, there is no point of the graph on the Z-axis, and it is also clear that no point could be chosen which would make the surface continuous there. The origin is an essential discontinuity for the function f.

The connection between limits and continuity is very close. Our definition of limit is based on continuity: The statement "$\lim_{p \to p_0} f(p) = L$" is true precisely when f can be made into a function continuous at p_0 by defining (or redefining) f at p_0 by $f(p_0) = L$. This approach to limits is also used in *Introduction to Calculus 1*, Willcox et al.[3]

[3] See note 1 above.

Exercises

1 Calculate

$$\lim_{\substack{p \to (0,0) \\ \text{on } \mathcal{C}}} \frac{x + y}{\sqrt{x^2 + y^2}}, \quad \text{where}$$

(a) $\mathcal{C}$ is the positive X-axis.

(b) $\mathcal{C}$ is the line $y = x$, where $x > 0$. What happens when $x < 0$?

(c) $\mathcal{C}$ is the curve $y = x^2$.

2 Calculate

$$\lim_{\substack{p \to \mathbf{O} \\ \text{on } \mathcal{C}}} \frac{xy + yz}{x^2 + y^2 + z^2}, \quad \text{where}$$

(a) $\mathcal{C}$ is the line from (1, 1, 1) to the origin,

(b) $\mathcal{C}$ is the line from (1, 2, 0) to the origin,

(c) $\mathcal{C}$ is the line from (a, b, c) to the origin,
(d) $\mathcal{C}$ is the curve $x = t^2$, $y = t$, $z = 2t$.

3 Calculate $\lim_{\substack{p \to (0,0) \\ \text{on } \mathcal{C}}} \sin(x/y)$, where $\mathcal{C}$ is

(a) the Y-axis, (b) the line $y = x$,
(c) the curve $y = \sqrt{x}$, $x > 0$, (d) the curve $x = \sqrt{y}$, $y > 0$.

4 Let $f(x, y) = xy + x \sin y$, $p_n = ((-1)^n, 1/n)$, and $q_n = (n, \pi/2)$. Calculate

(a) $\lim_{n\to\infty} f(p_n)$, (b) $\lim_{n\to\infty} f(q_n)$.

5 Let $f(x, y) = (xy - x + 1)/(y^2 + 1)$ and $p_n = (1/n, 2/n)$. Find $\lim_{n\to\infty} f(p_n)$.

6 Let $f(x, y) = x^2y + y$ and $p_n = \left(\frac{n-1}{n+1}, 1 + \frac{2}{n}\right)$. Find $\lim_{n\to\infty} f(p_n)$.

7 Let $f(x, y, z) = xyz/(x^2 + y^2 + z^2)$ and $p_n = (1/n, -2/n, 3/n)$. Find $\lim_{n\to\infty} f(p_n)$.

***8** Can you show that $\lim_{n\to\infty} f(p_n) = 0$ for every sequence $p_n \to 0$ when $f(x, y, z) = xyz/(x^2 + y^2 + z^2)$?

9 If $f(x, y) = x/(x + y)$, $g(x, y) = y/(x + y)$, and $p_n = (1/n, 2/n)$, what is (a) $\lim_{n\to\infty} f(p_n)$, (b) $\lim_{n\to\infty} g(p_n)$, and (c) $\lim_{n\to\infty} f(p_n)g(p_n)$?

***10** Is there a curve $\mathcal{C}$ that goes through the origin $(0, 0)$ such that

$$\lim_{\substack{p \to (0,0) \\ \text{on } \mathcal{C}}} \frac{x^2y^3}{x^4 + y^6} \neq 0?$$

***11** The function $f(x, y) = x^y$ is defined for all $x > 0$, $y > 0$. Look at the limiting behavior of $f(x, y)$ as $(x, y) \to (0, 0)$. Is there a "natural" choice for the value of 0^0? (*Hint:* Observe the limit on lines and on curves approaching the origin.)

2.5

Implications of Continuity

One reason for the importance of the concept of continuity is that continuity provides a single basic premise from which most of the intuitively desirable properties of well-behaved functions such as the intermediate value property can be derived. In this section, we shall examine these properties in the form they take for functions of several variables. We start with what has been called the inertia property.

Theorem 8 *If f is defined on a set D and is continuous at a point $p_0 \in D$ and $f(p_0) > 0$, then there is a neighborhood $\mathfrak{N}$ about p_0 such that $f(p) > 0$ for all points p in D that are also in $\mathfrak{N}$.*

Before proving this theorem, we list some of its consequences. First, it is easy to replace the hypothesis "$f(p_0) > 0$" by "$f(p_0) > C$" by applying the theorem to the function $g(p) = f(p) - C$, to obtain the conclusion that $f(p) > C$ must hold for all points p in D that are sufficiently near p_0. Another simple argument allows one to reverse the inequality, replacing "$>$" by "$<$." A further simple modification enables one to sharpen the conclusions, showing for example that under the hypothesis that $f(p_0) > 0$, there must exist a constant $d > 0$ such that $f(p) \geq d$ for all p in D near p_0. Finally, the general statement of the inertia property turns out to imply continuity itself. The precise statements of these theorems will be found in Exercises 1–3.

Proof Since f is assumed to be continuous at p_0, we know that $f(p)$ is nearly the same value as $f(p_0)$ for all points p in D that are near p_0. Choose the number $\frac{1}{2}f(p_0) > 0$ as ϵ. Then a neighborhood $\mathfrak{N}$ about p_0 can be selected so that $|f(p) - f(p_0)| < \epsilon$ for all p in D that are also in $\mathfrak{N}$. This inequality says that $f(p)$ must lie in the interval $[f(p_0) - \epsilon, f(p_0) + \epsilon]$ for all such points p. Recalling the value of ϵ, we see that the left endpoint of this interval is actually $f(p_0) - \frac{1}{2}(f(p_0)) = \frac{1}{2}(f(p_0))$. Calling this number d, we see that $d > 0$ and that $f(p) \geq d > 0$ for all p in D that are also in $\mathfrak{N}$.

Another fundamental result in the study of functions of one variable is also true in the several-variable case: Continuous functions take maximum and minimum values on certain kinds of sets. (Recall that this theorem was basic to all work on maxima and minima problems.) In one variable, the sets for which this result held were closed bounded intervals $[a, b]$, and examples were given to show that the property can fail if the interval is not closed or is unbounded. In the case of functions of several variables, the class of sets turns out to be all closed and bounded sets, in the plane for two variables, and in n-space if we are dealing with continuous functions of n variables. The general statement is Theorem 9. We shall make frequent use of the result, but we leave the proof to a more advanced course, since it involves a deeper analysis of the topological nature of the real numbers and of n-space than we are prepared for now. (However, see Exercise 13.)

Theorem 9 *Let f be defined and continuous on a closed and bounded set D. Then there are two numbers M and m such that $m \leq f(p) \leq M$ for all p in D, and points p_1, p_2, exist in D where $f(p_1) = m$ and $f(p_2) = M$.*

Part of the conclusion in this theorem is sometimes stated in the abbreviated form: *Continuous functions are always bounded on closed bounded sets.* This is to be interpreted as meaning that there is a number B such that $|f(p)| \leq B$ for all $p \in D$, but it does not necessarily mean that B need be the smallest such bound nor that B is actually a value achieved by the function f.

The points p_1 and p_2 mentioned in the theorem will be called minimum points for f and maximum points for f respectively. For a specific function f and a set D, it very often happens that these points are on the boundary of D. Such a function is the analogue of a one-variable function whose maximum occurs at an endpoint of an interval, rather than at an interior point.

What is the analogue of the intermediate value theorem? The intuitive picture is clear. Suppose f is a continuous function that gives the temperature at each point of a region D: At one point the temperature is 5, and at another point the temperature is 10. Then there must be a point somewhere in D at which the temperature is 7. But this can be false if you don't put some restrictions on the set D. For example, let D be the set of all points (x, y) where $x^2 - y^2 \geq 1$, and let $f(x, y) = x$. Then $(2, 1)$ and $(-2, 1)$ lie in D, $f(2, 1) = 2$ and $f(-2, 1) = -2$, but there is no point p in D where $f(p) = 0$. The reason is that the set D in this example is not a connected set: D consists of two separated components, on one of which f is always positive-valued, and on the other, negative. (See

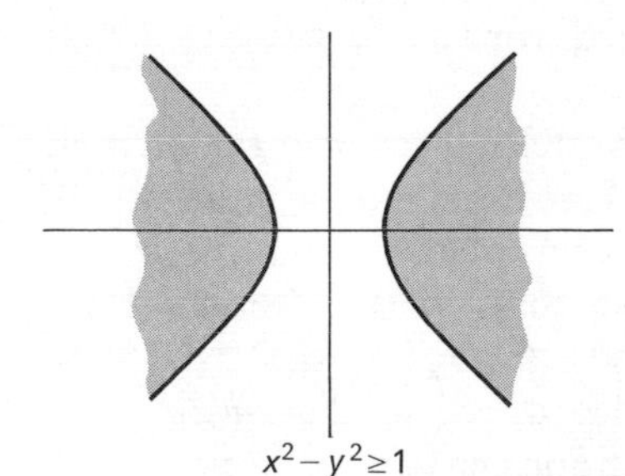

Figure 2–14

Figure 2–14.) A correct statement of the intermediate value theorem is the following.

Theorem 10 *Let f be defined and continuous on a connected set D. If f takes on the values A and C at two points in D and if $A < B < C$, then f must take on the value B at some point in D. In fact, if $\mathcal{C}$ is a continuous curve lying wholly in D and joining two points p_1 and p_2, where $f(p_1) = A$ and $f(p_2) = C$, then there must be a point p_0 on $\mathcal{C}$ where $f(p) = B$.*

Proof Suppose that the equation of the curve $\mathcal{C}$ is $x = X(t)$, $y = Y(t)$, and that the choice $t = t_1$ gives the point p_1, and $t = t_2$ the point p_2. (See Figure 2–15.) Now set

$$F(t) = f(X(t), Y(t)).$$

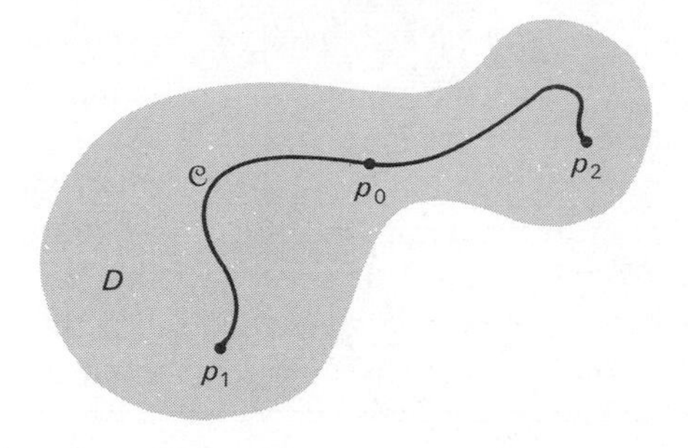

Figure 2–15

Then F is a continuous function of t for $t_1 \leq t \leq t_2$, and $F(t_1) = f(p_1) = A$ and $F(t_2) = f(p_2) = C$. By the intermediate value theorem for functions of one variable, we conclude that there is a value $t = t_0$ with $t_1 < t_0 < t_2$ such that $F(t_0) = B$. If p_0 is the point $(X(t_0), Y(t_0))$ on the curve $\mathcal{C}$, then we have obtained $f(p_0) = B$, as required.

In the exercises are a number of problems in which the intermediate value theorem can be used to prove geometric results, some of which may be surprising. The solution of each of these problems will require, as a first step, the continuity of certain functions. Rather than proving continuity, you should assume that each of the functions is continuous and proceed from there. Several illustrations may help to show how this is done.

Let f and g be continuous functions defined for x in the interval $[a, b]$, and suppose that $f(a) = 0$, $f(b) = 1$ and $g(a) = 1$, $g(b) = 0$. From the diagram in Figure 2–16, it seems plausible that the graphs of f and g must intersect. To prove this, consider the function $h = g - f$. Since h is continuous and $h(a) = 1 - 0 = 1$, $h(b) = 0 - 1 = -1$, there must be a choice of x_0 so that $h(x_0) = 0$; accordingly, $f(x_0) = g(x_0)$, and the graphs cross at $x = x_0$.

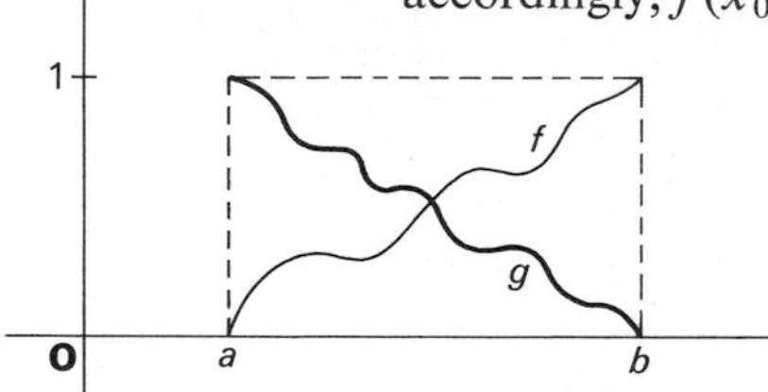

Figure 2–16

Next, consider the following assertion: *On any great circle on the earth, there must be two points at least five thousand miles apart that have the same temperature.* Our starting point is to *assume* that the temperature at each point on the given great circle depends continuously on the position. Choose any three points A, B, C equally spaced around the circle. (See Figure 2–17.) If any two of these points have the same temperature, we are done. We may therefore suppose that all three have distinct temperatures, and we shall assume that they have been labeled so that A is the hottest and C the coolest point. Then, by the intermediate value theorem, there

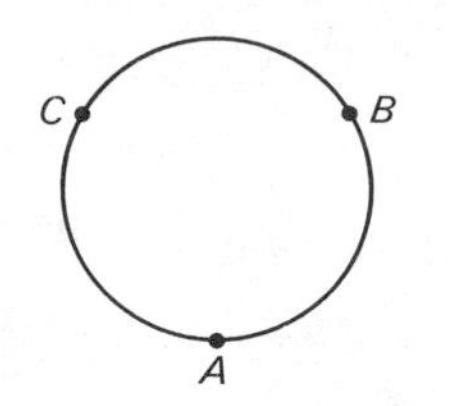

Figure 2–17

must be a point P somewhere on the 120° arc from A to C where the temperature is equal to that at B; it is clear that the distance from B to P is at least 5,000 miles.

Our last topic in this chapter is a sharpened form of the concept of continuity called *uniform continuity*. We shall see its importance later in integration of functions, in the important practical problem of preparing useful tables of function values (like logarithm tables), and in the approximation of complicated functions by simpler functions.

Let us return again to our intuitive picture of a function as prescribing a temperature distribution throughout a region D in the plane. Continuity means that about any point p_0 in D there is a neighborhood in which the temperature doesn't vary much from that at p_0. Suppose now that we invent a special instrument looking something like a compass or a caliper with two adjustable legs; if it is placed on the plane with one leg touching the point p_1 and the other touching p_2, then a meter will read the absolute difference of the temperatures at p_1 and p_2 (Figure 2–18). The value recorded by this instrument can be written as $|f(p_1) - f(p_2)|$.

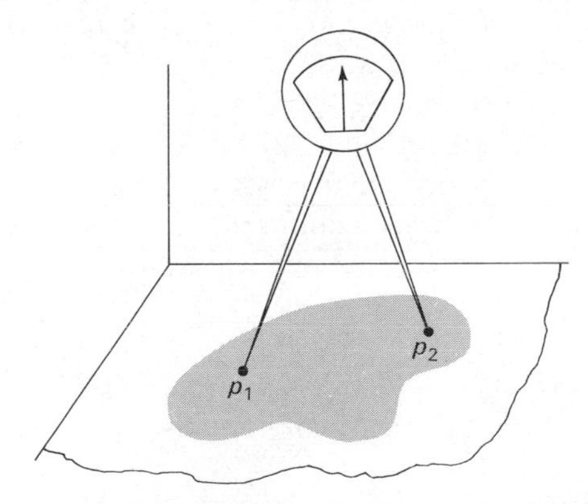

Figure 2–18

The meaning of continuity for f can be stated this way: Given any number ϵ and a point p_0, we can choose a setting of the caliper such that if one leg is set on p_0 and the other is placed anywhere with the opening of the caliper kept less than the allowed setting, the meter reading will be less than ϵ. (In terms of the formal definition in Section 2.3, the number δ corresponds to the allowed setting of the maximum opening of the caliper.)

Using this hypothetical instrument as an expository device, we explain the notion of uniform continuity.

INFORMAL DEFINITION The function f is said to be uniformly continuous on the set D if given any $\epsilon > 0$, it is possible to choose a setting of the caliper so that if the legs are not opened wider than this setting, and the caliper is placed anywhere at all with each leg resting on a point in D, then the meter reading is less than ϵ.

If this is restated in a formal mathematical way, we obtain the following.

Definition 6 The function f is uniformly continuous on D if given any $\epsilon > 0$, there is a constant $\delta > 0$ such that

$$|f(p_1) - f(p_2)| < \epsilon$$

for every pair of points p_1, p_2 in D that obey $|p_1 - p_2| < \delta$.

Many functions can easily be shown to be uniformly continuous on specific sets. For example, we shall show that the function $f(x, y) = x + y$ is uniformly continuous on the whole plane. Take any two points $p_1 = (x_1, y_1)$ and $p_2 = (x_2, y_2)$. Then,

$$\begin{aligned}|f(p_1) - f(p_2)| &= |(x_1 - x_2) + (y_1 - y_2)| \\ &\leq |x_1 - x_2| + |y_1 - y_2|.\end{aligned}$$

It is clear that $|x_1 - x_2|$ and $|y_1 - y_2|$ are each less than $|p_1 - p_2|$. Hence, for any points p_i, we have $|f(p_1) - f(p_2)| \leq 2|p_1 - p_2|$. Given $\epsilon > 0$, let us choose δ to be $(\frac{1}{2})\epsilon$. Then, if $|p_1 - p_2| < \delta$, $|f(p_1) - f(p_2)| < \epsilon$, and we have proved f to be uniformly continuous on the whole plane.

A key result in the theory of continuous functions is the following, which shows that on a familiar class of sets D, *every* function that is continuous on D is also uniformly continuous on D. This is another theorem whose proof we leave to a course in advanced calculus.

Theorem 11 *If D is a closed and bounded set in n-space and if f is defined and continuous on D, then f is uniformly continuous on D.*

For the reader interested in some of the theoretical aspects of mathematics, we point out that many of the results in this section have immediate extensions to continuous *vector*-valued functions. Sometimes the statement of a theorem must be changed considerably, as in the case of the intermediate value theorem: *If D is a connected set and f is a continuous vector-valued function defined on D, then the set of values $f(p)$ for p in D is a connected set of vectors.*

Exercises

1 (a) Use Theorem 8 to prove that if f is continuous on an open set D and $p_0 \in D$ with $f(p_0) > C$, then there is a neighborhood $\mathfrak{N} \subset D$ about p_0 such that $f(p) > C$ for all $p \in \mathfrak{N}$.

(b) Show that this implies that the set

$$\mathcal{O} = \{\text{all } p \in D \text{ where } f(p) > C\} \text{ is an open set.}$$

(c) State and prove a theorem similar to part (a) but with $f(p_0) < C$ as the hypothesis.

2 Prove that if f is continuous on an open set D, $p_0 \in D$, and $f(p_0) > c$, then there is a number $b > c$ such that $f(p) \geq b$ for all p near p_0. (*Hint:* First show that if $a > c$, then there is a number b with $a > b > c$.)

***3** Let f be a function defined on n-space $\mathbb{R}^n$ such that if $p_0 \in D$ and $a < f(p_0) < b$, then there is a neighborhood $\mathfrak{N}$ about p_0 in $\mathbb{R}^n$ such that $a < f(p) < b$ for all $p \in \mathfrak{N}$. Show that f must be continuous in $\mathbb{R}^n$.

4 Does the function $f(x, y) = xy/(x^2 + y^2 + 1)$ have a maximum value anywhere in the XY-plane? (*Hint:* Change to polar coordinates and see what happens as r increases.)

5 Let $f(x, y) = x^5y + y^3x^2 + xy^5 + 2$. Show that there is a point $p_0 = (x_0, y_0)$ with $|x_0| \leq 1$ and $|y_0| \leq 1$ where $f(p_0) = 0$. Can you tell anything about where p_0 is located in this square?

6 Suppose that a function f obeys the following inequality:

$$|f(p_1) - f(p_2)| \leq B|p_1 - p_2| \tag{2.6}$$

for all points p_1, p_2 on the domain D.

(a) Show that f is uniformly continuous on D.

(b) Use (a) to show that the function $f(p) = |p|$ is uniformly continuous in space.

7 (a) Show that the function $f(x) = x^2$ satisfies the relation (2.6) in Exercise 6, where D is the interval $1 \leqq x \leqq 4$ and $B = 8$.

(b) Show that the function $f(x, y) = x^2 - y^2$ also satisfies (2.6), where $D = \{\text{all } (x, y), -1 \leqq x \leqq 1, -1 \leqq y \leqq 1\}$ and $B = 4$.

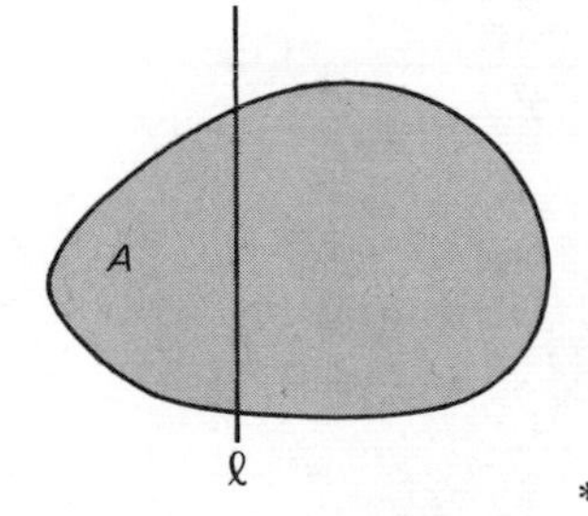

Figure 2–19

8 Let A be the region inside a smooth closed curve in the plane. (See Figure 2–19.) Give a mathematical argument to show that there is a *vertical* line ℓ which cuts the region in half so that each portion has the same area.

***9** Given two regions A and B (Figure 2–20), give a mathematical argument to show that there is a line l that cuts each in half at the same time. Note that l does not have to be vertical.

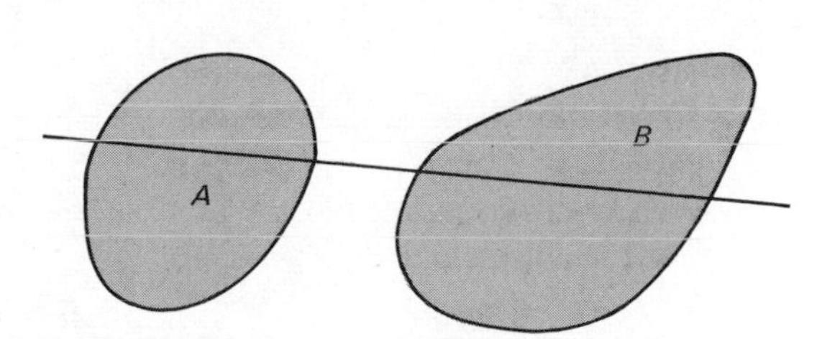

Figure 2–20

10 Give an argument to show that there must exist two points on the equator of a sphere that are diametrically opposite (antipodes) and have the same temperature (assuming temperature to be a continuous function of location).

11 Let D be a convex set bounded by a continuous curve $\mathcal{C}$. Give a mathematical argument to show that it is possible to draw a line ℓ that simultaneously divides the region in half and also cuts the curve $\mathcal{C}$ in half (Figure 2–21).

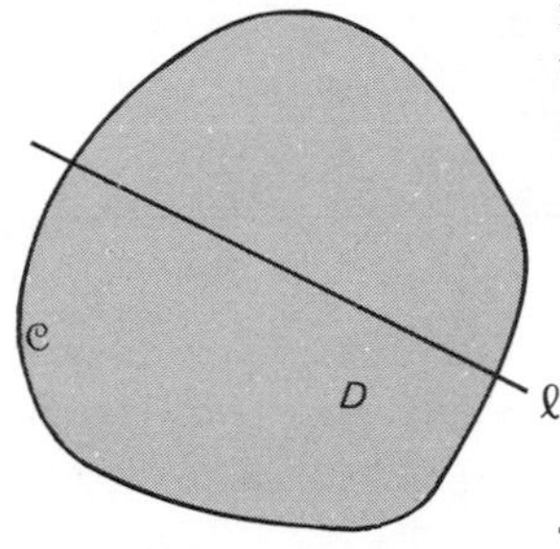

Figure 2–21

12 How many roots does the equation $5 \sin x - x = 0$ have? (Where did your argument use the intermediate value theorem?)

****13** Suppose that we take for granted the fact that any function f that is continuous on a closed bounded set D must be bounded there. With that supposition, prove that f must take on a maximum value someplace in D. (*Hint:* Look at $g(p) = 1/(B - f(p))$ for an appropriate choice of B.)

---

differentiation

chapter 3

3.1

Models for Motion

The basic motivation behind the creation and development of much of elementary calculus was the desire to achieve a usable mathematical model for the physical concept of motion. The first step was taken by Descartes, who realized that the idea of spatial position or location could be made concrete by the use of coordinate geometry; the second step was the realization that the subjective experience of time could be modeled by the real numbers.

From these steps follows the basic model for the motion of a point on a line—i.e., straight-line motion—as a function f from the time axis to one-dimensional space:

$$x = f(t), \tag{3.1}$$

and the model for the motion of a point in space as a function F from the time line to three-dimensional space,

$$(x, y, z) = F(t) = (f(t), g(t), h(t)), \tag{3.2}$$

which specifies the position of the moving point at each instant of time. The path or track of the moving point is a curve in space, and as a dividend, (3.2) becomes the parametric equation for a general space curve. Thus, at the start, we are operating in three contexts, subjective reality (motion), geometry (curve), and analysis (function).

The simplest motions are the straight-line motions, in which the point moves back and forth along a line and for which the model is (3.1). In this restricted framework, one asks for an interpretation within the model of the intuitive concept of speed or velocity and is led to adopt the definition:

$$\begin{aligned}\text{Velocity at time } t &= \lim_{h\to 0} \frac{f(t+h) - f(t)}{h} \\ &= f'(t) \\ &= \frac{dx}{dt}.\end{aligned}$$

At the same time that this interpretation of velocity was obtained, an unrelated study of the geometry of curves and the problem of defining and constructing tangents to curves led to the study of the graphs of functions and to a formula for the calculation of the slope of a curve at a given point identical with the velocity formula above.

At this stage, one might have summarized the situation by means of the following table:

Context	***Concept***	
subjective reality	straight-line motion	velocity
geometry	curve	slope of curve
analysis	function	derivative

Thus, looked at from the viewpoint of geometry, a straight-line motion becomes a particular curve in the plane (= space-time) which can also be viewed as the graph of a particular function, and the velocity of the moving point becomes the slope of the curve or the derivative of the function. (See Figure 3–1.)

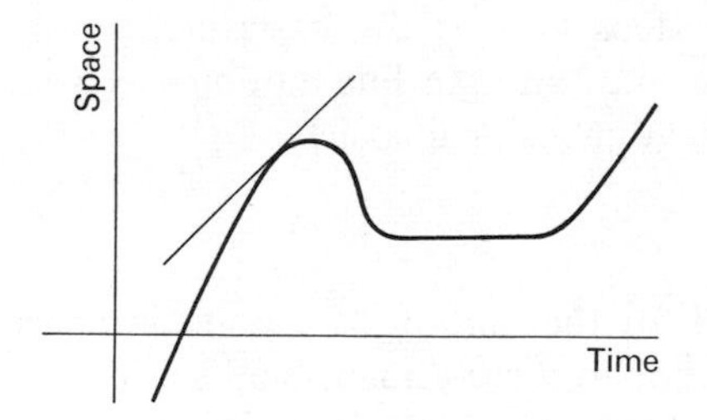

Figure 3–1

From the viewpoint of analysis, differentiation is a process which is applied to a function f to yield a new function f'. It is natural to examine the results of applying this process to f', obtaining f'', and then to wonder if f'' has some natural interpretation in the other two contexts. In the geometry of curves, f'' does not correspond to any immediately relevant concept, although it is connected with the idea of curvature. However, in the study of motion, f'' fits easily into the experimental facts; it may be identified with the concept of acceleration and gives immediate precision to the intuitive notion of "force."

Suppose now that we try to extend this model to cover the motion of a point in a plane, or more generally, in space. Since the motion itself is identified with the function F, it is natural to examine the graph of F to see if we can relate some of its geometric properties to aspects of the physical reality.

If we are concerned with a motion whose track is confined to the XY-plane, then the graph of this function can be regarded as a curve in a three-dimensional space, the Cartesian product of the XY-plane and the time line. In Figure 3–2, we have sketched two such curves corresponding to two particles moving around the same circular

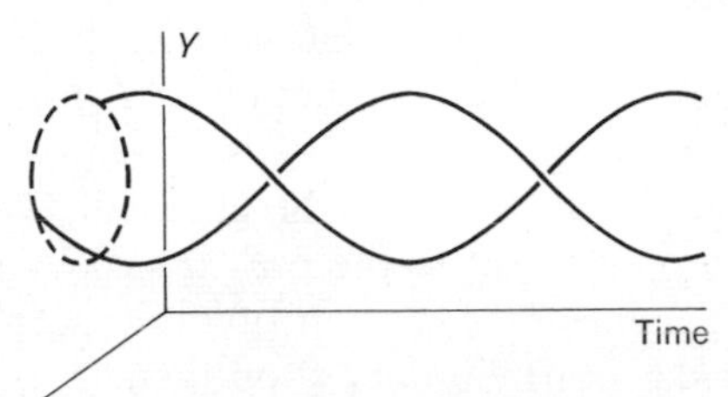

Figure 3–2

track. The corresponding picture for the motion of a point in space would be a curve in four-dimensional space-time. In physics, such a curve is called the "world line" of the particle and presents the complete history of the particle, past, present, and future. Although these pictures cannot be displayed, they are convenient mental images and help in certain branches of theoretical physics. Unfortunately, there is no simple relationship between geometric properties of the world line of a motion and concepts such as velocity or acceleration. Thus, the geometric approach which worked so well as a model for straight-line motion is only partly successful for the general motion of a point in space.

The analytic approach, however, is extremely successful. Suppose we mechanically apply to the general function

$$F(t) = (f(t), g(t), h(t)) = (x, y, z)$$

the same mathematical procedure that led to the derivative of a numerical-valued function. Define

$$F'(t) = \lim_{h \to 0} \frac{F(t+h) - F(t)}{h}.$$

Since

$$\frac{1}{h}\{F(t+h) - F(t)\}$$
$$= \left(\frac{f(t+h) - f(t)}{h}, \frac{g(t+h) - g(t)}{h}, \frac{h(t+h) - h(t)}{h}\right)$$

we may also write

$$F'(t) = (f'(t), g'(t), h'(t)) = \left(\frac{dx}{dt}, \frac{dy}{dt}, \frac{dz}{dt}\right). \tag{3.3}$$

Reasoning by analogy, we label $F'(t)$ the **velocity** of the moving point at time t and then observe that the value of F' is not a number but a triple of numbers and thus something which can be regarded as a vector. Intuitively, the instantaneous velocity of a point moving in a curved path can be thought of as a vector quantity, since we can speak of both the direction in which the point is moving at that instant and its speed, which we can think of as the magnitude of the velocity. In building our mathematical model, we treat $F'(t)$ as a free vector, and place its initial point at the point $F(t)$, the position of the moving point at time t, and we then define the direction of $F'(t)$ to be the direction of the motion and $|F'(t)|$ to be the speed. (See Figure 3–3.)

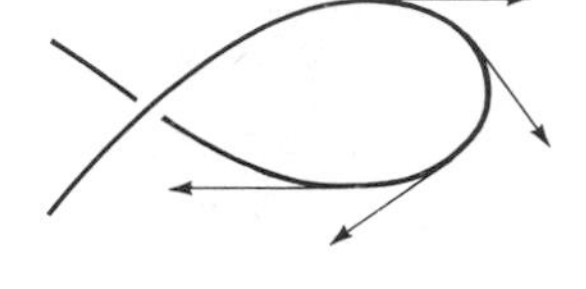

Figure 3–3

At this stage, something must be checked. Clearly, the velocity vector ought to be tangent to the path traced by the moving point. We recall the geometric definition of the tangent to a curve at a point P. Consider nearby points Q on the curve, and construct the line from P through Q. Then, calculate the limit of the direction of this line, as Q approaches P along the curve, and define this limit to be the direction of the tangent at P. (See Figure 3–4.)

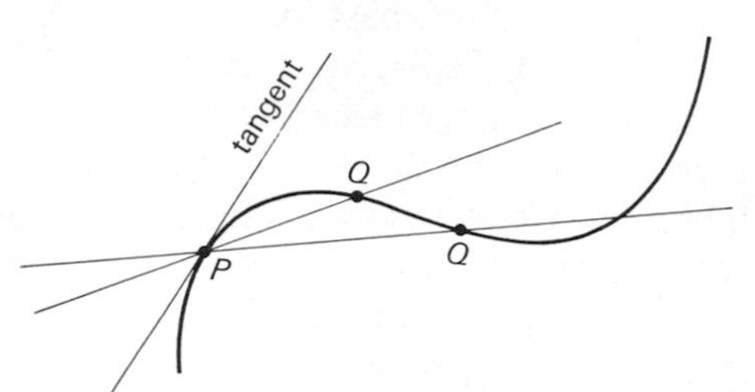

Figure 3–4

If $P = F(t)$, we may take Q to be $F(t + h)$, and the vector from P to Q is $F(t + h) - F(t)$. If $h > 0$, then $F(t + h) - F(t)$ and its nonnegative multiple $(1/h)\{F(t + h) - F(t)\}$ have the same direction. As h approaches 0, Q approaches P. Thus, the vector $F'(t)$ will have the same direction as the tangent vector T to the curve at the point $P = F'(t)$. (We must add the proviso that $F'(t) \neq 0$, since the vector $(0, 0, 0)$ does not have a direction.)

At this stage, we can summarize what we have learned under two headings:

GEOMETRY: If a curve in space is given by the parametric equation

$$(x, y, z) = F(t),$$

then the direction of the **tangent vector** to the curve at a point $P_0 = F(t_0)$ is that of the vector

$$T = F'(t_0) = \left(\frac{dx}{dt}, \frac{dy}{dt}, \frac{dz}{dt}\right)_{t_0}$$

(provided that $F'(t_0) \neq 0$).

PHYSICS: If the motion of a point in space is given by the formula $P = (x, y, z) = F(t)$, then the **velocity** of the point at time t_0 is given by $F'(t_0)$. The **speed** of the point is $|F'(t_0)|$, and the velocity vector is directed along the tangent to the path of the point.

If we continue this hypothetical retracing of the development of this important mathematical model, we would next proceed as before and repeat the process of differentiation, obtaining another vector-valued function

$$F''(t) = (f''(t), g''(t), h''(t)) = \left(\frac{d^2x}{dt^2}, \frac{d^2y}{dt^2}, \frac{d^2z}{dt^2}\right), \tag{3.4}$$

which, by analogy with the theory of straight-line motion, we label as the acceleration of the moving point. Again, we connect this vector-valued function with the intuitive physical concept of force, which we recognize as something which can also have a direction and a magnitude, and with the concept of mass, both via the equation $F = Ma$.

At this stage, we are able to propose a specific test of the model. Consider a particle of mass 1 moving in a circular path of radius R at a constant speed s. The equation for this motion is

$$F(t) = (R \cos \omega t, R \sin \omega t, 0),$$

where, as we shall see, $\omega R = s$. According to the model, the velocity is given by

$$F'(t) = (-\omega R \sin \omega t, \omega R \cos \omega t, 0).$$

We note that the velocity is not constant, but that—as required—the speed is $|F'(t)| = \sqrt{(-\omega R \sin t)^2 + (\omega R \cos t)^2} = \omega R = s$. The acceleration is then

$$\begin{aligned} F''(t) &= (-\omega^2 R \cos \omega t, -\omega^2 R \sin \omega t, 0) \\ &= -\omega^2 F(t) \\ &= \omega^2(-F(t)). \end{aligned}$$

The model thus predicts that the force needed to produce this circular path must have magnitude $|F''(t)| = \omega^2 R = s^2/R$ and must be directed toward the center of the circle, since it is opposite to $F(t)$.

The degree to which this prediction is confirmed by experiment is a measure of the success of this model. It should be noted that many of the steps in building the model might have been taken simply because they were mathematically elegant or motivated by analogy. This model conforms to the experimental evidence and predicts new patterns not previously revealed by experimentation: Thus we have an instance of what the physicist Eugene Wigner called the *Unreasonable effectiveness of mathematics in the natural sciences.*[1]

[1] *Comm. Pure Appl. Math.* 13 (1960), 1–14.

Exercises

1 The following space-time diagrams each describe the straight-line motion of one or more points. Give an equivalent verbal description of the events (Figure 3–5).

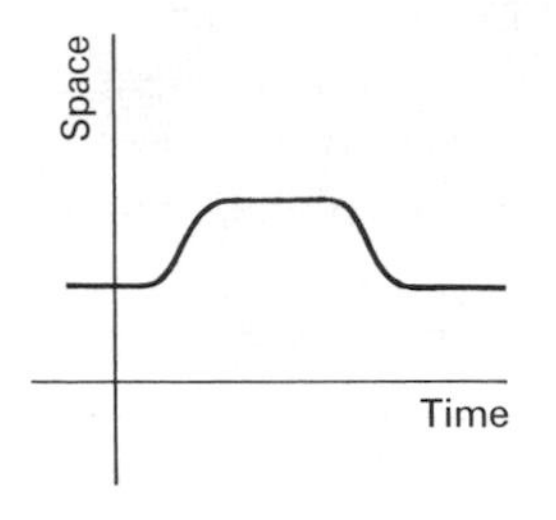

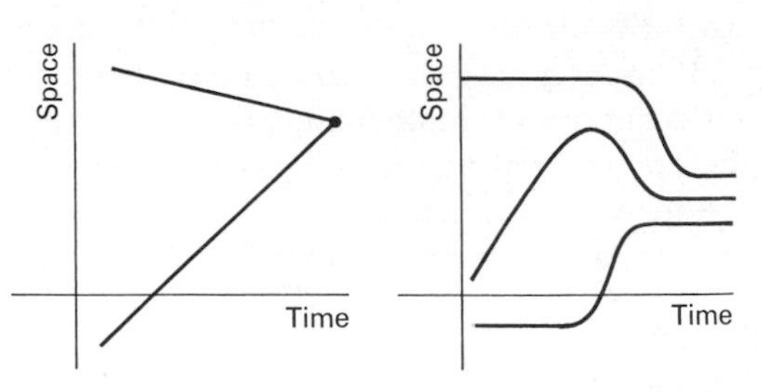

Figure 3–5

2 The graph of f' is shown in Figure 3-6. Given that $f(0) = 1$, draw the graph of f'' and f.

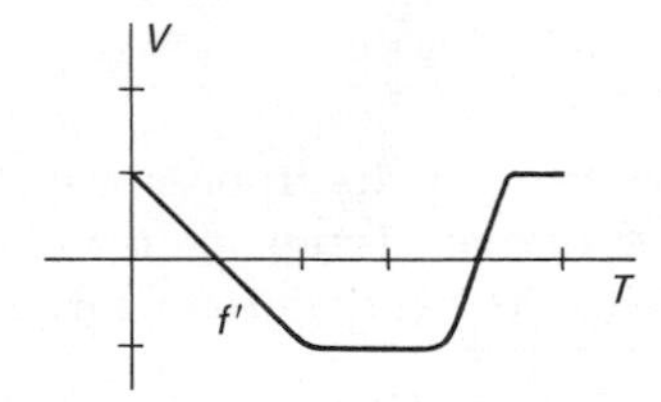

Figure 3–6

3 If $F(t) = (3t, t^3 - t^2, t^2 - 4t)$ describes a motion in space, find the velocity, speed, and acceleration at the points $(-3, -2, 5)$ and $(6, 4, -4)$.

For each of the following motions, sketch the path, and at a number of points draw the velocity and acceleration vectors:

4 $F(t) = (t, t^2, 0)$.

5 $F(t) = (t, t^3, 0)$.

6 $F(t) = (\cos t, \sin t, t)$.

7 Is there a point on the curve $F(t) = (t, t^2, t^3)$ where the tangent to the curve is

(a) parallel to the vector $V = (3, 3, 1)$,
(b) orthogonal to the vector $V = (3, 3, 1)$?

8 What is the cosine of the angle between the curves $(t, 2t^2 - 1, t + 1)$ and $(t^2, t + 2, 2t + 4)$ at the point $(1, 1, 2)$ where they meet?

9 Show that the following three curves meet at the point $(0, 0, 0)$ and are mutually perpendicular:

$$C_1\colon (t + t^2, -t, \sin 2t), \qquad C_2\colon (t^2 - t, \sin t, t),$$
$$C_3\colon (t - 1, 3t - t^2 - 2, 2t^2 - 4t + 2).$$

10 For the motion described by $F(t) = (t^2 - t, 2t + 1, t - t^2)$, is it ever true that the velocity and acceleration vectors are perpendicular at the same moment of time?

11 For the motion described by $F(t) = (t, t^2, t^3)$, is there ever a moment of time $t > 0$ at which the velocity and acceleration vectors are perpendicular?

12 A spiral curve is given by the parametric equation $x = \cos\theta$, $y = \sin\theta$, $z = \theta$. Are there a pair of points on this curve at which the tangent vectors are perpendicular?

3.2

Partial Derivatives

The fundamental concept that lies behind the differential calculus is that of rate of change. For functions of one variable, it is convenient to identify this variable conceptually with time, so that rate of change becomes synonymous with velocity, yielding the type of model outlined in the previous section.

The situation for functions of several variables differs greatly. If the domain of a function is a set in the plane, for example, if $f(x, y) = x^2 + 4xy + 2y^2$, there is little to be gained by interpreting one of the pair x and y to be time. As before, it is more useful to think of f as assigning a numerical value $f(x, y)$ to each point (x, y), regarding $f(x, y)$ as the temperature at that point. Then, the rate-of-change idea leads us to consider how the temperature changes according to the distance moved in units such as "degrees per centimeter."

A moment's reflection will show that this is inadequate, since the temperature change over a distance d will be very apt to depend upon the direction of motion as well as the distance traveled. In general, then, we would want to measure the rate of change of a function of several variables in each possible direction from a point, so that we do not obtain merely a single derivative of f but rather the set of *directional derivatives* of a function. When the directions are specialized to be those of the positive coordinate axes, the resulting directional derivatives are called the *partial derivatives*. As we will show, it is possible to calculate the directional derivative of a function of n variables in any given direction if the partial derivatives are known, or, in fact, if the directional derivatives are known in *any* n independent directions.

Suppose we start with functions of two variables. Let f be a continuous function defined in an open set D, and write $w = f(x, y)$, $p_0 = (x_0, y_0)$.

Definition 1 The partial derivatives of f at p_0 are given by

$$\frac{\partial f}{\partial x}(p_0) = f_x(x_0, y_0) = \lim_{h \to 0} \frac{f(x_0 + h, y_0) - f(x_0, y_0)}{h}, \tag{3.5}$$

(3.6) $$\frac{\partial f}{\partial y}(p_0) = f_y(x_0, y_0) = \lim_{h \to 0} \frac{f(x_0, y_0 + h) - f(x_0, y_0)}{h}$$

when these limits exist.

To see that these definitions are consistent with the intuitive notion of directional derivatives as discussed above, note that the point $(x_0 + h, y_0)$ is on the same horizontal line as p_0, at exactly a distance h from p_0 when h is positive. Thus, $f_x(p_0)$ is the directional derivative of f at p_0 in the horizontal direction determined by the unit vector $(1, 0)$, which is the direction of the X-axis. Likewise, $f_y(p_0)$ is the directional derivative at p_0 corresponding to the direction of the Y-axis. Other notations are often used for partial derivatives, such as w_x, $\partial w/\partial x$, f_1, and w_y, $\partial w/\partial y$, f_2; here, "1" and "2" refer to the *first* and *second* coordinates.

There is another very useful way to look at partial differentiation. In calculating $\partial f/\partial x$ at a point p_0, we carry out a familiar limit calculation on x, keeping $y = y_0$. This procedure is usually referred to as "differentiating with respect to x, holding y constant." In general, the partial derivative of a function with respect to one of the coordinate variables is calculated by holding all others constant and differentiating the resulting function of one variable in the usual way. The following examples should clarify this procedure and the use of various notations:

(i) If $f(x, y) = x^3y + 3x^2y^3 - 7xy^4$, then

$$f_x(x, y) = \frac{\partial f}{\partial x} = f_1(x, y) = 3x^2y + 6xy^3 - 7y^4,$$

$$f_y(x, y) = \frac{\partial f}{\partial y} = f_2(x, y) = x^3 + 9x^2y^2 - 28xy^3.$$

(ii) If $w = xy^4z^3$, then

$$\frac{\partial w}{\partial x} = y^4z^3, \quad \frac{\partial w}{\partial y} = 4xy^3z^3, \quad \text{and} \quad \frac{\partial w}{\partial z} = 3xy^4z^2.$$

(iii) If $T = 3A^2x - 4Axt^3$, then

$$\frac{\partial T}{\partial x} = T_x = 3A^2 - 4At^3, \quad \frac{\partial T}{\partial t} = -12Axt^2, \quad \text{and} \quad \frac{\partial T}{\partial A} = 6Ax - 4xt^3.$$

In passing, note that many problems in elementary calculus actually involve the calculation of partial derivatives, as in the example "If $y = ax^3 + bx^2$, find dy/dx." For we are in fact working with the function $F(a, b, x) = ax^3 + bx^2$, and the problem asks for the partial derivative F_x, whose value is $3ax^2 + 2bx$.

Any partial derivative of a numerical-valued function of several variables is again such a function. It is therefore possible to repeat the process of calculating a partial derivative, either differentiating with respect to the same variable or with respect to another. At first sight, the resulting notation for these higher partial derivatives looks complicated; however, the letters that appear tell which variables are involved in each successive differentiation. Suppose $w = x^3y^5$. Then, we have

$$\frac{\partial w}{\partial x} = 3x^2y^5,$$

$$\frac{\partial}{\partial y}\left(\frac{\partial w}{\partial x}\right) = \frac{\partial^2 w}{\partial y\,\partial x} = 15x^2y^4,$$

$$\frac{\partial}{\partial x}\left(\frac{\partial^2 w}{\partial y\,\partial x}\right) = \frac{\partial^3 w}{\partial x\,\partial y\,\partial x} = 30xy^4.$$

Something interesting happens if we carry out another succession of differentiations on the same function:

$$\frac{\partial w}{\partial y} = 5x^3y^4,$$

$$\frac{\partial}{\partial x}\left(\frac{\partial w}{\partial y}\right) = \frac{\partial^2 w}{\partial x\,\partial y} = 15x^2y^4,$$

$$\frac{\partial}{\partial x}\left(\frac{\partial^2 w}{\partial x\,\partial y}\right) = \frac{\partial^3 w}{\partial x^2\,\partial y} = 30xy^4.$$

We see that the result obtained apparently does not depend at all upon the order in which the various partial differentiations are performed, provided that in each case the number of differentiations in the X-axis direction and in the Y-axis direction are unchanged. Thus, in this experiment, we have seen that

$$\frac{\partial^2 w}{\partial x\,\partial y} = \frac{\partial^2 w}{\partial y\,\partial x},$$

$$\frac{\partial^3 w}{\partial x\,\partial y\,\partial x} = \frac{\partial^3 w}{\partial x^2\,\partial y}.$$

Based on the above, we might conjecture that each side of the last equation is equal to $\partial^3 w/\partial y\,\partial x^2$; you can verify this directly for our example. This is an instance of a general result, applying to all sufficiently smooth functions of several variables. We do not know any way to make it intuitively plausible, but a proof of a special case will be given later.

In the study of functions of one variable, the derivative of a function has an immediate geometric interpretation as the slope of a tangent to the graph of the function, and the second derivative can be related directly to the curvature of the graph. It is natural to hope for similar geometric interpretations for the partial derivatives of a function of several variables in connection with the graph of the function. Although we shall look at this topic in much more detail in a later section, we can make a useful start now.

Let us look at the graph of $w = f(x, y)$ in the vicinity of the point $p_0 = (x_0, y_0)$. In the calculation of the partial derivative f_x at p_0, we hold y constant, and differentiate with respect to x. Geometrically "hold y constant" means that we take a cross section of the graph in the vertical plane $y = y_0$. The intersection is a curve lying in this plane, and having the equation $w = f(x, y_0)$. (See Figure 3–7.)

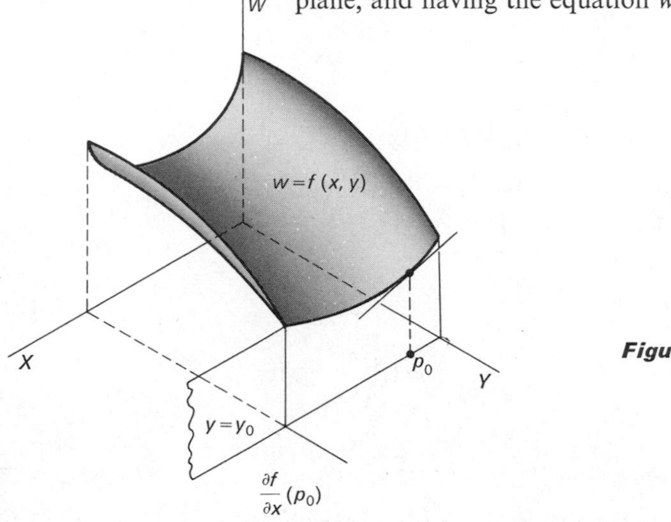

Figure 3–7

If we differentiate this equation with respect to x at x_0, obtaining $\partial w/\partial x = f_x(x_0, y_0)$, the geometric interpretation of this value is clearly the slope of the tangent to the curve of cross section, as shown in the diagram.

A similar analysis shows that the value of the partial derivative f_y at p_0 can be identified with the slope of the cross sectional curve obtained by slicing the graph of f with the plane $x = x_0$. (See Figure 3–8.)

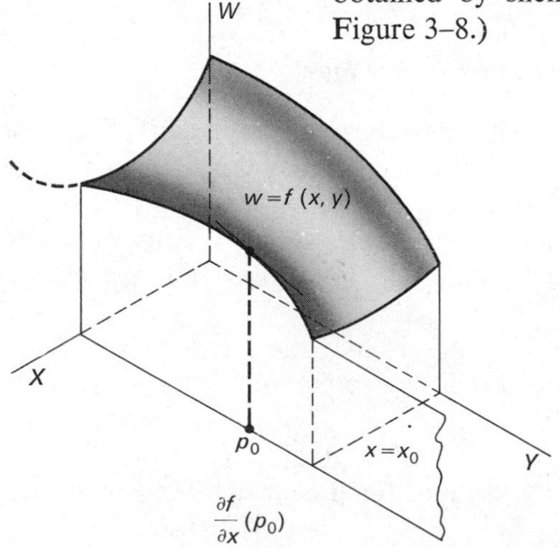

Figure 3–8

Using a similar picture, we can visualize the geometric meaning of the general directional derivative of a function of two variables. For any direction u in the XY-plane, we can construct a cross section of the graph of $w = f(x, y)$ in the direction u by passing a plane through the vector u perpendicular to the XY-plane. (See Figure 3–9.)

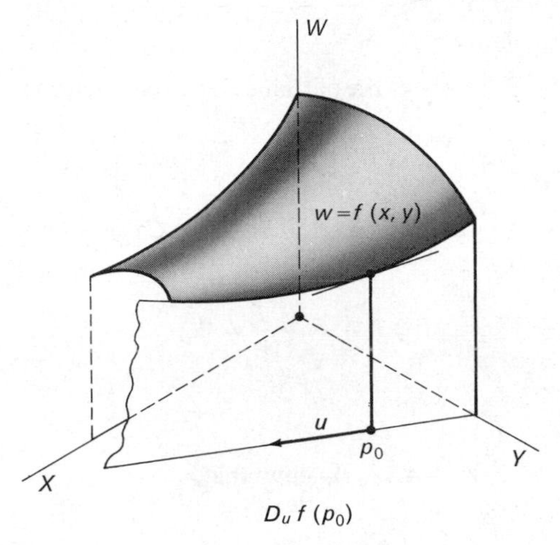

Figure 3–9

Then, the value of the directional derivative of f at p_0 in the direction u will be the slope of the curve of cross section at the point on the graph above the point $p_0 = (x_0, y_0)$.

Although it should be possible to show an analogous geometric interpretation of partial derivatives and directional derivatives for functions of three or more variables, we are severely limited by the fact that the graphs of such functions will be "surfaces" in 4-space or higher dimensional spaces, not something we can visualize with much success. In the next section, we shall find a better way to understand the directional derivative of a function of three variables.

Exercises

1 If $f(x, y) = x^3y - 2x^2y^2 + 5xy - 2x$, find $\partial f/\partial x$, $\partial f/\partial y$, $\partial^2 f/\partial x^2$, $\partial^2 f/\partial x\,\partial y$, $\partial^2 f/\partial y\,\partial x$, $\partial^2 f/\partial y^2$.

2 If $f(x, y) = x\cos(xy) - y\sin(xy)$, find $\partial f/\partial x$, $\partial f/\partial y$ and $\partial^2 f/\partial x^2$, $\partial^2 f/\partial x\,\partial y$, $\partial^2 f/\partial y^2$, $\partial^2 f/\partial y\,\partial x$.

3 If $f(x, y, z) = x^2z - 3xyz + 2xy^2$, find $\partial f/\partial x$, $\partial f/\partial y$, $\partial f/\partial z$ and

$$\frac{\partial^2 f}{\partial x\,\partial y},\ \frac{\partial^2 f}{\partial y\,\partial z},\ \frac{\partial^2 f}{\partial x\,\partial z},\ \frac{\partial^2 f}{\partial x^2},\ \frac{\partial^2 f}{\partial y^2},\ \frac{\partial^2 f}{\partial z^2}.$$

4 With $f(x, y, z) = x^2y^3z^4$, find

$$\frac{\partial^3 f}{\partial x\,\partial y\,\partial z},\ \frac{\partial^3 f}{\partial z\,\partial y\,\partial x},\ \frac{\partial^3 f}{\partial y\,\partial z\,\partial x}$$

and verify that these are all equal to $\partial^3 f/\partial z\, \partial x\, \partial y$.

5 Given $f(x, y, t) = xe^{-t} + ye^{xt}$, find $\partial^2 f/\partial x\, \partial t$ and $\partial^2 f/\partial t^2$.

6 Show that $f(x, y) = A(x) - B(y)$ obeys the partial differential equation

$$\frac{\partial^2 f}{\partial x\, \partial y} = 0.$$

7 Show that $f(x, y) = x^3 - 3xy^2$ obeys the partial differential equation

$$\frac{\partial^2 f}{\partial x^2} + \frac{\partial^2 f}{\partial y^2} = 0.$$

8 For what values of C does the function $f(x, y) = y^3 - Cx^2y$ satisfy the following differential equations?

(a) $\dfrac{\partial^2 f}{\partial x^2} + \dfrac{\partial^2 f}{\partial y^2} = 0,$ (b) $\dfrac{\partial^2 f}{\partial x^2} - \dfrac{\partial^2 f}{\partial y^2} = 0,$

(c) $\dfrac{\partial^2 f}{\partial x\, \partial y} = 0.$

9 If $f(x, y, z) = A(x, y) + B(y, z) + C(x, z)$, show that

$$\frac{\partial^3 f}{\partial x\, \partial y\, \partial z} = 0.$$

3.3 Directional Derivatives and Gradients

A function has as many first order partial derivatives as it has distinct variables. It also has as many directional derivatives as there are distinct directions. None of these can be called the principal derivative of f. It is natural to wonder if there is something which serves as *the* derivative of a function of several variables, as there was with functions of one variable. In this section, we shall find an answer in the vector-valued function called the gradient of f, from which the directional derivatives can be obtained. We first make the latter precise by a formal definition, the rationale for which can be seen from Figure 3–10. If u is a unit vector, the point $p_0 + hu$ is the point obtained by moving from the point p_0 in the direction u a distance

$$|hu| = |h|(1) = |h|.$$

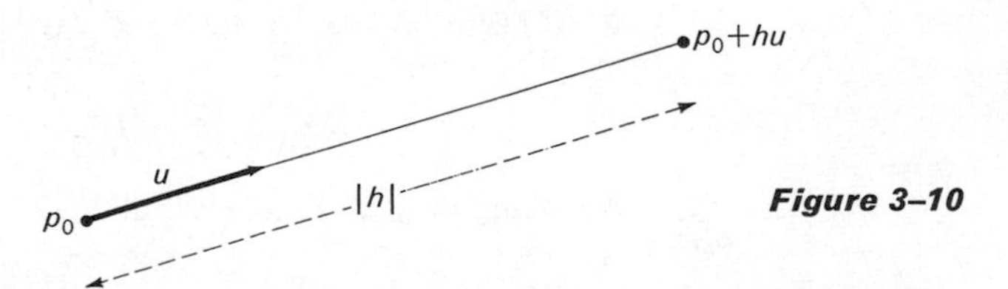

Figure 3–10

Definition 2 Let f be defined on a neighborhood of the point p_0, and choose any unit vector u. Then, the value of the directional derivative of f at p_0 in the direction u is the number $\mathbf{D}_u f(p_0)$ defined by

$$\lim_{h\to 0} \frac{f(p_0 + hu) - f(p_0)}{h} \tag{3.7}$$

when this limit exists.

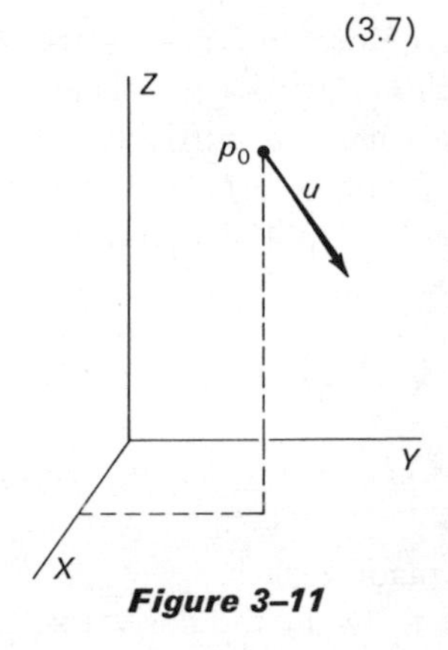

Figure 3–11

To illustrate the meaning of this definition, let us take the function $f(x, y, z) = 2x^2y + yz$. Let $p_0 = (1, 1, 2)$ and let u be the unit vector $(\frac{1}{3}, \frac{2}{3}, -\frac{2}{3})$, which points forward and downward from p_0. (See Figure 3–11.) Now, $f(p_0) = f(1, 1, 2) = 4$, and

$$\begin{aligned} p_0 + hu &= (1, 1, 2) + (\tfrac{1}{3}h, \tfrac{2}{3}h, -\tfrac{2}{3}h) \\ &= (1 + \tfrac{1}{3}h, 1 + \tfrac{2}{3}h, 2 - \tfrac{2}{3}h), \end{aligned}$$

so that

$$\begin{aligned} f(p_0 + hu) &= 2(1 + \tfrac{1}{3}h)^2(1 + \tfrac{2}{3}h) + (1 + \tfrac{2}{3}h)(2 - \tfrac{2}{3}h) \\ &= 4 + \tfrac{10}{3}h + \tfrac{2}{3}h^2 + \tfrac{4}{27}h^3. \end{aligned}$$

Accordingly,

$$\frac{f(p_0 + hu) - f(p_0)}{h} = \frac{10}{3} + \frac{2}{3}h + \frac{4}{27}h^2,$$

and

$$\mathbf{D}_u f(p_0) = \lim_{h\to 0} \tfrac{10}{3} + \tfrac{2}{3}h + \tfrac{4}{27}h^2 = \tfrac{10}{3}.$$

This is a tedious computation at best. One reason for the definition that follows and the accompanying general theorem is to make such computations much simpler. We state the theorem only for functions of three variables, but shall illustrate it for another case as well, since it is a general property of functions of n variables for any n.

Definition 3 Let f be a function which is defined and has continuous partial derivatives at all points (x, y, z) in a region D. Then, the gradient of f is the continuous vector-valued function ∇f that is defined on D by

$$\nabla f = (f_x, f_y, f_z) = \left(\frac{\partial f}{\partial x}, \frac{\partial f}{\partial y}, \frac{\partial f}{\partial z}\right). \tag{3.8}$$

Other names and notations are sometimes used for this function; in particular, it may be called the differential of f, and be denoted by df. There is also a growing tendency to call this vector-valued function simply the derivative of f.

Theorem 1 *The directional derivative of f at p_0 in the direction of the unit vector u can be calculated from the gradient of f at p_0 by the formula*

$$\mathbf{D}_u f(p_0) = u \cdot \nabla f(p_0). \tag{3.9}$$

We illustrate (3.9) by several examples before looking at the proof of the theorem. Suppose $f(x, y, z) = 2x^2y + yz$. Then, $\nabla f = (4xy, 2x^2 + z, y)$ for each (x, y, z) in space. In particular, if $p_0 = (1, 1, 2)$, then the gradient of f at p_0 will be $\nabla f(1, 1, 2) = (4, 4, 1)$. If we take the direction $u = (\frac{1}{3}, \frac{2}{3}, -\frac{2}{3})$, then the directional derivative of f at p_0 in the direction u will be

$$\begin{aligned}\mathbf{D}_u f(p_0) &= (\tfrac{1}{3}, \tfrac{2}{3}, -\tfrac{2}{3}) \cdot \nabla f(1, 1, 2)\\ &= (\tfrac{1}{3}, \tfrac{2}{3}, -\tfrac{2}{3}) \cdot (4, 4, 1)\\ &= \tfrac{4}{3} + \tfrac{8}{3} - \tfrac{2}{3} = \tfrac{10}{3},\end{aligned}$$

in agreement with the direct calculation we made earlier.

Again, with the same function f, suppose we want the derivative of f at $(3, -1, -2)$, in the direction of the vector $v = (2, 3, 6)$. We first note that v is not a unit vector, but that $|v| = \sqrt{4 + 9 + 36} = 7$. Accordingly, the unit vector in the direction of v is $v/|v| = (\frac{1}{7})v = (\frac{2}{7}, \frac{3}{7}, \frac{6}{7}) = u$. Using the fact that $\nabla f = (4xy, 2x^2 + z, y)$, we calculate the value of the gradient of f at $(3, -1, -2)$ to be $(-12, 16, -1)$. Finally, the directional derivative of f at $(3, -1, -2)$ in the direction u is

$$(\tfrac{2}{7}, \tfrac{3}{7}, \tfrac{6}{7}) \cdot (-12, 16, -1) = -\tfrac{24}{7} + \tfrac{48}{7} - \tfrac{6}{7} = \tfrac{18}{7}.$$

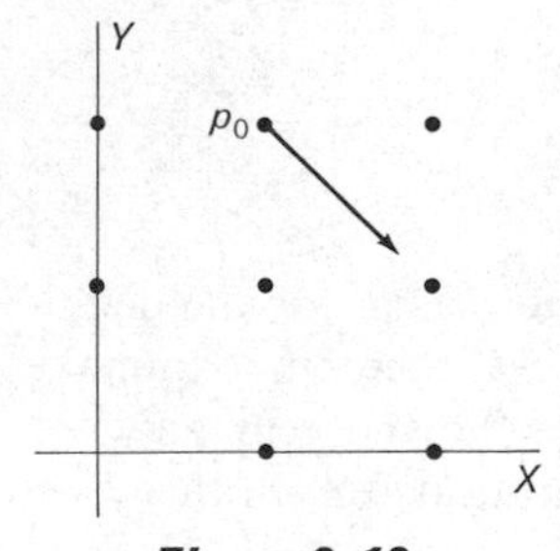

Figure 3–12

How do you apply such a procedure when dealing with a function of two variables? Suppose $F(x, y) = x^3y^2 - 2xy^3$, and that we want to find the derivative of F at the point $(1, 2)$ in the direction indicated in Figure 3–12, which is from $(1, 2)$ toward the point $(2, 1)$. This direction is that of the vector $v = (2, 1) - (1, 2) = (1, -1)$, and since $|v| = \sqrt{2}$, the corresponding unit vector is

$$u = \left(\frac{1}{\sqrt{2}}, -\frac{1}{\sqrt{2}}\right).$$

The gradient of F is $(F_x, F_y) = (3x^2y^2 - 2y^3, 2x^3y - 6xy^2)$, which at the point $(1, 2)$ has the value $(-4, -20)$. Thus, we calculate the value of $\mathbf{D}_u F$ to be $(1/\sqrt{2}, -1/\sqrt{2}) \cdot (-4, -20) = 16/\sqrt{2}$.

What is the meaning of the gradient itself? Recall a useful fact about vectors. Given any vector V and any unit vector u, we saw in formula (1.9) that the number $u \cdot V$ is the projection of V in the direction u. (In Figure 3–13 we have shown the projections of a vector V in three different directions.) It is clear that the longest projection of V is in the direction of V itself, and its value is $u \cdot V = |V|$, the magnitude of V.

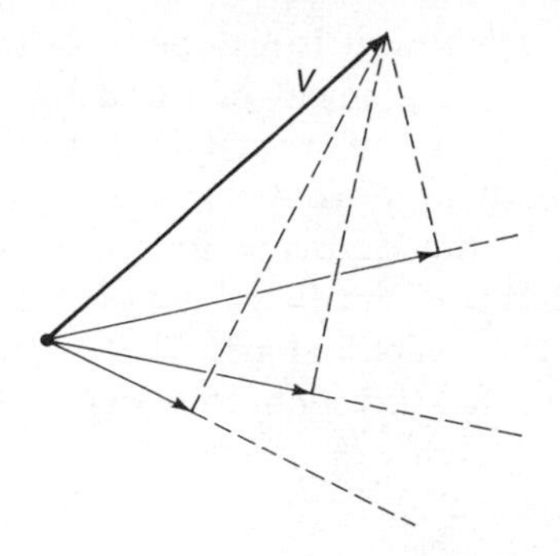

Figure 3–13

If we apply these remarks to Theorem 1, we see that the directional derivative or rate of change of a function f in the direction u is just the projection of ∇f, the gradient of f, in that direction. Moreover, we also see that the value obtained is greatest when the direction u is chosen to be the same as that of ∇f. This discussion has led us to the following important conclusion which we formulate as a theorem:

Theorem 2 *The gradient ∇f of a function f is a vector-valued function whose value at any point p_0 is a vector that points in the direction in which the function is increasing most rapidly, and whose length is the rate of change in that direction. Moreover, the rate of change of f in any other direction at p_0 is simply the projection of the gradient in that direction:*

$$\mathbf{D}_u f(p_0) = u \cdot \nabla f(p_0).$$

In Figure 3–14, we have shown part of the field of gradient vectors for the function $f(x, y) = \frac{1}{4}x^2y$. At each point p_0, we have drawn the vector $\nabla f(p_0)$, thus indicating the direction in which the value of f is increasing most rapidly.

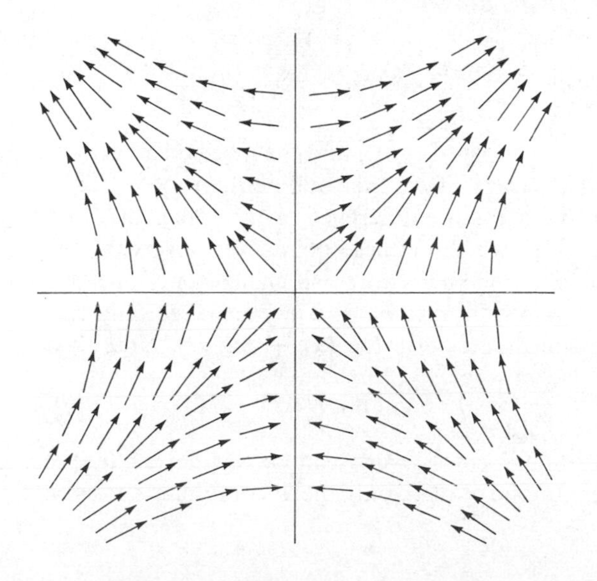

Gradient field for $f(x, y) = \frac{1}{4}x^2y$

Figure 3–14

For functions of two variables, it is also possible to find a geometric connection between the gradient of f and the shape of the surface that is the graph of f. If you are standing at the point $P_0 = (p_0, f(p_0))$ on the surface lying directly above the point p_0 in the XY-plane, and you look at the landscape immediately about you, then the vector $\nabla f(p_0)$ points in the direction you must move from P if you want to climb the steepest slope. The magnitude of this vector is the slope at P_0 in that direction. (See Figure 3–15.)

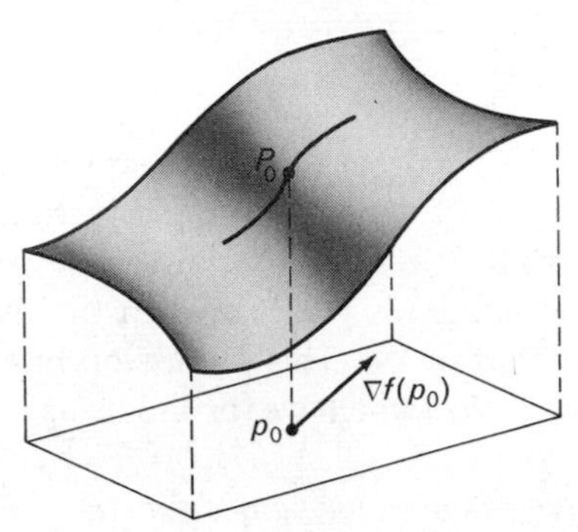

Figure 3–15

Having now seen some of the consequences of Theorem 1, we take up its proof. The proof depends on a special result called the **Approximation Lemma**, which is a several-variable analogue of the Mean Value Theorem. We state and prove it for functions of two variables.

Lemma *Let f be defined and have first order partial derivatives on a circular neighborhood of the point $p_0 = (x_0, y_0)$. Let p be another point in this neighborhood, and write $p = p_0 + \Delta p$, where $\Delta p = (\Delta x, \Delta y)$. Then, there are two points q_1 and q_2 with $|p_0 - q_i| \leq |\Delta p|$, $i = 1, 2$, such that*

$$f(p_0 + \Delta p) - f(p_0) = \Delta x \frac{\partial f}{\partial x}(q_1) + \Delta y \frac{\partial f}{\partial y}(q_2). \tag{3.10}$$

Proof The meaning of (3.10) and the mechanism of the proof may be somewhat clearer if we first outline the proof. Observe that the left side of (3.10) is the difference of the values of f at two points. In Figure 3–16, we show these points and an extra point q; note that if we move from p_0 to q we change only the X-coordinate, and if we move from q to $p_0 + \Delta p$, we change only the Y-coordinate. The coordinates of q are $(x_0 + \Delta x, y_0)$. Note also that

$$f(p_0 + \Delta p) - f(p_0) = \{f(p_0 + \Delta p) - f(q)\} + \{f(q) - f(p_0)\}. \tag{3.11}$$

Thus, the difference we want can be written as the sum of two differences: In one of these only the Y-coordinate varies; in the other only the X-coordinate changes. We shall apply the usual one-variable Mean Value Theorem to each of these separately.

Recall the ordinary one-variable Mean Value Theorem: If F is continuous on $[a, b]$ and $F'(t)$ exists for all t, $a < t < b$, then there is a number t_1 in $[a, b]$ such that

$$F(b) - F(a) = (b - a)F'(t_1).$$

We shall use this twice in proving the Approximation Lemma. We start by rewriting (3.11):

$$\begin{aligned} f(x_0 + \Delta x, y_0 + \Delta y) - f(x_0, y_0) \\ = \{f(x_0 + \Delta x, y_0 + \Delta y) - f(x_0 + \Delta x, y_0)\} \\ + \{f(x_0 + \Delta x, y_0) - f(x_0, y_0)\}. \end{aligned} \tag{3.12}$$

Examine the second of these two expressions, and note that it has the form $F(b) - F(a)$, where $F(t) = f(t, y_0)$, $a = x_0$, and $b = x_0 + \Delta x$. Also, note that $F'(t) = \partial f/\partial x\,(t, y_0)$, since in calculating the derivative of $f(t, y_0)$ with respect to t, we hold y_0 constant and differentiate with respect to the first variable. (This is an instance in which to indicate the partial derivative of f the notation f_1 would be better than $\partial f/\partial x$ or f_x.) Applying the Mean Value Theorem, we have

$$\begin{aligned} f(x_0 + \Delta x, y_0) - f(x_0, y_0) &= F(b) - F(a) \\ &= (b - a)F'(t_1) \\ &= (\Delta x)\frac{\partial f}{\partial x}(t_1, y_0) \\ &= \Delta x \frac{\partial f}{\partial x}(q_1), \end{aligned} \tag{3.13}$$

where q_1 is the point (t_1, y_0). Since t_1 lies between x_0 and $x_0 + \Delta x$, the point q_1 must lie somewhere on the lower edge of the triangle shown in Figure 3–16. Also, since the hypotenuse of this triangle has length Δp, $|p_0 - q_1| \leq |\Delta p|$.

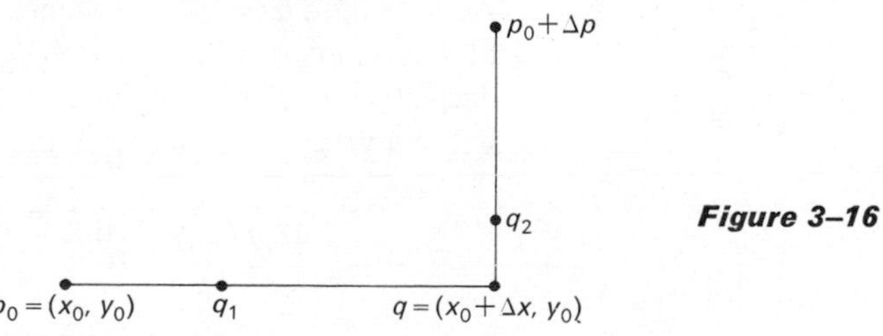

Figure 3–16

Return to (3.12), and observe that the first expression in brackets is also of the form $F(b) - F(a)$, where this time $F(t) = f(x_0 + \Delta x, t)$,

$a = y_0$, $b = y_0 + \Delta y$. Also, $F'(t) = \partial f/\partial y\,(x_0 + \Delta x, t)$, since we are differentiating with respect to the second variable, holding the first, $x_0 + \Delta x$, constant. Applying the Mean Value Theorem again, we have

$$\begin{aligned} f(x_0 + \Delta x, y_0 + \Delta y) - f(x_0 + \Delta x, y_0) &= F(b) - F(a) \\ &= (b - a)F'(t_2) \\ &= (\Delta y)\frac{\partial f}{\partial y}(x_0 + \Delta x, t_2) \qquad (3.14) \\ &= \Delta y \frac{\partial f}{\partial y}(q_2), \end{aligned}$$

where $q_2 = (x_0 + \Delta x, t_2)$. Since t_2 lies between y_0 and $y_0 + \Delta y$, the point q_2 lies on the vertical edge of the triangle, as shown in Figure 3–16, and again $|p_0 - q_2| \le |\Delta p|$.

Combining these two formulas (3.13) and (3.14) with (3.12), we have proved (3.10).

Proof (of Theorem 1) With the help of the lemma, we can now prove Theorem 1. Given p_0 and a unit vector $u = (\alpha, \beta)$, we want to compute the derivative of f at p_0 in the direction u. This requires that we calculate the limit as $h \to 0$ of the expression

$$\frac{f(p_0 + hu) - f(p_0)}{h}.$$

We use the lemma to calculate the top of this fraction, letting

$$\Delta p = hu = (h\alpha, h\beta) = (\Delta x, \Delta y).$$

Then

$$\begin{aligned} f(p_0 + hu) - f(p_0) &= (h\alpha)\frac{\partial f}{\partial x}(q_1) + (h\beta)\frac{\partial f}{\partial y}(q_2) \\ &= h\left\{\alpha\frac{\partial f}{\partial x}(q_1) + \beta\frac{\partial f}{\partial y}(q_2)\right\}. \end{aligned}$$

As $h \to 0$, $|\Delta p| = h$ approaches 0, and since $|p_0 - q_i| \le |\Delta p|$, each of the points q_i will approach p_0. By hypothesis, each of the partial derivatives of f are continuous at p_0, so that $\partial f/\partial x\,(q_1)$ and $\partial f/\partial y\,(q_1)$ will converge to $\partial f/\partial x\,(p_0)$ and $\partial f/\partial y\,(p_0)$ respectively. Thus,

$$\begin{aligned} \mathbf{D}_u f(p_0) &= \lim_{h\to 0}\frac{f(p_0 + hu) - f(p_0)}{h} \\ &= \lim_{|\Delta p|\to 0}\left\{\alpha\frac{\partial f}{\partial x}(q_1) + \beta\frac{\partial f}{\partial y}(q_2)\right\} \\ &= \alpha\frac{\partial f}{\partial x}(p_0) + \beta\frac{\partial f}{\partial y}(p_0). \end{aligned}$$

Since $u = (\alpha, \beta)$ and $\nabla f = (\partial f/\partial x, \partial f/\partial y)$, we have proved (for functions of two variables) that $\mathbf{D}_u f = u \cdot \nabla f$.

To prove this theorem for functions of three or more variables requires that one start from a general form of the Approximation Lemma. The picture to illustrate the three-variable form showing where the needed points q_1, q_2, q_3 will lie is given in Figure 3–17.

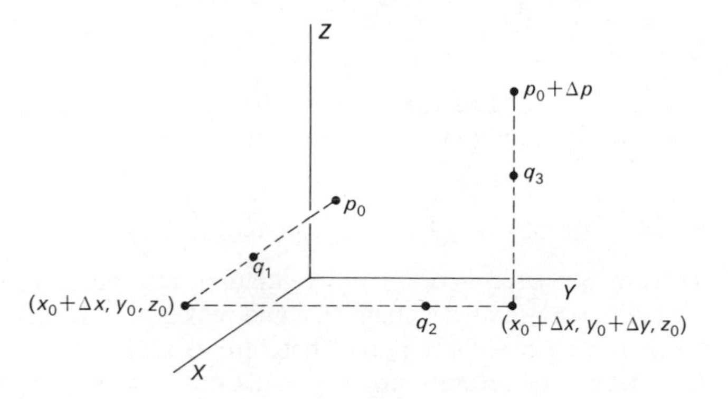

Figure 3–17

Exercises

1 Calculate the directional derivative $\mathbf{D}_u f$ of $f(x, y) = 2xy^2 - 3x^2 + 5y$ at the point $p_0 = (1, 2)$ in the direction of the following unit vectors:

(a) $u = \left(\dfrac{1}{\sqrt{2}}, -\dfrac{1}{\sqrt{2}}\right)$, (b) $u = (\frac{3}{5}, \frac{4}{5})$,

(c) $u = (\frac{5}{13}, -\frac{12}{13})$.

2 For $f(x, y) = 2xy^2 - 3x^2y$, calculate ∇f at the points $(1, 2)$, $(-1, 1)$, and $(2, -1)$, and draw the vectors.

3 Calculate the directional derivative $\mathbf{D}_u f$ for $f(x, y, z) = x^2y + 2yz - 3xz^2$ at the point $(1, 2, -1)$ in the direction of the vectors:

(a) $u = \left(\dfrac{1}{3}, \dfrac{-2}{3}, \dfrac{2}{3}\right)$, (b) $u = (-1, 4, 8)$,

(c) $u = (4, -4, -7)$.

4 For $f(x, y, z) = 2x^2y - yz^2 + xyz$, calculate ∇f and evaluate it at $(1, 1, 0)$, $(1, 2, -1)$, and $(-1, 2, 1)$.

5 For $f(x, y, z, w) = 2xy + zw - xyzw$, calculate ∇f and evaluate it at $(1, 1, -1, 2)$, $(1, 2, -1, 1)$, and $(2, 0, -1, 2)$.

6 Sketch a gradient field using at least 5 different points for the function given by $f(x, y) = 2x - y$.

7 Sketch a gradient field using at least 10 points for the function given by $f(x, y) = xy$.

8 Use the Approximation Lemma to prove that if f is continuous and has continuous partial derivatives in a disc D and $f_x = f_y = 0$ at all points of D, then f is *constant* in D.

9 (a) State the Approximation Lemma for 3 variables.
(b) Prove it (see Figure 3–17).

3.4

The Chain Rule

For functions of one variable, the rule for differentiating the composite of two functions takes on a simple form:

$$\frac{dw}{dx} = \frac{dw}{du}\frac{du}{dx}.$$

The notation used here for differentiation, introduced by Leibnitz, simplifies the use of the chain rule and makes it easier to remember by suggesting cancellation and therefore its validity. When we turn to functions of several variables and formulate the corresponding chain rules for differentiating functions of the form

$$H(x, y) = F(f(x, y), g(x, y)),$$

the results are complicated and seem much less reasonable. For example, if we were to use the numerical subscript notation for partial derivatives (according to which, for example, H_1 is the partial derivative of the function H with respect to the first variable), we would write

$$H_1(x, y) = F_1(f(x, y), g(x, y))f_1(x, y) + F_2(f(x, y), g(x, y))g_1(x, y).$$

It helps immensely to use a Leibnitz-like notation for partial derivatives, just as for functions of one variable, but this will not eliminate all of the complexity inherent in the situation. In particular, different combinations of functions and numbers of variables lead to apparently different forms of the chain rule. By considering a small number of typical cases, we hope that the general pattern will become clear.

We start by introducing a special term for an important class of functions.

Definition 4 A function f is said to be of class C^1 (or to be continuously differentiable) on an open set D if f is continuous and has continuous first partial derivatives everywhere in D.

There are also classes C^n, consisting of functions all of whose partial derivatives up to order n are continuous. Most of the functions that you will meet in this book are at least of class C^1 on their domains of definition, and many are of class C^∞. The importance of the class C^1 is that all the functions in this class satisfy the requirements for the Approximation Lemma and thus have a continuous gradient ∇f and many other useful properties.

We take up first a simple case of the chain rule. Let F be a function of one variable and f a function of three variables, and define a function H of three variables by

$$H(x, y, z) = F(f(x, y, z)). \tag{3.15}$$

An example is

$$H(x, y, z) = \sin(x^2y + yz).$$

Suppose we ask for the partial derivatives of H. You should have no difficulty finding these if you merely do what seems natural. To find $\partial H/\partial x$, we want to differentiate with respect to x holding y and z constant. If we do this, using the standard chain rule for functions of one variable (since we are now treating H as a function of x alone), we have

$$\begin{aligned}\frac{\partial H}{\partial x} &= \cos(x^2y + yz)\frac{\partial}{\partial x}(x^2y + yz)\\ &= (2xy)\cos(x^2y + yz).\end{aligned}$$

In the same way,

$$\frac{\partial H}{\partial y} = (x^2 + z)\cos(x^2y + yz).$$

In order to formulate this chain rule conveniently in general terms, we adopt a different notation. Another way to describe the functional relationship in (3.15) is to introduce some auxiliary variables, writing

$$\begin{aligned}w &= F(u),\\ u &= f(x, y, z),\\ w &= H(x, y, z).\end{aligned} \tag{3.16}$$

It is clear that (3.15) is obtained at once if you replace u in the first equation with its value from the second. Instead of asking for the partial derivatives of H, we can ask for $\partial w/\partial x$, $\partial w/\partial y$, $\partial w/\partial z$, and the same process applied in the example above leads to the general solution

$$\begin{aligned}\frac{\partial H}{\partial x} &= \frac{\partial w}{\partial x} = \frac{dw}{du}\frac{\partial u}{\partial x},\\ \frac{\partial H}{\partial y} &= \frac{\partial w}{\partial y} = \frac{dw}{du}\frac{\partial u}{\partial y},\\ \frac{\partial H}{\partial z} &= \frac{\partial w}{\partial z} = \frac{dw}{du}\frac{\partial u}{\partial z}.\end{aligned} \tag{3.17}$$

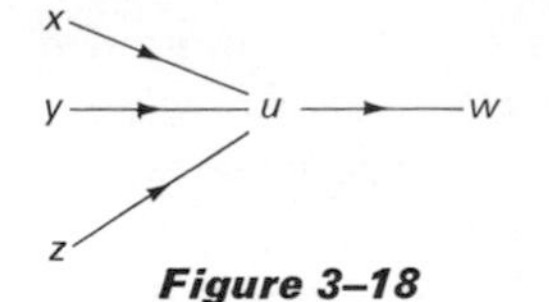

Figure 3–18

More complicated relationships between variables can often be presented more clearly by means of special diagrams. In this simple case, the diagram in Figure 3–18 shows the dependencies expressed in (3.16).

The next case is quite different, and its solution requires a new result. Consider the equations

$$w = f(x, y, z)$$
$$x = X(t),\ y = Y(t),\ z = Z(t). \tag{3.18}$$

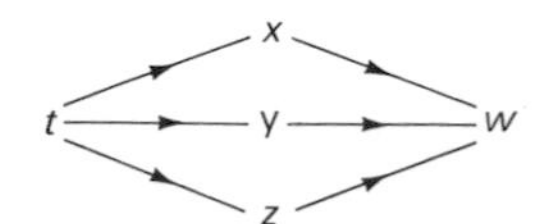

Figure 3–19

The corresponding diagram is in Figure 3–19. The effect of (3.18) is to make w a function of t alone; w is $G(t)$, where

$$G(t) = f(X(t), Y(t), Z(t)).$$

We want to find a way to express $dw/dt = G'(t)$ in terms of the partial derivatives of f, namely $\partial w/\partial x$, $\partial w/\partial y$, $\partial w/\partial z$, and the ordinary derivatives of the functions X, Y, and Z.

Theorem 3 *If* (3.18) *holds, with the functions* f, X, Y, Z *of class* C^1, *then*

$$\frac{dw}{dt} = \frac{\partial w}{\partial x}\frac{dx}{dt} + \frac{\partial w}{\partial y}\frac{dy}{dt} + \frac{\partial w}{\partial z}\frac{dz}{dt}\,. \tag{3.19}$$

For convenience, we restate Theorem 3 for the case where w is a function of x and y alone, and x and y are functions of t. Then the corresponding chain rule is Theorem 3′.

Theorem 3′ *If* $w = f(x, y)$, $x = X(t)$, $y = Y(t)$, *all of class* C^1, *then*

$$\frac{dw}{dt} = \frac{\partial w}{\partial x}\frac{dx}{dt} + \frac{\partial w}{\partial y}\frac{dy}{dt}\,. \tag{3.20}$$

It is possible to use the diagram in Figure 3–19 to make the formula (3.19) look plausible. We want to calculate dw/dt, which is the rate of change in w due to changes in t. Since w is a function of x, y, and z, and each of these in turn is a function of t, we must take account of three different ways in which w depends upon t. Each of these ways corresponds to a path joining t and w in the diagram.

Suppose it were possible to look at one of these paths and to ignore the other two. In effect, this would be assuming that x alone depends on t, while y and z are constant. If this were true, we would have $w = f(X(t), y_0, z_0)$. We could then find dw/dt at once by the elementary chain rule, obtaining $dw/dt = f_x(X(t), y_0, z_0)X'(t)$. Note that this can also be written $(\partial w/\partial x)(dx/dt)$. Again, if it should happen that x and z were constants, and that y alone depended on t, then the same reasoning would yield for dw/dt the result $(\partial w/\partial y)(dy/dt)$.

In fact, we cannot expect that any of the functions X, Y, or Z is constant. A change in t is likely to change each of the values of the variables x, y and z, and the reasoning above would not apply. How then can we take into account the interaction of separate changes in each of these variables in order to compute the rate of change in w? The surprising fact, expressed in formula (3.19), is that the correct answer is merely the sum of these fragmentary special terms $(\partial w/\partial x)(dx/dt)$, $(\partial w/\partial y)(dy/dt)$, and $(\partial w/\partial z)(dz/dt)$, and that no more complicated expression is needed.

In effect, each path in the functional relationship diagram illustrates one way in which w depends upon t, and each path thus gives rise to a fragmentary rate of change which could be interpreted as the portion of the total rate of change that corresponds to this path, assuming that no other dependency holds. Theorem 3 asserts that the total rate of change is simply the sum of all these fragmentary rates of change, one corresponding to each path between t and w. A formal proof of Theorem 3′, based upon the Approximation Lemma that was proved in Section 3.3, will be given later in the present section.

To illustrate the use of these results, suppose $w = x^2y^3$, where $x = \sin t$ and $y = \cos t$. Then, according to (3.20),

$$\begin{aligned}\frac{dw}{dt} &= \frac{\partial w}{\partial x}\frac{dx}{dt} + \frac{\partial w}{\partial y}\frac{dy}{dt}\\ &= (2xy^3)(\cos t) + (3x^2y^2)(-\sin t).\end{aligned}$$

(This can be checked by replacing x and y by their values in terms of t to obtain $w = (\sin t)^2(\cos t)^3$, and then calculating dw/dt directly.)

The transition to the most general situation should now be clear, and it should be possible to arrive at the correct form of the chain rule that is appropriate to any special case. For example, returning to the original problem that was given at the start of this section, consider the relations

$$\begin{aligned} w &= F(u, v),\\ u &= f(x, y),\\ v &= g(x, y). \end{aligned} \tag{3.21}$$

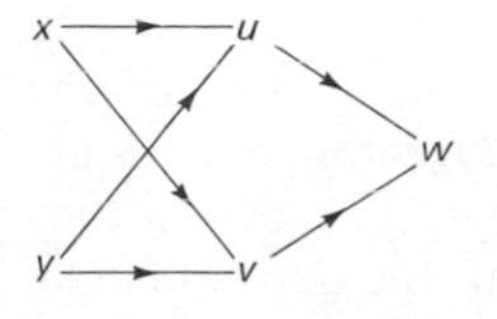

Figure 3–20

The corresponding dependency diagram is given in Figure 3–20.

Clearly, as we see either from this diagram or from the equations (3.21), w can be regarded as a function of x and y, and we can therefore seek a way to find the partial derivatives $\partial w/\partial x$ and $\partial w/\partial y$. To find the former, we look at the paths that join x and w. There are two of these, and one corresponds to the fragmentary rate of change $(\partial w/\partial u)(\partial u/\partial x)$ and the other to $(\partial w/\partial v)(\partial v/\partial x)$. The sum of these will be $\partial w/\partial x$. Similar reasoning leads us to expect that the correct form of the chain rule in this example would be

Theorem 4 *If $w = F(u, v)$, where $u = f(x, y)$ and $v = g(x, y)$, and if f, g, F are continuously differentiable, then*

$$\frac{\partial w}{\partial x} = \frac{\partial w}{\partial u}\frac{\partial u}{\partial x} + \frac{\partial w}{\partial v}\frac{\partial v}{\partial x}, \tag{3.22}$$

$$\frac{\partial w}{\partial y} = \frac{\partial w}{\partial u}\frac{\partial u}{\partial y} + \frac{\partial w}{\partial v}\frac{\partial v}{\partial y}. \tag{3.23}$$

As an illustration, if $w = 3u^2v$, where $u = 2x - 3y$ and $v = xy$, then

$$\frac{\partial w}{\partial x} = (6uv)(2) + (3u^2)(y) = 12uv + 3u^2y,$$

$$\frac{\partial w}{\partial y} = (6uv)(-3) + (3u^2)(x) = -18uv + 3u^2x.$$

A word of explanation may clear up a frequently asked question, namely, is there a difference between the notations $\partial w/\partial x$ and dw/dx? The best answer we can give is that these notations are convenient signals: $\partial w/\partial x$ reminds us that there are other variables that help determine the values of w and we are interested in the rate of change of w as x alone is altered; dw/dx suggests that x alone is the determining or independent variable. There are cases where a single variable, say "time," enters into a function both by itself and also through other variables that in turn depend upon time. In such complicated cases, both $\partial w/\partial t$ and dw/dt may occur. You may meet such cases in connection with subjects such as economics, physics, or psychology.

Once a formula has been obtained for a derivative or a partial derivative of a variable which is a complicated function of other variables, it is possible to repeat the process and find formulas for higher derivatives. This can become quite complicated, since the number of terms that are involved can be large. We will carry out one such example to illustrate the method.

Suppose that $w = T(x, y, z)$ gives the temperature w at each point (x, y, z) in a region of space. If $x = f(t)$, $y = g(t)$, and $z = h(t)$, then these equations describe a point which is moving in a certain curved path, and $w = T(f(t), g(t), h(t))$ is the temperature of this moving point at each moment of time. The dependency diagram is the same as that in Figure 3–19, and the corresponding formula for the chain rule gives

$$\frac{dw}{dt} = \frac{\partial w}{\partial x}\frac{dx}{dt} + \frac{\partial w}{\partial y}\frac{dy}{dt} + \frac{\partial w}{\partial z}\frac{dz}{dt}, \tag{3.19}$$

which describes the rate of change of the temperature of the moving point. Can we now differentiate again to find a formula for d^2w/dt^2? Let us first consider a special case.

Suppose that the moving point is described by

$$x = \cos t,$$
$$y = \sin t,$$
$$z = 2t.$$

Then (3.19) becomes

$$\frac{dw}{dt} = -\sin t\frac{\partial w}{\partial x} + \cos t\frac{\partial w}{\partial y} + 2\frac{\partial w}{\partial z}\cdot$$

We differentiate this with respect to t, recalling that each of the terms $\partial w/\partial x$, $\partial w/\partial y$, $\partial w/\partial z$ are functions of t, obtaining

$$\begin{aligned}\frac{d^2w}{dt^2} &= (-\cos t)\frac{\partial w}{\partial x} - (\sin t)\frac{d}{dt}\left\{\frac{\partial w}{\partial x}\right\} + (-\sin t)\frac{\partial w}{\partial y}\\ &+ (\cos t)\frac{d}{dt}\left\{\frac{\partial w}{\partial y}\right\} + 2\frac{d}{dt}\left\{\frac{\partial w}{\partial z}\right\}\cdot\end{aligned} \tag{3.24}$$

Let us look separately at

$$\frac{d}{dt}\left\{\frac{\partial w}{\partial x}\right\}.$$

Since $w = f(x, y, z)$, $\partial w/\partial x$ is also a function of x, y and z, namely $f_x(x, y, z)$ or $f_1(x, y, z)$. We can therefore apply the rule in Theorem 3 to carry out the desired differentiation, and we obtain

$$\begin{aligned}\frac{d}{dt}\left\{\frac{\partial w}{\partial x}\right\} &= \frac{\partial}{\partial x}\left(\frac{\partial w}{\partial x}\right)\frac{dx}{dt} + \frac{\partial}{\partial y}\left(\frac{\partial w}{\partial x}\right)\frac{dy}{dt} + \frac{\partial}{\partial z}\left(\frac{\partial w}{\partial x}\right)\frac{dz}{dt}\\ &= (-\sin t)\frac{\partial^2 w}{\partial x^2} + (\cos t)\frac{\partial^2 w}{\partial y\,\partial x} + 2\frac{\partial^2 w}{\partial z\,\partial x}\cdot\end{aligned}$$

In the same way, we obtain

$$\frac{d}{dt}\left\{\frac{\partial w}{\partial y}\right\} = (-\sin t)\frac{\partial^2 w}{\partial x\,\partial y} + (\cos t)\frac{\partial^2 w}{\partial y^2} + 2\frac{\partial^2 w}{\partial z\,\partial y},$$

$$\frac{d}{dt}\left\{\frac{\partial w}{\partial z}\right\} = (-\sin t)\frac{\partial^2 w}{\partial x\,\partial z} + (\cos t)\frac{\partial^2 w}{\partial y\,\partial z} + 2\frac{\partial^2 w}{\partial z^2}\cdot$$

If we substitute the above into (3.24) and then use the equality of mixed double derivatives such as $\partial^2 w/\partial x\,\partial y$ and $\partial^2 w/\partial y\,\partial x$, we arrive at the final answer:

$$\frac{d^2w}{dt^2} = -\cos t \frac{\partial w}{\partial x} - \sin t \frac{\partial w}{\partial y}$$

$$+ (\sin t)^2 \frac{\partial^2 w}{\partial x^2} + (\cos t)^2 \frac{\partial^2 w}{\partial y^2}$$

$$+ 4 \frac{\partial^2 w}{\partial z^2} + (-2 \sin t \cos t) \frac{\partial^2 w}{\partial x \, \partial y}$$

$$+ (4 \cos t) \frac{\partial^2 w}{\partial y \, \partial z} + (-4 \sin t) \frac{\partial^2 w}{\partial z \, \partial x} \cdot$$

If the same sequence of steps is carried out for the general functions $x = X(t)$, $y = Y(t)$, $z = Z(t)$, the result is

$$\frac{d^2w}{dt^2} = \frac{\partial^2 w}{\partial x^2}\left(\frac{dx}{dt}\right)^2 + \frac{\partial^2 w}{\partial y^2}\left(\frac{dy}{dt}\right)^2 + \frac{\partial^2 w}{\partial z^2}\left(\frac{dz}{dt}\right)^2$$

$$+ 2 \frac{\partial^2 w}{\partial x \, \partial y} \frac{dx}{dt} \frac{dy}{dt} + 2 \frac{\partial^2 w}{\partial y \, \partial z} \frac{dy}{dt} \frac{dz}{dt} + 2 \frac{\partial^2 w}{\partial z \, \partial x} \frac{dz}{dt} \frac{dx}{dt} \qquad (3.25)$$

$$+ \frac{\partial w}{\partial x} \frac{d^2x}{dt^2} + \frac{\partial w}{\partial y} \frac{d^2y}{dt^2} + \frac{\partial w}{\partial z} \frac{d^2z}{dt^2} \cdot$$

In most cases it is usually easier to carry out the process completed above than to remember and apply this complex formula.

Proof (of (3.20)) We finally return to the chain rules themselves and give a proof of (3.20) in Theorem 3′, which is typical of the proof of (3.19) or any of the other variations of the chain rule. Restating the problem, we have $w = f(x, y)$ with $x = X(t)$, $y = Y(t)$, and we want to express dw/dt in terms of the partial derivatives of f and the ordinary derivatives of X and Y. According to (3.20), we seek to prove the formula

$$\frac{dw}{dt} = \frac{\partial w}{\partial x} \frac{dx}{dt} + \frac{\partial w}{\partial y} \frac{dy}{dt} \cdot$$

We have $w = f(X(t), Y(t)) = F(t)$, and we assume that f, X, Y are functions of class C^1.

We start with the basic definition

$$\frac{dw}{dt} = \lim_{h \to 0} \frac{F(t + h) - F(t)}{h} \cdot$$

Set $\Delta x = X(t+h) - X(t)$ and $\Delta y = Y(t+h) - Y(t)$, with $(x, y) = (X(t), Y(t)) = p$ and $\Delta p = (\Delta x, \Delta y)$. Then

$$\begin{aligned} F(t+h) - F(t) &= f(x + \Delta x, y + \Delta y) - f(x, y) \\ &= f(p + \Delta p) - f(p). \end{aligned}$$

To this, we apply the Approximation Lemma from the last section, obtaining

$$F(t+h) - F(t) = \Delta x \frac{\partial f}{\partial x}(q_1) + \Delta y \frac{\partial f}{\partial y}(q_2),$$

where q_1 and q_2 are points that satisfy $|p - q_i| \leq |\Delta p|$. We next note that

$$\frac{\Delta x}{h} = \frac{X(t+h) - X(t)}{h},$$

so that

$$\lim_{h \to 0} \frac{\Delta x}{h} = X'(t) = \frac{dx}{dt}.$$

Similarly, $\lim_{h \to 0} \Delta y/h = Y'(t) = dy/dt$. Also, since X and Y are continuous functions, $\Delta p = (\Delta x, \Delta y)$ will approach $(0, 0)$ as $h \to 0$, so that q_1 and q_2 must approach p. Putting all this together, we have

$$\begin{aligned} F'(t) &= \lim_{h \to 0} \frac{F(t+h) - F(t)}{h} \\ &= \lim_{h \to 0} \frac{\Delta x}{h} \frac{\partial f}{\partial x}(q_1) + \frac{\Delta y}{h} \frac{\partial f}{\partial y}(q_2) \\ &= \frac{dx}{dt} \frac{\partial f}{\partial x}(p) + \frac{dy}{dt} \frac{\partial f}{\partial y}(p) \\ &= \frac{\partial w}{\partial x} \frac{dx}{dt} + \frac{\partial w}{\partial y} \frac{dy}{dt}. \end{aligned}$$

The chain rule in Theorem 4 can be obtained from the result we have just obtained. We assume that $w = F(u, v)$, where $u = f(x, y)$ and $v = g(x, y)$, and we want $\partial w/\partial x$ and $\partial w/\partial y$. To find the first, we regard y as constant and differentiate w with respect to x, using the rule given in (3.20),

$$\frac{\partial w}{\partial x} = \frac{\partial w}{\partial u} \frac{\partial u}{\partial x} + \frac{\partial w}{\partial v} \frac{\partial v}{\partial x}.$$

This is formula (3.22); a similar procedure yields (3.23).

The most general form of the chain rule can be stated thus: Let w be a C^1 function of $u_1, u_2, \ldots, u_n$, and let each u_i be a C^1 function of the variables $x_1, x_2, \ldots, x_m$. Then w is a C^1 function of the $x_1, x_2, \ldots, x_m$, and

$$\frac{\partial w}{\partial x_j} = \frac{\partial w}{\partial u_1}\frac{\partial u_1}{\partial x_j} + \frac{\partial w}{\partial u_2}\frac{\partial u_2}{\partial x_j} + \frac{\partial w}{\partial u_3}\frac{\partial u_3}{\partial x_j} + \cdots + \frac{\partial w}{\partial u_n}\frac{\partial u_n}{\partial x_j}.$$

Figure 3–21 is the corresponding diagram.

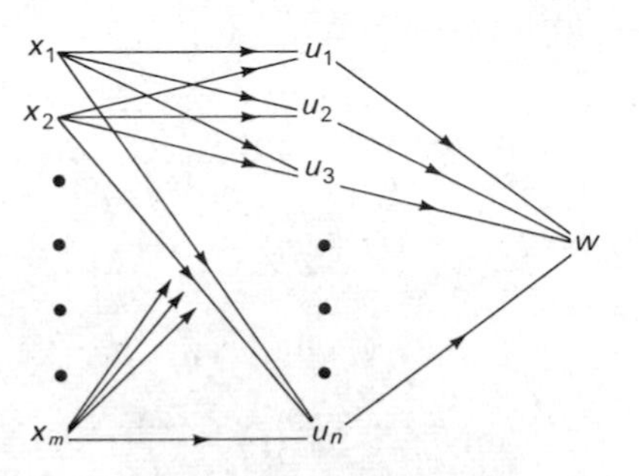

Figure 3–21

Exercises

1 Find $\partial w/\partial x$ and $\partial^2 w/\partial x^2$ when

(a) $w = \sin(x^2 y)$, (b) $w = (x^2 - y^2)^3$, (c) $w = \dfrac{x}{y}$.

2 If $w = F(u) = u^4$ and $u = f(x, y, z) = x^2y - y^2z + 3xyz$, find $\partial w/\partial x$, $\partial w/\partial y$, $\partial w/\partial z$.

3 If $w = f(x, y)$ and $x = t^2 - t$ and $y = 3t + t^2$, find dw/dt and then check for $f(x, y) = xy$.

4 Apply the chain rule when $w = F(u, v)$, $u = x^2 - y^2$, $v = 2xy$ to find $\partial w/\partial x$, $\partial w/\partial y$.

5 Apply the chain rule to find $\partial w/\partial x$, $\partial w/\partial y$, $\partial w/\partial z$ when $w = F(u, v, t)$ and $u = x - y$, $v = yz$, $t = x^2 - z^2$.

6 (a) Draw the functional relation diagram to represent

$$w = \rho(x, y, z),\ x = \varphi(s, t),\ y = \psi(s, t),\ z = \theta(s, t).$$

(b) In this case formulate the chain rule to obtain $\partial w/\partial s$ and $\partial w/\partial t$.

7 Apply the general formula (3.25) to find d^2w/dt^2 for $w = f(at, bt, ct)$.

8 Let $w = F(u)$, $u = f(x, y, z)$. Find a formula for $\partial^2 w/\partial x^2$ and for $\partial^2 w/\partial x\,\partial y$.

9 Which of the following formulas correspond to each of these diagrams (Figure 3–22)?

(a) $w = f(x, y, t)$, $x = \varphi(t)$, $y = \psi(t)$;
(b) $w = f(x, y)$, $x = \varphi(t)$, $y = \psi(t)$;
(c) $w = f(x, y)$, $x = g(y, t)$, $y = h(t)$.

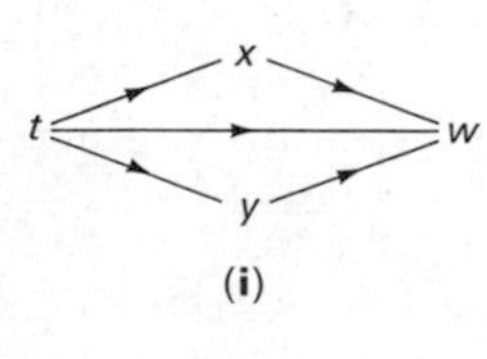

(i)

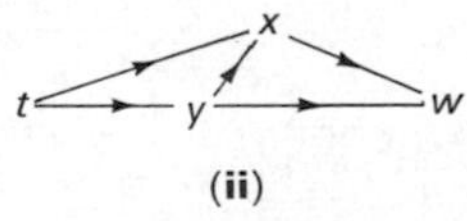

(ii)

Figure 3–22

10 (a) For $w = f(u, v)$, $u = \varphi(x, y)$, $v = \psi(x, y)$, $x = A(t)$, $y = B(t)$, draw the functional relation diagram.

(b) Guess the chain rule applicable to find dw/dt.

11 Consider the following set of functional dependencies: $w = F(x, y, z)$, $x = f(u, v)$, $y = g(u, v)$, $z = h(u, v)$, $u = \alpha(t)$, $v = \beta(t)$.

(a) Explain how w can be regarded as a function of t alone.

(b) Construct a diagram to display all the functional relationships between w, x, y, z, u, v, t.

(c) Show that there are 6 different paths connecting w and t in the diagram.

(d) Complete the formula for dw/dt which starts

$$\frac{dw}{dt} = \frac{\partial w}{\partial x}\frac{\partial x}{\partial u}\frac{du}{dt} + \frac{\partial w}{\partial x}\frac{\partial x}{\partial v}\frac{dv}{dt} + \cdots.$$

3.5

Applications: Extremal Problems and Geometry

In this section, we shall look at some of the ways in which partial derivatives can be used to examine the shape of graphs of functions and to solve maximum-minimum problems in several variables. In order to be able to use pictures to illustrate the results, we limit most of the discussion to functions of two variables; some statements will appear without proofs, leaving a fuller treatment to a course in advanced calculus.

In the study of the graph of a function f of one variable, the critical points were of special importance; these points x_0 where $f'(x_0) = 0$ comprised the local maxima, local minima, and points of inflection. We can hope for something similar, although more complicated, for functions of several variables.

Definition 5 Let f be of class C^1 in an open set D. Then, the critical points for f in D are those points $p \in D$ at which $\nabla f(p) = 0$.

If f is a function of two variables, $\nabla f = (f_x, f_y)$, so that a critical point is one where both f_x and f_y are zero. For example, if $f(x, y) = x^2 - 2xy + 3y^2 - x$, then (x, y) will be a critical point if and only if

$$\begin{aligned} 2x - 2y - 1 &= 0, \\ -2x + 6y &= 0. \end{aligned}$$

Solving, we find that this function has exactly one critical point $(\frac{3}{4}, \frac{1}{4})$.

We next want to learn something about the varieties of possible behavior for a function near one of its critical points. Recall first that a point p_0 is said to be a **maximum** point for f on the set S if $f(p_0) \geq f(p)$ for all points p in S. The number $M = f(p_0)$ is said to be the **maximum value** of f on S. The terms "minimum point" and "minimum value" are defined in a similar way, and the terms "extremal point" and "extremal value" are used to cover both maxima and minima. A point p_0 not on the boundary of the domain f is called a local maximum (minimum) point for f if there is a neighborhood $\mathfrak{N}$ about p_0 such that p_0 is a maximum (minimum) point for f on $\mathfrak{N}$. The value $f(p_0)$ is then called a local maximum or minimum value for f.

These definitions are also easily pictured by means of the graph of f. If p_0 is a local maximum point for f and $f(p_0) = M$, then the point $(p_0, M) = P_0$ is on the graph of f and is the top of a local mountain peak. (See Figure 3–23.) It is natural to call the point P_0

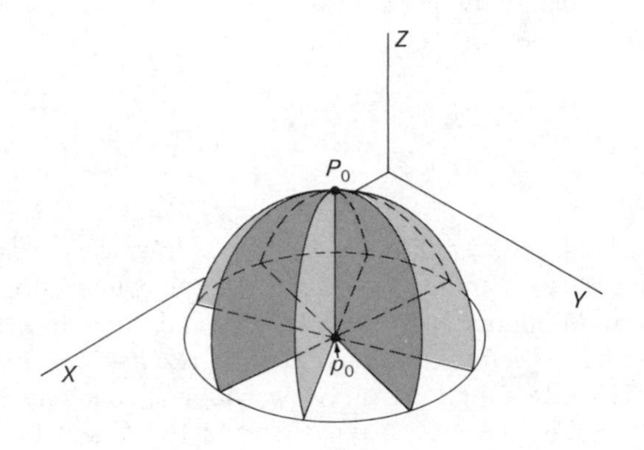

Figure 3–23

a local maximum point on the graph of f; there may be other points on the graph of f that are higher than P_0, but no point *near* P_0 is higher. The three terms—the maximum point p_0, which is in the domain of the function, the maximum value M, which is a number, and the point P_0, which lies on the graph of f, often are all called the maximum of f and used interchangeably. To illustrate the distinction, the function $f(x) = 2 - x^2$ has maximum value 2, maximum point $x = 0$, and (0, 2) for the maximum point on its graph. We shall try to make these distinctions visible in discussions of extremal problems, but there will be times when the context must be used to clarify the meaning. Finally, we emphasize that the maximum or minimum of a continuous function f on a closed bounded region D need not be a local maximum or minimum; it may occur on the boundary of D.

Our first result is the following.

Theorem 5 *Let f be of class C^1 in an open set D. Then any points of D that are local extremal points for f are critical points for f.*

Proof When f is a function of two variables, we can give a simple geometric proof for this theorem based on our knowledge of elementary calculus. Let p_0 be a local maximum for f, and look at the corresponding point $P_0 = (p_0, f(p_0))$ on the graph of f. As explained above, P_0 will be the highest point of the surface $z = f(x, y)$ in a neighborhood around P_0. (See Figure 3–23.) Thus, any vertical plane cross section of this surface through P_0 is a curve in that plane which also has as its highest point P_0, where it therefore must have zero slope. Since the slope of such a cross-sectional curve is the value of the corresponding directional derivative of f at the point p_0, which lies directly below P_0, we conclude that every directional derivative of f at p_0 must have the value zero. In particular, $f_x(p_0) = 0$ and $f_y(p_0) = 0$, so that p_0 is a critical point for f.

If f is a function of three variables, its graph is in 4-space, something like a surface, and the proof given above would not be very convincing. A direct analytic argument can be given which is quite similar in spirit to the above one, but which works if f is a function of n variables for any n. We do not think about the graph of f at all, but look only at the set D which is the domain of f. Take any unit vector u, and look at the function of one variable defined by $F(t) = f(p_0 + tu)$. If p_0 is a local maximum for f, then $f(p_0) = F(0) \geq F(t)$ for all values of t in a neighborhood of 0. Accordingly, $t = 0$ is a maximum point for the function F on an interval $-\delta < t < \delta$, so that $F'(0) = 0$. To make use of this fact, we must find $F'(t)$ in terms of f, using the methods of the last section. To see how this procedure goes, we carry it out for $n = 3$. We have $p_0 = (x_0, y_0, z_0)$, and we let $u = (\alpha, \beta, \gamma)$, so that

$$p_0 + tu = (x_0 + \alpha t, y_0 + \beta t, z_0 + \gamma t).$$

The function F is then given by

$$F(t) = f(x_0 + \alpha t, y_0 + \beta t, z_0 + \gamma t)$$

and we want to find $F'(t)$. To use the chain rule, we put $w = f(x, y, z)$ where $x = x_0 + \alpha t$, $y = y_0 + \beta t$, $z = z_0 + \gamma t$, so that by Theorem 3 of the last section

$$\frac{dw}{dt} = F'(t) = \frac{\partial w}{\partial x}\frac{dx}{dt} + \frac{\partial w}{\partial y}\frac{dy}{dt} + \frac{\partial w}{\partial z}\frac{dz}{dt}$$

$$= f_x\alpha + f_y\beta + f_z\gamma.$$

At $t = 0$, $p_0 + ut = p_0$, and (f_x, f_y, f_z) becomes ∇f at p_0, so that

$$F'(0) = u \cdot \nabla f(p_0) = \mathbf{D}_u f(p_0).$$

We now return to the fact that $F'(0) = 0$—true because F had a local maximum at $t = 0$—to conclude that every directional derivative of f at p_0 is zero. As before, this implies that $\nabla f = 0$ at p_0, and p_0 is therefore a critical point for f.

This result shows that some of the critical points of f can be accounted for by local extrema. Simple examples show that there can be other critical points. For example, the function $f(x) = x^3 + 1$ has 0 as its only critical point, and it is neither a maximum nor a minimum for f. Indeed, the point (0, 1) is a point of inflection on the graph of f. For functions of several variables, the term "saddle point" is used for critical points that are not maxima or minima, and is supposed to suggest that the graph of the function near this point resembles that of the function $x^2 - y^2 + 4$, near the origin, as shown in Figure 3–24. Here, the origin is the critical point, but every neighborhood about the origin contains points where the value of the function is greater than, and points where the value of the function is less than, that of the function at (0, 0).

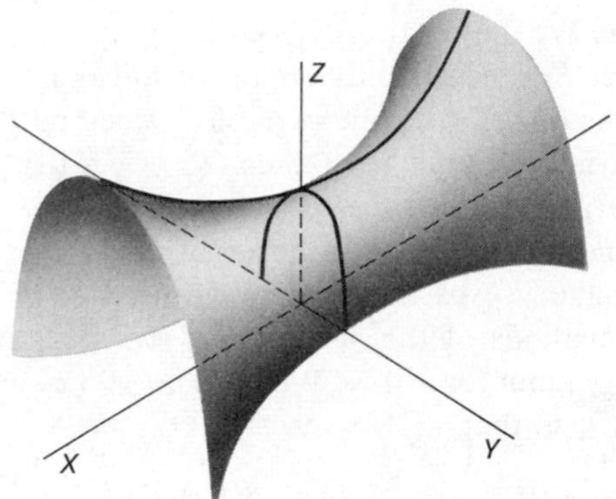

Figure 3–24

In the study of functions of one variable, the second derivative was used to distinguish maximum points from minimum points and also to recognize points of inflection. Something similar but more complicated is true for functions of several variables; we give only part of the general theory.

If the point p_0 is a local maximum point for f, as shown in Figure 3–23, then all the curves of cross section must be concave downward. Applying this fact to the slices made by the planes $x = x_0$ and $y = y_0$, we see that $f(x, y_0)$ and $f(x_0, y)$ must be functions of one variable with graphs that are concave downward. Hence, their second derivatives must be negative or zero at p_0, so that we have both $f_{xx}(x_0, y_0) \leq 0$ and $f_{yy}(x_0, y_0) \leq 0$. If the point p_0 is a minimum point for f, the same analysis ends with $f_{xx}(p_0) \geq 0$ and $f_{yy}(p_0) \geq 0$. This argument yields the following simple test for distinguishing maxima from minima.

Theorem 6 *If f is of class C^2 in an open set D in the plane and p_0 is a point that is known to be a local maximum for f, then it is necessary that $\partial^2 f/\partial x^2 \leq 0$ and $\partial^2 f/\partial y^2 \leq 0$ at p_0. Likewise, if it is known that p_0 is a local minimum, then $\partial^2 f/\partial x^2 \geq 0$ and $\partial^2 f/\partial y^2 \geq 0$ there.*

Observe that at a maximum or minimum point, the two second partial derivatives of f must have the same sign. This leads to the following result, which is more useful than the theorem itself.

Corollary *If f is of class C^2 in D and p_0 is a critical point for f at which $((\partial^2 f/\partial x^2)(\partial^2 f/\partial y^2)) < 0$, then p_0 cannot be either a maximum or minimum point and is therefore a saddle point.*

For example, in the illustration used above, $f(x, y) = x^2 - y^2 + 4$, $\partial^2 f/\partial x^2 = 2$, and $\partial^2 f/\partial y^2 = -2$, so that their product is -4. The origin is therefore a saddle point. We can improve this criterion by looking at *all* the cross sections through the graph of f, instead of only two. If p_0 is a maximum point for f, then every curve of cross section must be concave downward there. By setting $F(t) = f(p_0 + tu)$, we see that this condition is equivalent to requiring that $F''(0) \leq 0$ for every choice of the unit vector u. By an argument that will be outlined in the exercises, this approach yields the following results.

Theorem 7 *If f is of class C^2 in an open region D of the plane and p_0 is a local extremal point for f, then it is necessary that at p_0*

$$f_{xx} f_{yy} - (f_{xy})^2 \geq 0.$$

Corollary *If p_0 is a critical point for f at which*

$$f_{xx} f_{yy} - (f_{xy})^2 < 0,$$

then p_0 is a saddle point for f.

As an illustration, let $f(x, y) = xy$. Then, $f_x = y$ and $f_y = x$, so the origin $(0, 0)$ is the only critical point. There, $f_{xx} = f_{yy} = 0$, and $f_{xy} = 1$. Since $(0)(0) - (1)^2 = -1$, the origin is a saddle point.

It is harder to find conditions that identify a critical point as an extremal point. It is not enough to have $f_{xx} < 0$ and $f_{yy} < 0$, even though this means that the corresponding cross sections in planes parallel to the XZ- and YZ-planes yield curves that are concave downward, since a cross section in a different direction might be concave upward. (See Exercise 3.) In fact, it is possible to produce a function $f(x, y)$ which has a critical point at the origin and whose graph is such that *every* plane cross section through the origin is a curve having a minimum at the origin, but such that the origin is *not* a local minimum for the function.[2] The best result that can be easily stated is the following, whose proof is left to a more advanced text.

[2] *See R.C. Buck, Advanced Calculus, 2nd ed. (New York; McGraw-Hill, 1965) p. 363.*

Theorem 8 *If f is of class C^2 in an open set D in the plane and $p_0 \in D$ is a critical point for f at which*

$$f_{xx}f_{yy} - (f_{xy})^2 > 0,$$

then p_0 is either a local maximum point or a local minimum point for f, as determined by the sign of f_{xx} (or f_{yy}).

If this result is compared with that in the corollary to Theorem 7, it will be seen that neither says anything about the nature of a critical point p_0 at which $f_{xx}f_{yy} - (f_{xy})^2 = 0$. The reason is simply that in this case no conclusion can be drawn. It is similar to the ambiguous case in one-variable calculus which arises when $f'(x_0) = 0$ and $f''(x_0) = 0$, where x_0 can be a maximum point, a minimum point, or a point of inflection. More complicated tests, involving higher derivatives, can be derived which help to determine which sort of critical point one is dealing with.

The above results make it possible to solve maximum-minimum problems for functions of several variables in much the same way as for functions of one variable. Suppose that f is a function defined on a closed bounded set D and we want to find its maximum value on D. This must exist and be attained at some point P in D. (There might be several points P at which the same maximum value is attained.) Either the point P lies on the boundary of D, or else it is somewhere in the interior of D. In the latter case, it is a local maximum for f, and if f is of class C^1, P must be one of the critical points for f. This leads to the following solution strategy:

Step 1 List all the critical points for f interior to D.

Step 2 Eliminate any that cannot be local maxima for f.

Step 3 Find the value of f at each of the remaining points in the list, and choose the largest.

Step 4 Compare the value obtained in Step 3 with the values that f takes on the boundary of D, and again take the largest.

The result will of course be the maximum value of f on D, and you will know where the point P must be.

We illustrate this with several examples. In each case, D is the square in the plane with vertices $(0, 0)$, $(1, 0)$, $(1, 1)$, $(0, 1)$.

First, take $f(x, y) = x^2 - 2xy + 3y^2 - x$. There is one critical point for f in the plane at the point $p_0 = (\frac{3}{4}, \frac{1}{4})$, which happens to lie in the interior of D. At p_0, we find $f_{xx} = 2, f_{yy} = 6$. Accordingly, we know that p_0 cannot be a maximum point for f, since the sign of f_{xx} is wrong. If we seek the maximum of f on D, we know at this stage that we must search for it on the boundary of f. If we are looking for the minimum of f on D, we proceed to find f_{xy} at p_0, note that it turns out to be -2, and then calculate $f_{xx}f_{yy} - (f_{xy})^2 = 12 - 4 = 8 > 0$. By Theorem 8, we now know that p_0 is certainly a *local* minimum for f. It is still possible that p_0 is not *the* minimum of f, that a smaller value is achieved on the boundary of D. We calculate $f(p_0)$, obtaining $-\frac{3}{8}$.

We next examine f on the boundary of D, finding its formula and extremal values on each edge of the square. In Figure 3–25 we indicate both the appropriate formula for f and the maximum and minimum values of f on that edge, obtained in turn by looking at the corresponding one-variable maximum-minimum problem.

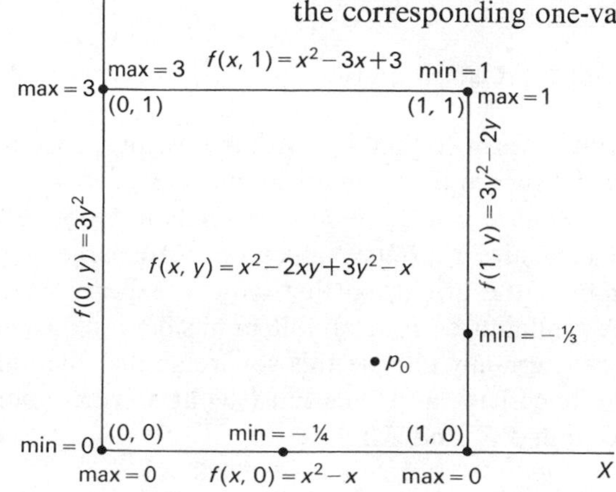

Figure 3–25

Comparing values, we see that the maximum of f on D is 3, and it is attained at $(1, 0)$. Similarly, the minimum of f on the boundary of D is $-\frac{1}{3}$, attained at $(1, \frac{1}{3})$. However, $-\frac{3}{8}$ is smaller than $-\frac{1}{3}$, so the absolute minimum of f on D is $-\frac{3}{8}$ and is attained at the interior point $p_0 = (\frac{3}{4}, \frac{1}{4})$.

Now consider a second function $g(x, y) = x^2 - 3xy + y^2 + 2x - y$. Again, this has only one critical point $(\frac{1}{5}, \frac{4}{5})$, and again it happens to be in the interior of D. However, when we check the higher derivatives of g, we find that $g_{xx}g_{yy} - (g_{xy})^2 = -5$. We know at once that the point $(\frac{1}{5}, \frac{4}{5})$ is a saddle point for g, so that both the minimum and maximum of g on D must occur on the boundary of D. Looking at g on the boundary of D yields the information displayed in Figure 3–26.

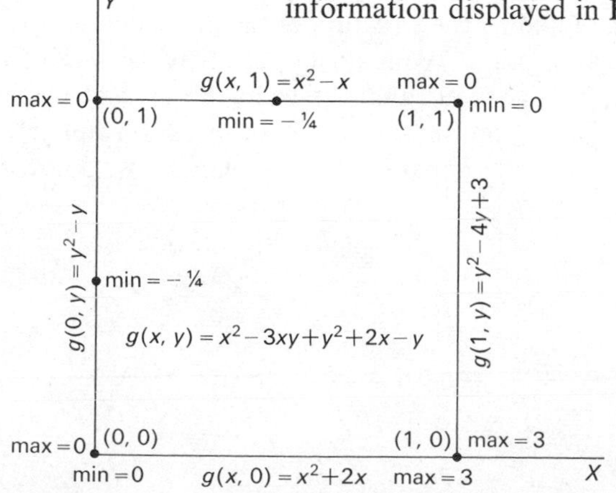

Figure 3–26

We conclude that the maximum of g is 3, attained at (1, 0), and the minimum value is $-\frac{1}{4}$, attained both at $(0, \frac{1}{2})$ and at $(\frac{1}{2}, 1)$.

There are many times when the nature of the problem makes it possible to discard the boundary immediately, since it is clear that the desired extremal will not be there. In such cases, all that one has to do is to locate the critical points, and compare the functional values there. For example, let us ask what point (x, y) in the plane will minimize the function

$$F(x, y) = (x + y - 1)^2 + (x + 3y - 2)^2 + (2x - y + 1)^2.$$

This type of problem arises in connection with the use of the method of least squares in statistics and in numerical analysis.

The region D is the entire plane, which is certainly not a bounded set. However, we recognize that the values of F are large if the point (x, y) is far from the origin, so that we can expect that the choice of (x, y) that minimizes $F(x, y)$ will be inside a sufficiently large square. We can certainly choose this square so that the minimum is not on the boundary, and thus must be at a critical point for F. We proceed to find F_x and F_y:

$$F_x = 2(x + y - 1) + 2(x + 3y - 2) + 2(2x - y + 1)(2),$$
$$F_y = 2(x + y - 1) + 2(x + 3y - 2)(3) + 2(2x - y + 1)(-1).$$

Setting $F_x = 0$ and $F_y = 0$, we have the pair of equations

$$6x + 2y = 1,$$
$$2x + 11y = 8.$$

Solving, we obtain the only critical point $(-5/62, 23/31)$. We do not have to test that this is in fact a local minimum. The function F, being bounded below by 0, *must* have a local minimum somewhere; since $(-5/62, 23/31)$ is the only critical point, it must be the minimum point.

We end this section by returning to the geometry of surfaces, and solve the problem of finding the equation of the plane that is tangent to a surface at a given point. We may suppose that the surface has the equation $F(x, y, z) = 0$ and that we want the tangent plane at a point $p_0 = (x_0, y_0, z_0)$ on the surface. We can use formula (1.14) in Section 1.5 to write the equation of this plane if we know the

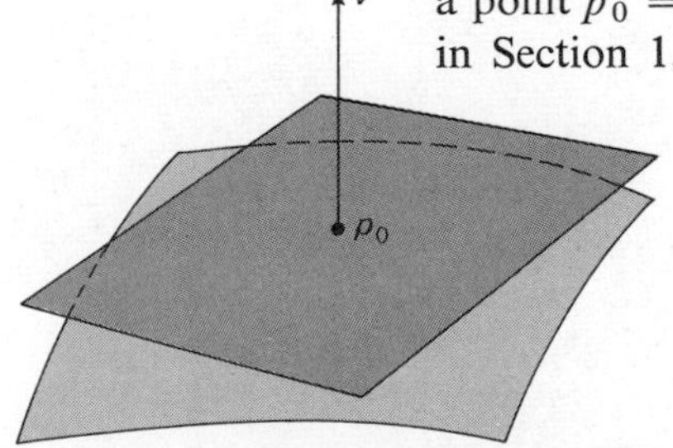

Figure 3–27

direction of the normal to the plane. Since the plane is to be tangent to the surface at p_0, we should require that the normal to the plane coincide with the normal to the surface at p_0. (See Figure 3–27.)

What property ought to be used to define the normal vector to a surface? Recall that the normal to a plane was chosen as a vector orthogonal to all lines in the plane. This suggests that an appropriate requirement would be that the vector u normal to a surface at a point p_0 ought to be orthogonal to *all* curves on the surface that pass through p_0. What is not at all obvious is that there exists such a vector u. The validity of this definition comes from the following result.

Theorem 9 *Let $\mathcal{S}$ be the surface whose equation is $F(x, y, z) = C$, and let p_0 lie on $\mathcal{S}$. Let $v = \nabla F(p_0)$, the gradient of the function F at p_0, and assume that p_0 is not a critical point of F, so that $|v| \neq 0$. Then, the direction of the vector v is normal to the surface $\mathcal{S}$ at p_0, and v is orthogonal to every smooth curve $\mathcal{C}$ on $\mathcal{S}$ that passes through p_0.*

Proof The parametric equations for a general smooth curve $\mathcal{C}$ in space can be written

$$\mathcal{C}: \begin{cases} x = X(t), \\ y = Y(t), \\ z = Z(t), \end{cases} \qquad 0 \leq t \leq 1. \tag{3.26}$$

If this curve is to go through p_0, we can assume that it does so when $t = t_0$, so that $p_0 = (X(t_0), Y(t_0), Z(t_0))$. If the curve $\mathcal{C}$ is to lie entirely on $\mathcal{S}$, then each point on the curve must satisfy the equation for the surface; thus, we must have

$$F(X(t), Y(t), Z(t)) = C$$

holding for all choices of t in $[0, 1]$. Treating this equation as an identity, we differentiate it with respect to t, using the chain rule developed in Theorem 3, to obtain

$$\frac{\partial F}{\partial x}\frac{dx}{dt} + \frac{\partial F}{\partial y}\frac{dy}{dt} + \frac{\partial F}{\partial z}\frac{dz}{dt} = \frac{d}{dt}C = 0. \tag{3.27}$$

This too must hold for all t, in particular for $t = t_0$. With this substitution, we have

$$F_x(p_0)X'(t_0) + F_y(p_0)Y'(t_0) + F_z(p_0)Z'(t_0) = 0.$$

However, the vector $(F_x(p_0), F_y(p_0), F_z(p_0))$ is the vector v, and the vector $(X'(t_0), Y'(t_0), Z'(t_0))$ is the tangent vector T to the curve (3.26) at the point p_0, so the relation (3.27) merely says that $v \cdot T = 0$. Hence v is orthogonal to the curve $\mathcal{C}$. (As explained in Section 3.1, the vector $T = (X'(t), Y'(t), Z'(t))$ describes the tangents to the curve $\mathcal{C}$ for points where $T \neq O$; this is the reason for the restriction that $\mathcal{C}$ be a smooth curve.)

Corollary *If $p_0 = (x_0, y_0, z_0)$ is a point on the surface whose equation is $F(x, y, z) = C$, and p_0 is not a critical point for the function F, then an equation for the tangent plane to the surface through p_0 is*

$$F_x(p_0)(x - x_0) + F_y(p_0)(y - y_0) + F_z(p_0)(z - z_0) = 0. \tag{3.28}$$

We shall give several illustrations of the above corollary. First, consider the sphere $x^2 + y^2 + z^2 = 9$ and the point $(1, -2, 2)$ which lies on it. With $F(x, y, z) = x^2 + y^2 + z^2$ and $C = 9$, $\nabla F = (2x, 2y, 2z)$, so that the normal vector at $(1, -2, 2)$ is in the direction of the vector $(2, -4, 4)$. The unit normal vector will then be $(\frac{1}{3}, -\frac{2}{3}, \frac{2}{3})$. (See Figure 3–28.) One may verify easily that the normal vector at any point of this sphere points away from the center of the sphere.

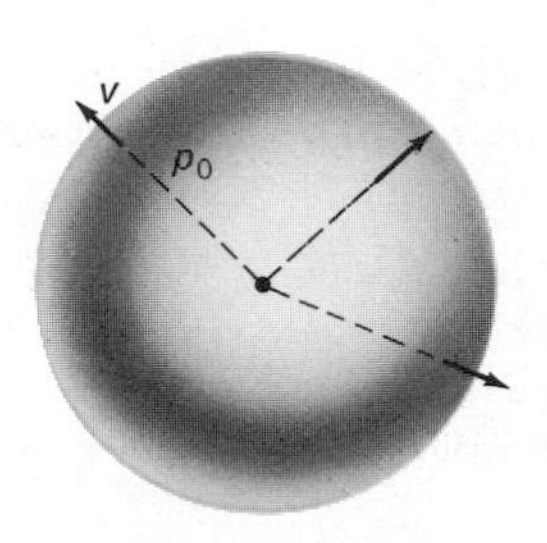

Figure 3–28

Again, consider the surface (hyperbolic paraboloid) $x^2 - y^2 = 3z$, and the point $p_0 = (2, 1, 1)$ on it. When F is defined by $F(x, y, z) = x^2 - y^2 - 3z$, the equation of the surface becomes $F(x, y, z) = 0$. Since $\nabla F = (2x, -2y, -3)$, the normal to the surface at p_0 is $(4, -2, -3)$, and the equation of the tangent plane at p_0 is

$$4(x - 2) - 2(y - 1) - 3(z - 1) = 0.$$

Finally, we consider the surface with equation $x^2 + y^2 - z^2 = 0$ and ask for its normal at the origin. Since $\nabla F = (2x, 2y, -2z)$, we find that this point is a critical point for F, and the above method fails. Is there some evident geometric reason to expect that we shall be unable to find a well-defined normal vector to this surface at the origin? An answer is found by looking at this surface, which turns out to be a cone with the origin as its vertex; there cannot exist a well-defined normal at this point. A similar situation will be found in Exercise 13, where the surface turns out to be a pair of intersecting planes such that no normal is defined at any point on their line of intersection.

There is another way to interpret the result of Theorem 9. Given the function $F(x, y, z)$, the surfaces whose equations are $F(x, y, z) = C$ for each possible value of C are called the *level surfaces* for F. If $F(p)$ is thought of as the temperature at the point p, then the level surfaces of F are the equithermal surfaces. The fact that $\nabla F(p_0)$ is the normal at the point p_0 to the surface $F(x, y, z) = C$ which passes through p_0 can be restated thus: *At each point of space, the gradient vector of F is orthogonal to the level* *surface through the point.* If you know the shapes of the level surfaces of F, then it is possible to draw the directions of the gradient field of F, although the magnitudes will have to be calculated.

Exercises

1 Find all critical points in the domain of definition of each of these functions:

(a) $f(x, y) = 3x^2 - 4xy + 5y^2 - 2x + 3y$,

(b) $f(x, y) = (x^2 - 1)(y^2 - 4)$, (c) $f(x, y) = x^2y - 8x - 4y$,

(d) $f(x, y) = \dfrac{x - y}{x + y}$, (e) $f(x, y) = \dfrac{xy}{x^2 + y^2}$.

2 Find all critical points in the domain of definition of each of these functions:

(a) $f(x, y, z) = x^2 + yz + z^2 - xz$,

(b) $f(x, y, z) = xy - 3yz + 4xz - x + y - 2z$,

(c) $f(x, y, z) = 4x^2 + y^2 + z^2 - xyz$.

3 (a) Show that the origin is the only critical point for $f(x, y) = 3xy - x^2 - y^2$ and that both f_{xx}, f_{yy} are negative there.

(b) By taking a suitable plane cross section, show that (0, 0) is not a maximum point for f.

4 Classify the critical points of the following functions:

(a) $f(x, y) = 4xy - 3x^2 + 4y^2$,

(b) $f(x, y) = 2x^2 + y^2 - 3xy$,

(c) $f(x, y) = 5x^2 + y^2 - 4xy$.

5 Classify the critical points for f where

$$f(x, y) = (x^2 - 1)(y^2 - 4). \text{ (Recall Exercise 1(b).)}$$

On the square S with opposite vertices at (0, 0) and (1, 1), find the maximum and minimum values of the following functions:

6 $f(x, y) = x^2 + 2y^2 - x - 3y$,

7 $f(x, y) = x^2 - y^2 - 6xy + x + 2y$,

8 $f(x, y) = x^2 + y^2 - 3x + 4y$,

9 $f(x, y) = 3x^2 + y^2 - 3xy - y$.

10 Show that the absolute minimum (for all x and y) of

$$F(x, y) = (2x + 3y)^2 + (x + y - 1)^2 + (x + 2y - 2)^2$$

is achieved for $x = -1$, $y = 1$.

11 Find a normal vector at the point $(3, -1, -2)$ on the surface $xyz = 6$.

12 Find a normal vector at the point $(1, -1, 2)$ on the surface

$$2xy + z^2 - y^2 = 1.$$

13 Attempt to find a normal to the surface given by $y^2 - x^2 + x - y = 0$ at the points $(2, 2, 1)$, $(3, -2, -1)$, and $(\frac{1}{2}, \frac{1}{2}, 2)$. Explain.

14 Attempt to find a normal to the surface given by $x^2 + y^2 - z^2 = 0$ at the points $(3, 4, 5)$ and $(0, 0, 0)$. Explain.

***15** Show that the line

$$x = 1 - t,\ y = 2t,\ z = -1 - 2t$$

meets the surface $4x^2 + y^2 + z^2 - 2x + 2z = 7$ at two points and is orthogonal to the surface at one of them.

***16** Invent and prove a two-variable analogue for Rolle's Theorem.

17 Show that critical points p_0 for a function $f(x, y)$ correspond to points P on the graph of f where the normal is vertical.

****18** Suppose that the origin $(0, 0)$ is a local minimum point for the function $f(x, y)$. Take any vector $u = (\alpha, \beta)$ and form the function

$$F(t) = f(ut) = f(\alpha t, \beta t).$$

(a) Use the chain rules of the last section to find that

$$F'(t) = \alpha f_x(ut) + \beta f_y(ut),$$
$$F''(t) = \alpha^2 f_{xx}(ut) + 2\alpha\beta f_{xy}(ut) + \beta^2 f_{yy}(ut).$$

(b) Since the cross section function F must have a local minimum at $t = 0$, show that

$$\alpha^2 f_{xx}(0, 0) + 2\alpha\beta f_{xy}(0, 0) + \beta^2 f_{yy}(0, 0) \geq 0$$

for all choices of α, β.

(c) By taking $\alpha = \sqrt{f_{yy}}$ and $\beta = -\sqrt{f_{xx}}$, show that $f_{xx}f_{yy} \geq (f_{xy})^2$ at the minimum point, thereby proving part of Theorem 7.

3.6

Functions Defined Implicitly

In the initial stages of the study of calculus, it is usual to say that an equation of the form $G(x, y) = 0$ determines "y as a function of x," or that it can be "solved for y in terms of x." It is now necessary to look at these and other similar statements a little more critically and to extend them to more than two variables.

First, there are equations that *cannot* be solved for y for various reasons. Since we are limited to real numbers, there may be no choices of x and y that satisfy the equation, as for $x^2 + y^2 + 1 = 0$. There are equations which do not admit solutions for y because y is not really present—i.e., $G(x, y)$ is independent of y. An example to illustrate this is the equation

$$G(x, y) = (x - y)(x + y) + (y - 2x)(y + 2x) = 0.$$

Second, there must be ground rules regarding what is to be considered a solution. If we have the equation $y^2 - x = 0$, what are the solutions expressing y in terms of x? Is there just one, $y = \sqrt{x}$? What about $y = -\sqrt{x}$? What about the function defined for all $x \geq 0$ by

$$y = \begin{cases} \sqrt{x} & \text{when } x \text{ is rational,} \\ -\sqrt{x} & \text{when } x \text{ is irrational?} \end{cases}$$

Finally, in discussing the solution of equations, we must be careful to make a clear distinction between the questions of whether there exists a solution and whether we know a process for finding a solution. (For example, every polynomial equation of odd degree of the form $x^n + a_1x^{n-1} + \cdots + a_n = 0$ has a real solution, but there is no general algebraic method for solving such equations when $n \geq 5$.)

In general, when we speak of solving $G(x, y) = 0$ for y, we are asking for a continuous function f defined on some interval I such that $y = f(x)$ satisfies the equation for all $x \in I$. This means that the following identity holds:

$$G(x, f(x)) = 0 \text{ for all } x \in I.$$

There is then an obvious geometric relationship between the graph of f and the graph of the equation $G(x, y) = 0$. Each point (x, y) on the graph of f, being of the form $y = f(x)$, must satisfy the equation $G(x, y) = 0$ and must therefore lie on its graph; thus, the graph of f must be a subset of the graph of the equation.

We can recast the above discussion to provide us with a general way of dealing with the problem of solving equations. Given an equation $G(x, y) = 0$, we look at its graph, and then look for continuous functions f whose graphs are subsets of the graph of the equation. Each such function yields a solution of the equation for y. It is also customary to choose each such function f to have as large a domain of definition as is allowed by the nature of the graph of the equation.

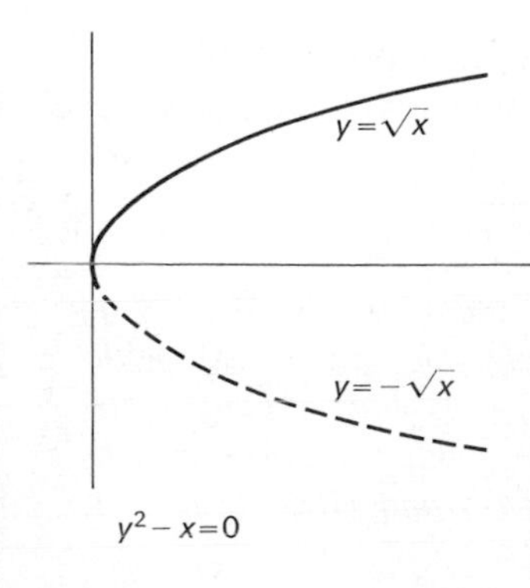

Figure 3–29

An easy illustration is the familiar equation $y^2 - x = 0$. Its graph is given in Figure 3–29, and we see at once that it can be divided into two pieces, each of which is the graph of a continuous function. This equation therefore has two (continuous) solutions,

$y = \sqrt{x}$ given by the top half of the graph, and $y = -\sqrt{x}$ given by the bottom half. (Recall again that the graph of a function must be such that no vertical line cuts it more than once; thus the portion of the parabola in a neighborhood of the vertex at (0, 0) cannot be the graph of a function.)

It is now possible to guess correctly the main theorem dealing with the solution of equations of the form $G(x, y) = 0$. If (x_0, y_0) is a point on its graph and if the slope of the curve is finite there (i.e., the tangent there is not vertical), then there is a function f such that $y_0 = f(x_0)$, and $y = f(x)$ is a solution of the equation $G(x, y) = 0$. We say that this equation "has a solution for y near the point (x_0, y_0)."

This conclusion can be converted into a condition on the partial derivatives of the function G. Let $x = X(t)$, $y = Y(t)$ be parametric equations for the curve $\mathcal{C}$ which is that part of the graph of the equation $G(x, y) = 0$ going through the point $p_0 = (x_0, y_0)$. We want to find the slope of the curve $\mathcal{C}$ at p_0. From elementary calculus, we have

$$\text{slope} = \left.\frac{dy}{dx}\right|_{p_0} = \frac{Y'(t_0)}{X'(t_0)}.$$

Since all points on the curve $\mathcal{C}$ satisfy the equation $G(x, y) = 0$, we have $G(X(t), Y(t)) = 0$ for all t, and differentiating as in the last section, we have, for all t,

$$G_x \frac{dx}{dt} + G_y \frac{dy}{dt} = 0.$$

Setting $t = t_0$ so that $(x, y) = p_0$, we have

$$G_x(p_0)X'(t_0) + G_y(p_0)Y'(t_0) = 0,$$

and

$$\text{slope of } \mathcal{C} = -\frac{G_x(p_0)}{G_y(p_0)}.$$

(This presupposes that the point p_0 is not a critical point for G, since then we could not carry out this algebra.) When will this slope be finite? Clearly, we will need $G_y(p_0) \neq 0$. Thus, we are led to formulate the following (conjectured) theorem.

Theorem 10 *If G is a function of two variables of class C^1 on an open set D, and if $p_0 = (x_0, y_0)$ is a point of D satisfying $G(x_0, y_0) = 0$ with $G_y(x_0, y_0) \neq 0$, then the equation $G(x, y) = 0$ has a continuous solution for y near p_0—that is, there is a continuous function f defined and continuous near x_0 such that $y_0 = f(x_0)$ and $G(x, f(x)) = 0$ for all x in a neighborhood of x_0.*

This important result is called the implicit function theorem. Its proof is not easy, and we leave it for a more advanced course, hoping that the preceding discussion has at least made it plausible. Notice that the theorem states only that such a function f must exist; it gives no instructions for finding a formula for f.

Let us see how it applies to some of our examples. First, if $G(x, y) = y^2 - x$, then $G_y(x, y) = 2y$, and we must avoid points p_0 for which $y_0 = 0$. Hence, the equation $y^2 - x = 0$ can be solved near any point on the graph of the equation, with the exception of the origin.

What happens if we choose the function

$$G(x, y) = (x - y)(x + y) + (y - 2x)(y + 2x)?$$

Since this reduces to $G(x, y) = -3x^2$, we have $G_y(x, y) = 0$ for all (x, y). It is therefore not surprising that the equation $G(x, y) = 0$ cannot be solved anywhere for y.

Let us turn next to the study of equations of the form $G(x, y, z) = 0$, and look for solutions for z of the form $z = f(x, y)$. The same general analysis can be made. The graph of the equation can be thought of as a surface, and we look for portions of that surface that are the graphs of functions f. When is part of a surface the graph of a function? Clearly, since we want vertical lines to meet the surface only once, it is sufficient to have the surface connected and without a vertical tangent plane—i.e., a tangent plane perpendicular to the XY-plane. In terms of the normals to the surface, this means that none shall be parallel to the XY-plane and thus none orthogonal to the Z-axis. Since the normal is a multiple of $\nabla G = (G_x, G_y, G_z)$, this is equivalent to the requirement that G_z be nonzero. We are thus led to state an implicit function theorem for three variables:

Theorem 11 *If G is a C^1 function of three variables and $G(x_0, y_0, z_0) = 0$, $G_z(x_0, y_0, z_0) \neq 0$, then the equation $G(x, y, z) = 0$ can be solved for z in the form $z = f(x, y)$ in a neighborhood of (x_0, y_0, z_0).*

As with the earlier result, we leave the proof of this and its n-variable generalization to a more advanced course. The general theorem, as one would guess, says that an equation of the form $F(x_1, x_2, \ldots, x_n) = 0$ can be solved for x_j as a function of all the remaining variables in a neighborhood of any point that satisfies the equation and at which $\partial F/\partial x_j \neq 0$.

How can we use a theorem like the implicit function theorem, which tells us that an equation *has* a solution, but not how to find it? Perhaps the examples that follow will show how it is useful.

The equation $x^2 + y^2 + z^2 = 9$ is satisfied by $x = 1$, $y = -2$, $z = -2$. Also, $\partial/\partial z\,(x^2 + y^2 + z^2 - 9) = 2z \neq 0$ when $z = -2$.

Hence, there is a solution of the equation for z in the form $z = f(x, y)$ which gives $f(1, -2) = -2$ and is valid in a neighborhood of $x = 1$, $y = -2$. (In this elementary case, we can give f explicitly, $f(x, y) = -\sqrt{9 - x^2 - y^2}$.) Without finding the function f, can we evaluate $\partial z/\partial x$ and $\partial z/\partial y$ at $x = 1$, $y = -2$? If we return to the original equation $x^2 + y^2 + z^2 = 9$ and think of z as replaceable by a certain C^1 function $f(x, y)$, we can treat the equation as an identity in x and in y, and differentiate it separately with respect to x and with respect to y. If we do so, we have

$$
\begin{aligned}
&2x + 0 + 2z\frac{\partial z}{\partial x} = 0, \\
&0 + 2y + 2z\frac{\partial z}{\partial y} = 0.
\end{aligned}
\tag{3.29}
$$

From (3.29), we obtain formulas for the partial derivatives of the function f:

$$
\begin{aligned}
f_x &= \frac{\partial z}{\partial x} = -\frac{2x}{2z} = -\frac{x}{z}, \\
f_y &= \frac{\partial z}{\partial y} = -\frac{2y}{2z} = -\frac{y}{z},
\end{aligned}
$$

and at the point $(1, -2, -2)$, we have $\partial z/\partial x = \frac{1}{2}$, $\partial z/\partial y = -1$.

For a more complicated example, consider the equation $2x^5y + 3y^5z + z^5x = 6$, which is satisfied by $x = 1$, $y = 1$, $z = 1$. Since $G_z = 3y^5 + 5z^4x$, which is nonzero at $(1, 1, 1)$, we conclude that the equation can be solved for $z = f(x, y)$ in a neighborhood of $x = 1$, $y = 1$. This time, no explicit algebraic solution can be presented, since this is a polynomial of degree 5 in z. However, the implicit function theorem guarantees that there *is* such a function f. To find $f_x = \partial z/\partial x$ and $f_y = \partial z/\partial y$ at the point $(1, 1)$, we take the original equation $2x^5y + 3y^5z + z^5x = 6$ and differentiate it, regarding z as $f(x, y)$. Thus, holding y constant, we obtain

$$10x^4y + 3y^5\frac{\partial z}{\partial x} + 5z^4\frac{\partial z}{\partial x}x + z^5 = 0.$$

Setting $x = 1$, $y = 1$, $z = 1$, we have

$$10 + 3\frac{\partial z}{\partial x} + 5\frac{\partial z}{\partial x} + 1 = 0,$$

so that $\partial z/\partial x = -11/8$. Likewise, differentiating the equation with respect to y, we have

$$2x^5 + 15y^4z + 3y^5\frac{\partial z}{\partial y} + 5z^4\frac{\partial z}{\partial y}x = 0,$$

and at (1, 1, 1),

$$2 + 15 + 3\frac{\partial z}{\partial y} + 5\frac{\partial z}{\partial y} = 0,$$

so that $\partial z/\partial y = -17/8$.

The same equation $2x^5y + 3y^5z + z^5x = 6$ can also be looked at as defining x as a function of y and z, since

$$G_x = 10xy + 0 + z^5,$$

which is nonzero at p_0. If we put $x = h(y, z)$ and want to evaluate h_y and h_z at $y = 1$, $z = 1$, we can once more return to the original equation, and—regarding x as $h(y, z)$—differentiate it as an identity in y and z. Thus, one first would get, differentiating with respect to y with z constant,

$$10x^4\frac{\partial x}{\partial y}y + 2x^5 + 15y^4z + z^5\frac{\partial x}{\partial y} = 0$$

and

$$\frac{\partial x}{\partial y} = -\frac{2x^5 + 15y^4z}{10x^4y + z^5}.$$

At the point (1, 1, 1), we would therefore have $\partial x/\partial y = -17/11$. Differentiating the original equation with respect to z, with x again regarded as $h(y, z)$ and y constant, we have

$$10x^4\frac{\partial x}{\partial z}y + 3y^5 + 5z^4x + z^5\frac{\partial x}{\partial z} = 0.$$

From this, we have

$$\frac{\partial x}{\partial z} = -\frac{3y^5 + 5z^4x}{10x^4y + z^5},$$

and at $x = 1$, $y = 1$, $z = 1$, we have $\partial x/\partial z = -8/11$.

Implicit differentiation was used in beginning calculus to handle certain types of maxima-minima problems. It is also useful for the same purpose in several variables. For example, suppose we want to find the rectangular box of greatest volume that can be placed with one corner at the origin, and the opposite corner on the ellipsoid $x^2 + y^2/4 + z^2/9 = 1$. (See Figure 3–30.)

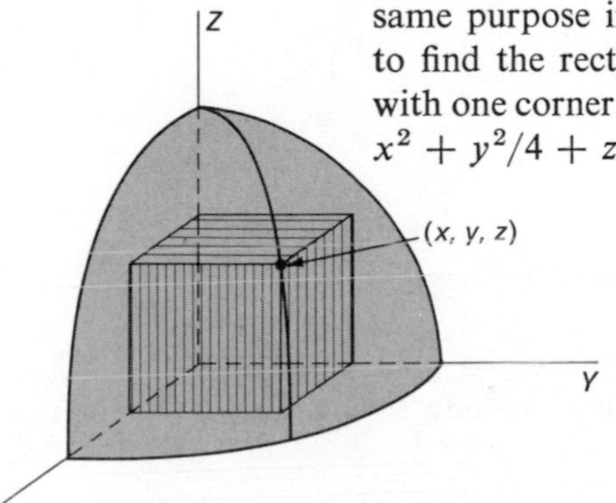

Figure 3–30

If the dimensions of the box are x, y, z, then we can restate the problem as follows: Among the points (x, y, z) which obey $x^2 + y^2/4 + z^2/9 = 1$, which will maximize the function $V(x, y, z) = xyz$? It is clear how we might reduce this to the type of extremal problem we solved in the last section. We can solve the ellipsoid equation to express z as $f(x, y)$, replace z by $f(x, y)$ in V, and obtain the problem of maximizing a function of two variables $F(x, y) = V(x, y, f(x, y))$. However, by using implicit differentiation, we can solve this problem without having to express z explicitly in terms of x and y. Recognizing that we *could* find f so that $z = f(x, y)$, we can treat the equations as though this had been done. The requirement that the extremal point must be a critical point for the function which expresses the volume in terms of x and y leads to the pair of equations $\partial V/\partial x = 0$, $\partial V/\partial y = 0$. Regarding z as some function of x and y, and V as xyz, we have

$$(3.30)\qquad \begin{aligned} 0 &= \frac{\partial V}{\partial x} = yz + xy\frac{\partial z}{\partial x}, \\ 0 &= \frac{\partial V}{\partial y} = xz + xy\frac{\partial z}{\partial y}. \end{aligned}$$

We need expressions for $\partial z/\partial x$ and $\partial z/\partial y$, which we obtain by differentiating the equation $x^2 + y^2/4 + z^2/9 = 1$:

$$2x + 0 + \frac{2}{9}z\frac{\partial z}{\partial x} = 0,$$

$$0 + \frac{2y}{4} + \frac{2}{9}z\frac{\partial z}{\partial y} = 0.$$

Solving for $\partial z/\partial x$ and $\partial z/\partial y$ and putting these into (3.30), we obtain a pair of algebraic equations

$$0 = yz + xy\left(-9\frac{x}{z}\right),$$

$$0 = xz + xy\left(-\frac{9}{4}\frac{y}{z}\right),$$

which simplify to

$$0 = y(z^2 - 9x^2),$$
$$0 = x(z^2 - \tfrac{9}{4}y^2).$$

The choices $x = 0$ or $y = 0$ would give $V = 0$, which is clearly not a maximum. Hence, we conclude that the desired maximum must occur for (x, y, z) obeying $z = 3x$, $z = \frac{3}{2}y$. Since the point is also required to satisfy the equation of the ellipsoid, we substitute $(1/3)z$ for x and $(2/3)z$ for y in the equation. We obtain $3z^2 = 9$, and thus $z = \sqrt{3}$, $x = \sqrt{3}/3$, and $y = \frac{2}{3}\sqrt{3}$.

A different illustration of the use of implicit differentiation is the following. The surface $xyz = 6$ and the plane $5x - 3y + z = 2$ intersect in a curve that passes through the point $P = (1, 2, 3)$, since this satisfies both equations. What is the direction of this curve at P? If we knew parametric equations for the curve, we could answer this question by finding the direction of the vector $(dx/dt, dy/dt, dz/dt)$. Can we find it without knowing the parametric equations?

If we assume that there are equations $x = X(t)$, $y = Y(t)$, $z = Z(t)$ which satisfy the pair of given equations identically, then we can differentiate both of the given equations with respect to t as though the substitutions had in fact been made. Thus, differentiating $xyz = 6$ yields

$$yz\frac{dx}{dt} + xz\frac{dy}{dt} + xy\frac{dz}{dt} = \frac{d}{dt}(6) = 0,$$

and differentiating $5x - 3y + z = 2$ yields

$$5\frac{dx}{dt} - 3\frac{dy}{dt} + \frac{dz}{dt} = 0.$$

If we now put in the known values of x, y, z corresponding to the point (1, 2, 3) in which we are interested, we arrive at a pair of linear equations in three unknowns

$$\begin{aligned} 6\frac{dx}{dt} + 3\frac{dy}{dt} + 2\frac{dz}{dt} &= 0, \\ 5\frac{dx}{dt} - 3\frac{dy}{dt} + \frac{dz}{dt} &= 0. \end{aligned} \tag{3.31}$$

Since there are more unknowns than equations, we do not expect to find unique numerical values for dx/dt, dy/dt, dz/dt. Indeed, we cannot expect to, since there is no unique parametric representation for a curve. However, we can find one solution of (3.31) such that every solution is a multiple of it. To find this solution, set $dz/dt = c$, and solve (3.31), obtaining

$$\frac{dx}{dt} = -\frac{3}{11}c \quad \text{and} \quad \frac{dy}{dt} = -\frac{4}{33}c.$$

Accordingly, the general solution of (3.30) is

$$\left(\frac{dx}{dt}, \frac{dy}{dt}, \frac{dz}{dt}\right) = \left(-\frac{3}{11}, -\frac{4}{33}, 1\right)c.$$

Different choices of c will alter the magnitude or reverse the direction of these vectors. Thus, the tangent to the curve at the point P will be in the direction of the vector $(-3/11, -4/33, 1)$ (or equivalently, in the direction of the vector $(9, 4, -33)$, which we get by taking $c = -33$.).

Exercises

1 Why cannot the following equation be "solved for y":

$$\frac{x^2}{x+y} + y = 1 + \frac{y^2}{x+y}?$$

2 Would you expect to be able to solve the equation

$$3z^2 - 2xyz + x^2 + 2 = 0$$

for z in a neighborhood of the point $(1, 3, 1)$? The point $(1, 5, 3)$?

3 The point $(1, 1, 1)$ lies on the surface described by $xy + 2yz^2 + xz^3 = 4$. What are the values of $\partial z/\partial x$ and $\partial z/\partial y$ there?

4 The point $(1, 1, 0)$ lies on the surface described by $xe^y + ye^z + ze^x = e + 1$. What are the values of $\partial z/\partial x$ and $\partial z/\partial y$ there?

5 Suppose $x^2y + 2yz^3 + xyz = 8$.

(a) Find formulas for $\partial z/\partial x$ and $\partial z/\partial y$.

(b) Find a formula for $\partial x/\partial z$ and for $\partial x/\partial y$.

(c) Evaluate these at the point $(1, 2, 1)$.

6 What is the volume of the largest rectangular box that can be placed as shown in Figure 3–31 with one vertex on the plane

$$3x + 4y + 5z = 12?$$

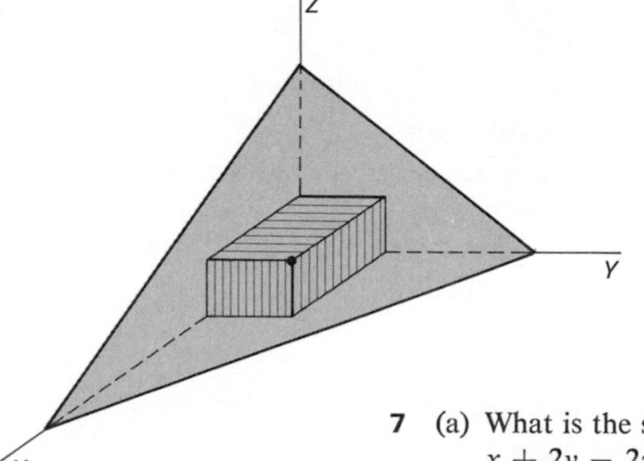

Figure 3–31

7 (a) What is the shortest distance from the point $(1, 2, 3)$ to the plane $x + 2y - 2z = 5$? Solve this using the methods of the present section by minimizing

$$(x - 1)^2 + (y - 2)^2 + (z - 3)^2.$$

(b) Compare your answer with that found by the method of Exercise 13, Section 1.5.

8 It is intuitively clear that the most economical closed rectangular box to hold a given volume should be cubical. Prove this by minimizing $A = 2xy + 2yz + 2xz$ when $xyz = V$.

9 The cylinders $x^2 + y^2 = 5$ and $x^2 + z^2 = 13$ intersect to form two closed curves. One curve passes through $(2, 1, 3)$. Find the direction of the tangent to this curve there.

10 Find the distance between the following nonintersecting lines in space:

$$(x, y, z) = (t, 1 + t, 2 - t),$$
$$(x, y, z) = (t + 2, 2t - 1, t).$$

(*Hint:* Change the parameter in the second equation to s and then minimize the distance between a point on one line and a point on the other.)

***11** An area next to a wall is to be fenced in on three sides as shown in Figure 3–32 using a fixed amount of material, 1000 square feet. What design will maximize the volume enclosed? (*Hint:* After using the implicit function method, analyze the function for the volume by finding z explicitly in terms of x and y.)

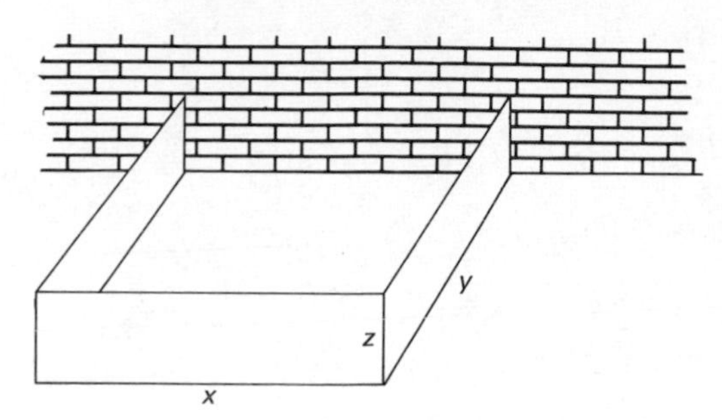

Figure 3–32

multiple integrals

chapter 4

4.1

Introduction

In your previous experience with calculus, you have learned to calculate the integrals of a wide variety of functions of one variable and to use integrals to find areas, to compute the work done by a variable force, to find the volumes of certain simple solids, and perhaps to solve certain very simple differential equations. When the functions involved were especially nice, you could obtain "exact" numerical answers by evaluating the antiderivative of the given function, and by using the so-called Fundamental Theorem of Calculus, which relates the definite integral of a function to its antiderivative; for other cases —including most of those likely to arise outside a calculus textbook —you know that you can return to the basic definition of the integral and use any of several simple numerical procedures such as the trapezoidal formula to obtain the answer to any degree of accuracy desired.

When we turn to functions of several variables, some things remain the same, while others alter radically. There is nothing quite like the Fundamental Theorem of Calculus, nor is there anything that plays the role of an antiderivative for a function of several variables. Furthermore, there are two different sorts of integration which are so closely associated that it is sometimes hard to keep them distinct; for functions of two variables, we call them the **double integral** and the **iterated integral**. As we shall see, a double integral is most frequently evaluated by converting it into an equivalent iterated integral; nevertheless, the concepts are distinct, and applications may give rise to one or to the other. The *iterated integral* is something akin to the inverse of a mixed partial derivative; in evaluating an iterated integral of a function of two variables, one first integrates with respect to one of the variables and then with respect to the second. Here the order in which the integrations are performed is important. In contrast, the concept of the *double integral* (and more generally that of the multiple integral) is a direct extension of the concept of the usual definite integral and has its roots in the problem of defining area, volume, or mass for regions of arbitrary shape.

In the study of integration of functions of one variable, it is easy to produce examples such as

$$\int_0^1 e^{-x^2}\,dx \quad \text{or} \quad \int_0^1 \frac{1}{\sqrt{x^3+1}}\,dx$$

which cannot be evaluated by the Fundamental Theorem, since the antiderivative of the integrand is a function which cannot be expressed in terms of any of the so-called elementary functions. In such cases, one must resort to a numerical approximation procedure. This is even more likely to occur with integrals of functions of several variables. So in order to know how to use approximate numerical integration well it is important for a student to grasp the basic definition of the multiple integral and to understand its connection with practical aspects of approximate numerical integration.

To return to the contrasts between integration in one variable and in several variables, we note the role of change of variable or substitution. This is a technique of great usefulness in evaluating single integrals; recall, for example, the results of the substitution $x = \sin\theta$ for evaluating integrals of functions containing the term $\sqrt{1-x^2}$. In multiple integrals, however, substitution is only of limited usefulness, and because of the complexity of the technique, we shall illustrate only a few of the most standard changes of variable.

4.2

The Iterated Integral

An iterated integral is merely an ordinary single definite integral in which the integrand is itself a function defined by an integral. Each of the following is an example:

$$I_1 = \int_0^2 \left\{\int_1^u 5u^2v\,dv\right\} du, \tag{4.1}$$

$$I_2 = \int_{-1}^2 \left\{\int_0^1 (x^2+y^2)\,dx\right\} dy, \tag{4.2}$$

(4.3)
$$I_3 = \int_0^1 \left\{ \int_{-1}^2 (x^2 + y^2)\, dy \right\} dx,$$

(4.4)
$$I_4 = \int_0^1 \left\{ \int_0^1 \left\{ \int_0^1 (x^2 - yz)\, dz \right\} dy \right\} dx,$$

(4.5)
$$I_5 = \int_{-1}^1 \left\{ \int_0^x \left\{ \int_x^y (x^2 - yz)\, dz \right\} dy \right\} dx.$$

To evaluate any of the above, one works from the inside out. For example, if we try (4.1), we first compute the integral that defines the integrand

$$f(u) = \int_1^u 5u^2 v\, dv = \frac{5u^2v^2}{2}\bigg|_{v=1}^{v=u}$$
$$= \tfrac{5}{2}u^2u^2 - \tfrac{5}{2}u^2 = \tfrac{5}{2}(u^4 - u^2).$$

The iterated integral presented in (4.1) is therefore exactly $I_1 = \int_0^2 \frac{5}{2}(u^4 - u^2)\, du$, which in the usual way becomes

$$\frac{5}{2}\left(\frac{u^5}{5} - \frac{u^3}{3}\right)\bigg|_0^2 = \frac{28}{3}.$$

Before proceeding to the computation of some of the remaining iterated integrals, we should mention that several other notations are often used for such integrals. Perhaps the most common modification is omission of the braces. Then (4.1) appears as

(4.1)′
$$I_1 = \int_0^2 \int_1^u 5u^2 v\, dv\, du$$

and (4.4) as

(4.4)′
$$I_4 = \int_0^1 \int_0^1 \int_0^1 (x^2 - yz)\, dz\, dy\, dx.$$

Conventionally, the order of the differentials dz, dy, dx from left to right indicates the order in which the successive integration is to be done; the braces are implicit.

A better notation is obtained when we omit the braces but place the differential adjacent to the corresponding integral sign. To illustrate, the integral in (4.1) then appears as

(4.1)″
$$\int_0^2 du \int_1^u 5u^2 v\, dv,$$

and (4.5) as

$$\int_{-1}^{1} dx \int_{0}^{x} dy \int_{x}^{y} (x^2 - yz)\, dz. \tag{4.5''}$$

Using this form, one merely works from right to left and follows the instructions given in the limits of integration. For example, the sequence of steps involved in doing (4.5)″ is

$$\begin{aligned}
I_5 &= \int_{-1}^{1} dx \int_{0}^{x} dy \{x^2 z - \tfrac{1}{2}yz^2\}\Big|_{z=x}^{z=y} \\
&= \int_{-1}^{1} dx \int_{0}^{x} dy \{(x^2 y - \tfrac{1}{2}yy^2) - (x^2 x - \tfrac{1}{2}yx^2)\} \\
&= \int_{-1}^{1} dx \int_{0}^{x} \{\tfrac{3}{2}x^2 y - \tfrac{1}{2}y^3 - x^3\}\, dy \\
&= \int_{-1}^{1} dx \{\tfrac{3}{4}x^2 y^2 - \tfrac{1}{8}y^4 - x^3 y\}\Big|_{y=0}^{y=x} \\
&= \int_{-1}^{1} dx \{(\tfrac{3}{4}x^2 x^2 - \tfrac{1}{8}x^4 - x^3 x) - (0)\} \\
&= \int_{-1}^{1} -\tfrac{3}{8}x^4\, dx = -\tfrac{3}{40}x^5\Big|_{x=-1}^{x=1} \\
&= -\tfrac{3}{40}\{1 - (-1)\} = -\tfrac{3}{20}.
\end{aligned}$$

For comparison, let us return to (4.4) and evaluate it, noting the effect of having simpler limits of integration. Switching to the alternate notation, we have

$$I_4 = \int_{0}^{1} dx \int_{0}^{1} dy \int_{0}^{1} (x^2 - yz)\, dz,$$

and proceeding as outlined above, we obtain in turn

$$\begin{aligned}
I_4 &= \int_{0}^{1} dx \int_{0}^{1} dy \{x^2 z - \tfrac{1}{2}yz^2\}\Big|_{z=0}^{z=1} \\
&= \int_{0}^{1} dx \int_{0}^{1} dy \{x^2 - \tfrac{1}{2}y\} \\
&= \int_{0}^{1} dx \{x^2 y - \tfrac{1}{4}y^2\}\Big|_{y=0}^{y=1} \\
&= \int_{0}^{1} dx \{x^2 - \tfrac{1}{4}\} = \{\tfrac{1}{3}x^3 - \tfrac{1}{4}x\}\Big|_{x=0}^{x=1} \\
&= \tfrac{1}{3} - \tfrac{1}{4} = \tfrac{1}{12}.
\end{aligned}$$

Many of the applications of elementary integration which you have encountered previously have in fact been concealed instances of iterated integrals, because the integrand originated as an integral. This is the case, for example, in the formula for the volume of a solid that is often given in elementary calculus and is called Cavalieri's principle. If $A(z)$ is the area of a plane cross section of a solid at height z, then the volume of the solid is given by

$$V = \int_a^b A(z)\, dz.$$

(See Figure 4–1.)

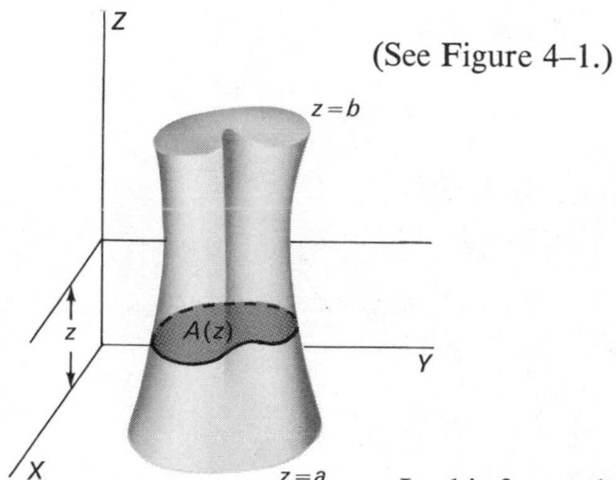

Figure 4–1

In this form, the right side of the above equation is not an iterated integral. However, in most cases, the cross-sectional area $A(z)$ must itself be calculated by means of integration. For example, we may have

$$A(z) = \int_{\alpha(z)}^{\beta(z)} F(x, z)\, dx,$$

where the limits of integration and the integrand depend upon which cross section we are taking and thus upon the value of z.

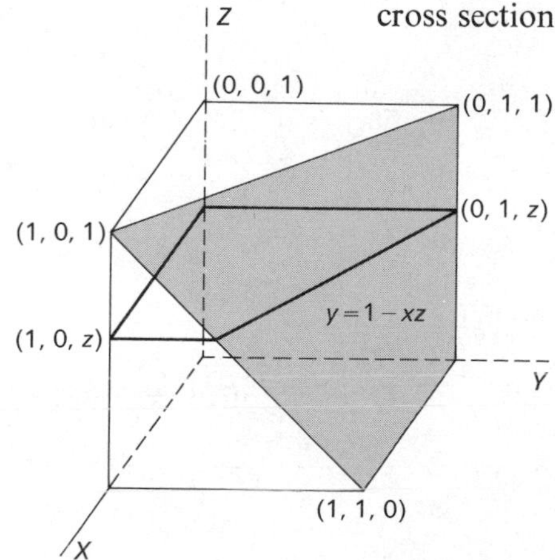

Figure 4–2

To illustrate, consider the solid shown in Figure 4–2. We have shown a typical cross section at height z. It is a trapezoid whose base is against the XZ-plane and whose top is the intersection of the horizontal plane at height z with the defining surface $y = 1 - xz$. If we transport this cross section to the XY-plane as in Figure 4–3, we see that its area is given by

$$A(z) = \int_0^1 (1 - xz)\,dx.$$

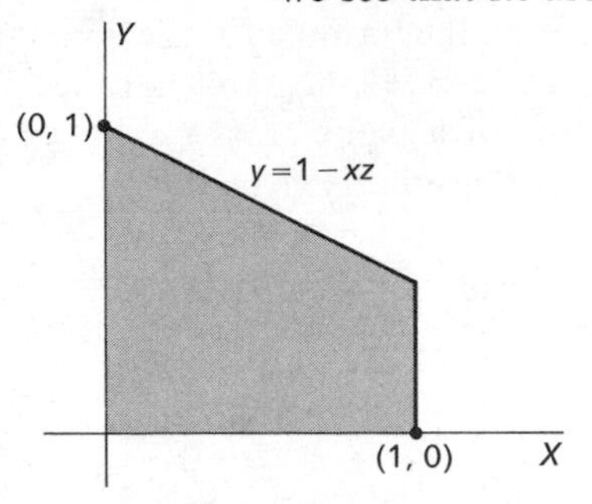

Figure 4–3

Thus, the volume of the solid is given by

$$V = \int_0^1 dz \int_0^1 (1 - xz)\,dx.$$

Evaluating, we have

$$V = \int_0^1 dz\,\{x - \tfrac{1}{2}x^2 z\}\Big|_{x=0}^{x=1}$$

$$= \int_0^1 (1 - \tfrac{1}{2}z)\,dz = (z - \tfrac{1}{4}z^2)\Big|_0^1 = \tfrac{3}{4}.$$

Exercises

1 Evaluate $\int_1^2 F(x)\,dx$, where

(a) $F(x) = \int_{-1}^{1} (y^2x - yx^2)\,dy$, (b) $F(x) = \int_0^x (y^2 + y)\,dy$,

(c) $F(x) = \int_0^x (y^2x - yx^2)\,dy$.

2 Evaluate

(a) $\int_0^1 du \int_0^u (u - t)(u + t)\,dt$, (b) $\int_{-1}^1 du \int_{-u}^{2u} (t^2u - t)\,dt$.

3 Evaluate

(a) $\int_{-1}^1 dy \int_y^{2y} (xy - x - y)\,dx$, (b) $\int_0^2 dx \int_{1-x}^{1+x} (1 + xt)\,dt$.

4 Evaluate

(a) $\displaystyle\int_0^1 \int_0^2 (x^2 + y)\,dx\,dy,$ (b) $\displaystyle\int_0^2 \int_0^1 (x^2 + y)\,dx\,dy.$

5 Evaluate

(a) $\displaystyle\int_0^2 \int_1^x (xy + 4y)\,dy\,dx,$ (b) $\displaystyle\int_{-1}^1 \int_{-v}^v (4u - 3v)\,du\,dv.$

6 Evaluate

$$\int_0^1 dx \int_1^x dy \int_0^{2y} (xyz)\,dz.$$

7 Evaluate

$$\int_1^2 du \int_0^u dv \int_0^{u+v} (u - v + w)\,dw.$$

8 Evaluate

$$\int_0^1 dx \int_0^x dy \int_0^y (z - 1)\,dz.$$

9 Evaluate

$$\int_0^1 dx \int_0^x dy \int_x^y (xy + yz)\,dz.$$

10 (a) Use elementary geometry to show that the area of a plane section of the solid shown in Figure 4–4 is

$$A(z) = \tfrac{1}{25}(10 - z)(15 - 2z).$$

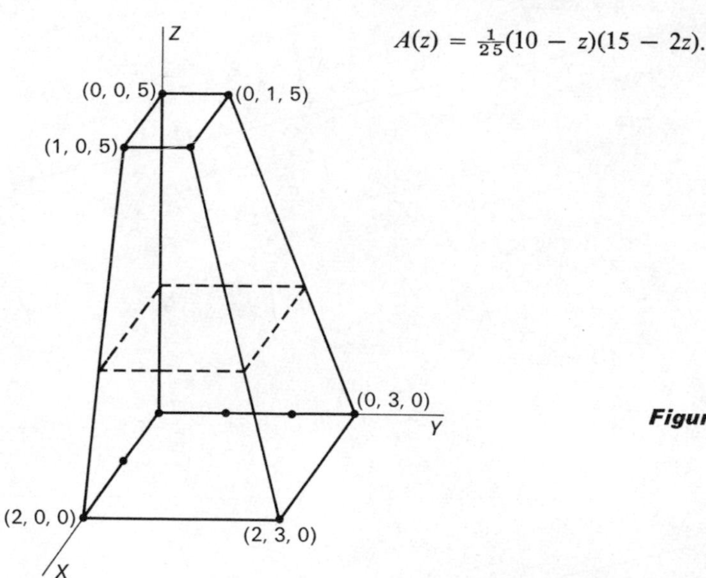

Figure 4–4

(b) Find the volume of the solid using Cavalieri's principle.

11 Given a solid bounded by the coordinate planes, the plane $y = 3$, and the surface $z = (y + 1) - x^2$, as shown in Figure 4–5:

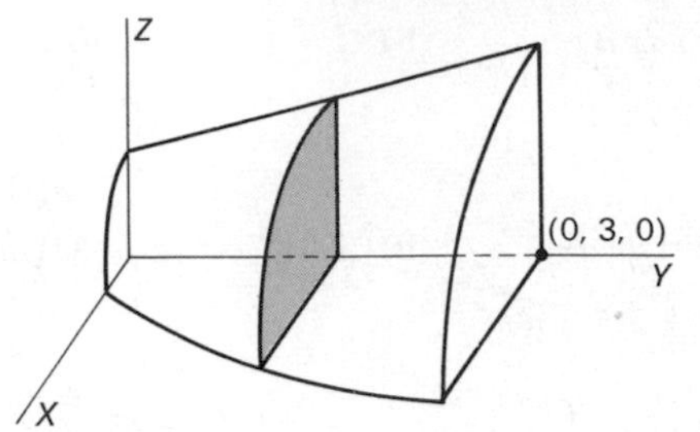

Figure 4–5

(a) Show that if $A(y)$ is the area of a vertical cross section at y, then

$$A(y) = \int_0^{\sqrt{y+1}} (y + 1 - x^2)\, dx.$$

(b) Show that the volume of the solid is given by

$$V = \int_0^3 \tfrac{2}{3}(y + 1)\sqrt{y + 1}\, dy.$$

(c) Evaluate the integral to obtain $V = (4/15)(31)$.

12 Evaluate

$$\int_0^1 dy \int_1^2 ye^{xy}\, dx.$$

13 Evaluate

$$\int_{-\pi/8}^{\pi/8} dy \int_{\sin y}^{\cos y} 4x \cos 2y\, dx.$$

14 Evaluate

$$\int_1^e dx \int_1^{\sqrt{x}} \frac{y}{x}\, dy.$$

15 Evaluate

$$\int_1^e dx \int_1^x \frac{\log y}{xy}\, dy.$$

16 Evaluate

$$\int_2^4 dt \int_0^{\sqrt{t}} \frac{s}{\sqrt{t + s^2}}\, ds.$$

4.3

Area and Volume

The Greek mathematician Eudoxus was probably the first person to build a satisfactory theory for the concept of length and to make an intelligent start on a corresponding theory for area and volume. The most naive approach is to measure the length of a line segment in terms of a standard unit segment, so that the length of a segment is simply the number of copies of the unit segment needed to produce the given segment. When the unit segment doesn't fit evenly into the given segment, one subdivides the unit segment into smaller equal unit segments and uses them to measure the given segment. The success of this process depends on the assumption that given any two line segments, one can find a suitable common unit segment such

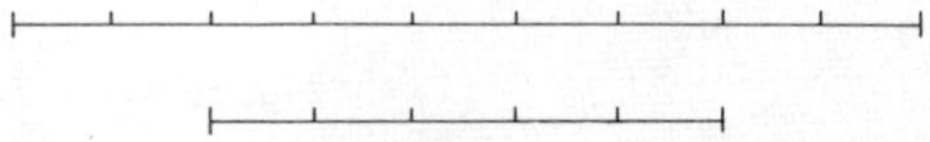

Figure 4–6

that each of the segments can be partitioned into a whole number of copies of the unit. (See Figure 4–6.) However, the discovery by Pythagoras (circa 540 B.C.) that the side and the diagonal of a square were incommensurable forever destroyed this simpleminded approach to length.

To restate the original approach in more modern terms, it sought to measure all line segments by means of rational numbers, and the discovery of Pythagoras showed that there were line segments whose length could not be measured by rational numbers. Having nothing comparable to the real number system, the Greeks could not restate their discovery in the simple form: The real number $\sqrt{2}$ is irrational.

Eudoxus, coming approximately 150 years after Pythagoras, overcame the existence of incommensurables and the limitations of the rational number system by a systematic process that used sequences of rational numbers, which in effect was equivalent to the sophisticated approach used by Dedekind more than two thousand years later. In modern language, an irrational number is identified by means of one or more sequences of rational numbers that converge to it. For a specific problem, Eudoxus frequently produced two sequences of rational numbers both converging to the desired quantity, one monotone increasing and the other monotone decreasing.

This idea is most transparent in the proofs of some of the classical theorems about volumes to be found in Books V and XII of Euclid. Little of this approach is found in our expurgated modern versions of Euclid except in the calculation of the area of a circle by "exhaustion," that is, by looking at the areas of a sequence of circumscribed and inscribed polygons. In a more general form, it is this approach that we shall use in this section to discuss the notions of area and volume.

First, let us formulate some of the properties which we might expect the notion of area to exhibit. If S is any set in the plane, it should have an area which we will denote by $A(S)$. Thus, we are dealing with a function A whose values are nonnegative real numbers and whose domain consists of sets S in the plane. If a set S_0 is a subset of S, i.e., $S_0 \subset S$, then we expect to have $A(S_0) \leq A(S)$. If a set S is divided into two disjoint pieces S_1 and S_2, then we expect that $A(S) = A(S_1) + A(S_2)$. If two sets S_1 and S_2 are congruent, so that one can be placed on top of the other by a rigid motion, then we expect that $A(S_1) = A(S_2)$. We also want a thin set such as a line segment to have zero area and a rectangle with length ℓ and width w to have area ℓw.

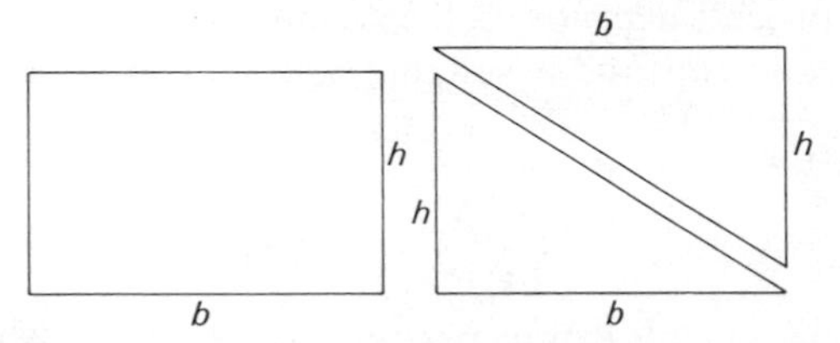

Figure 4–7

Based on these assumed properties, can we calculate the area of an arbitrary set? In elementary geometry, a simple argument (suggested in Figure 4–7) was used to show that the area of a right-angled triangle was given by $A(T) = bh/2$, and another argument (see Figure 4–8) showed that this formula works for any triangle. (Which properties of the area function do these proofs use?) Using this formula, we can then find the area of any polygonal set that can be partitioned into triangles, as in Figure 4–9.

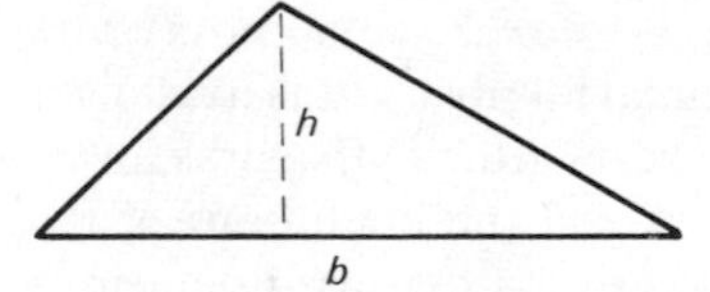

Figure 4–8

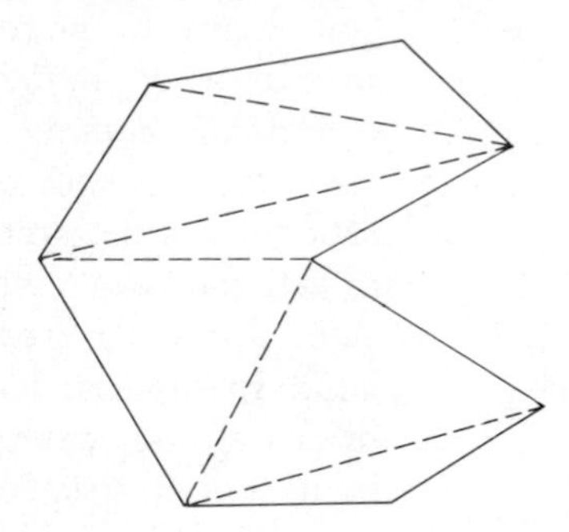

Figure 4–9

The area of a region with a curved boundary can be studied by finding polygonal sets which approximate it from inside and from outside and seeing what happens as the fit of the approximating sets is improved. For the circular disk, this was the technique called the method of exhaustion used by Eudoxus and his successors (see Figure 4–10).

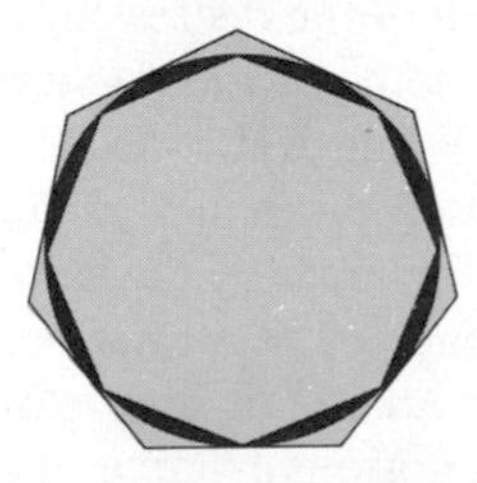

Figure 4–10

However, this entire approach can be questioned. For example, how do we know that we will arrive at the same numerical answer if we partition an irregular region such as that in Figure 4–9 into triangles in two different ways? Indeed, how do we know that there *is* such a function as the area function? Why does it have all the properties we listed? And is the concept of area applicable to all sets in the plane?

A complete treatment of the subject of area answering all these objections is best left to an advanced analysis course. However, we can give a useful partial treatment that simplifies the discussion of the double integral in the next section.

Let us begin again. This time, we assume only that we have defined area for rectangles and see how we might proceed to define an area function A for sets in the plane. If we were to give a complete treatment, we would then check to see that our function A has all the desirable properties that we listed above, and in the process we would answer the various questions posed. We shall not carry through this whole program, but we will sketch several of the important steps.

Suppose that D is a set whose area we wish to find. We assume that it is a bounded set, and we enclose D in a large rectangle (Figure 4–11).

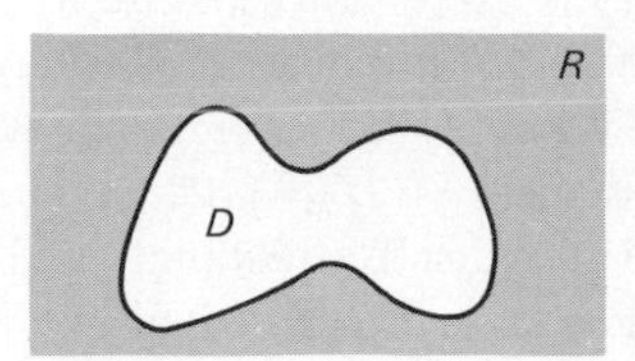

Figure 4–11

We choose a large integer N, and then divide each edge of the rectangle into N equal parts, constructing a grid work of lines dividing the original rectangle into N^2 small rectangles. (See Figure 4–12.) We next look at all of these small rectangles that lie entirely within the set D so that none of their points touches the edge of the set D. (These have been shaded in Figure 4–12.) We add up the areas of each of these rectangles and call the sum the inner approximation, denoting its value by $s(N)$. (The lower case s indicates "inner approximation.")

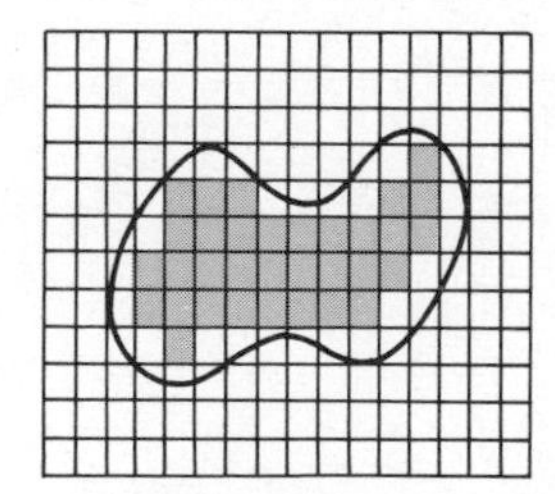

Figure 4–12

We return to the picture and this time we look at all the small rectangles formed by the grid which contain any part of the set D (see Figure 4–13, where these rectangles have been shaded). We add up the areas of these rectangles, call the sum the outer approximation, and denote its value by $S(N)$. Clearly, for any integer N, we have $s(N) \leq S(N)$.

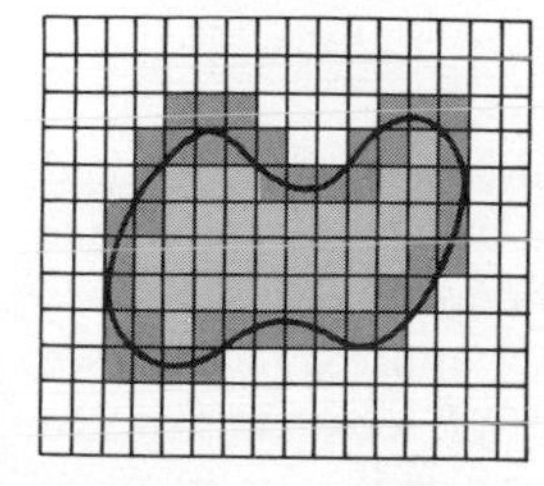

Figure 4–13

Definition 1 A set D is said to have an area α if and only if each of the sequences $\{s(N)\}$ and $\{S(N)\}$ converge to α as N increases; in this case, we write $A(D) = \alpha$.

The first question raised by this definition is whether it is possible to have $\lim_{N\to\infty} s(N)$ and $\lim_{N\to\infty} S(N)$ both exist, but be different. This is possible, and in such a case we say that the set D does not have an area. Do not confuse this with the somewhat similar statement "D has area zero," which means that both sequences $\{s(N)\}$, $\{S(N)\}$ converge to 0. (A linguistic analogy would be the words "free" and "priceless"; an item is free if its price is zero, but something (e.g., liberty) is priceless if it is impossible to define a price for it.) An example of a set D which does not have an area defined for it is given in the exercises (Exercise 8).

A second question of importance is whether this definition is consistent. Suppose we apply it to a rectangle with length a and width b. Will the process yield $A(D) = ab$, as it should? This is easy to check. If the set D is itself a rectangle with sides of length a and b and we subdivide it as described above, then each of the small rectangles has sides a/N and b/N and area ab/N^2. Calculating the inner approximation is easy, since we have exactly $(N - 2)^2$ of these smaller rectangles, so that $s(N) = (N - 2)^2(ab/N^2)$. (See Figure 4–14.) Likewise, there are N^2 rectangles in the outer approximation, and $S(N) = N^2(ab/N^2)$. It is clear that $\lim S(N) = ab$; since we have

$$s(N) = \frac{(N-2)^2}{N^2}\,ab = \left(1 - \frac{2}{N}\right)^2 ab,$$

it is also clear that $\lim s(N) = ab$. Hence, D has area; and $A(D) = ab$, as we hoped.

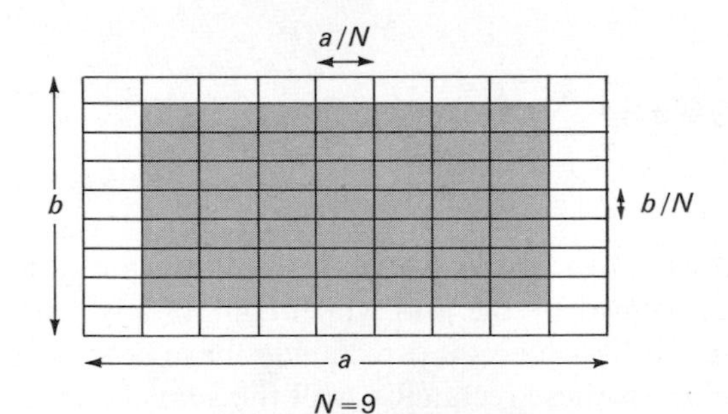

Figure 4–14

Several other properties that we listed earlier can be checked easily. For example, one may show directly that any line segment has zero area. Since no rectangles can fit inside a line segment, the inner approximation is just zero. Looking at the outer approximation (see Figure 4–15), we see that a line segment of length L located as shown will not require more than $3N$ rectangles to cover it, for at most

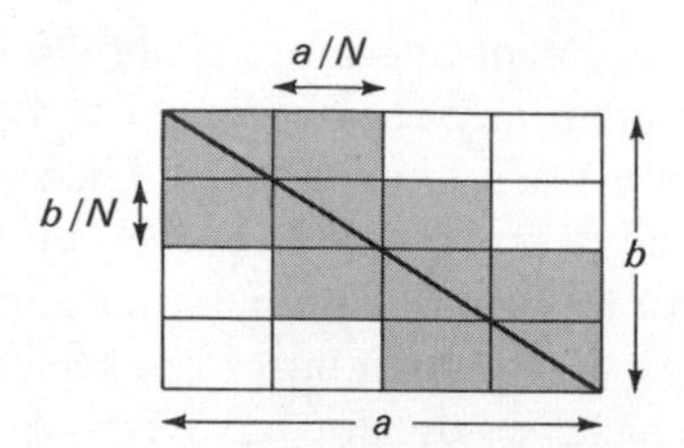

Figure 4–15

three rectangles in either column or row touch the line segment. Each of these rectangles will have area ab/N^2. Thus,

$$S(N) \leq 3(N)(ab/N^2) = 3ab/N, \quad \text{and} \quad \lim S(N) = 0.$$

Again, if $D_0 \subset D$, then consideration of a picture will show that any inner approximation for D_0 is smaller than or equal to the inner approximation for D, with the same value for N. A similar statement holds for outer approximations, so that one easily concludes that $A(D_0) \leq A(D)$, provided that each of these sets is of the sort for which area is defined. (In the exercises, we have suggested a few examples which will help you to see this.)

Is there a way to determine which sets D have an area? From the definition of area given above, it is clear that what is crucial is that $\lim_{N\to\infty} s(N) = \lim_{N\to\infty} S(N)$, or what amounts to the same thing, $\lim_{N\to\infty} \{S(N) - s(N)\} = 0$. Since each of the numbers $s(N)$ and $S(N)$ comes from adding the areas of a collection of small rectangles, the difference $S(N) - s(N)$ is merely the sum of the areas of those rectangles used in the outer approximation but not in the inner approximation. In Figure 4–16 we have shaded these rectangles;

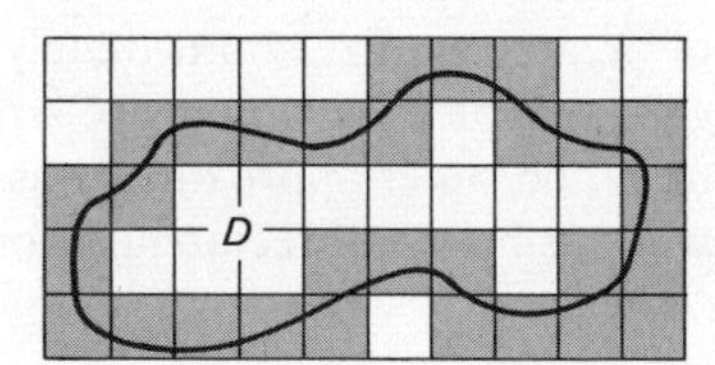

Figure 4–16

you will see that they are exactly those rectangles that cover up the edge of the set D. The number $S(N) - s(N)$ is therefore the outer approximation for the chosen value of N that would arise in the process of finding the area of the set of points consisting exactly of the edge or boundary of the set D. The statement

$$\lim_{N\to\infty} \{S(N) - s(N)\} = 0$$

asserts that the boundary of the set D has zero area. The important result that we have shown by this argument is the following:

Theorem 1 *Let D be a bounded set. Then the area of D is defined if and only if its boundary is a set of zero area.*

Since a line segment and more generally any polygonal curve is a set with zero area, any region D whose boundary is one or more polygonal curves is a set having area. The same is true for any region bounded by one or more smooth curves, and thus for any region that we are going to meet in this book. (Mathematicians have discovered strange regions whose boundaries are such pathological curves that the curves themselves do not have zero area, and as a result, the concept of area does not apply to these regions.)

It is a difficult task to complete the program of checking that the area function created by Definition 1 obeys all the desired properties. In particular, it is nontrivial to show that two congruent regions always have the same area. We shall not complete the program, but ask you to accept our word that it can be accomplished, and that the above concept of area rests on firm foundations.

The ability to find the area of a given region D is an important tool in mathematics. For many regions, particularly those of complex shape, there is nothing better than the naively simple procedure of subdividing and counting squares. In terms of Definition 1, this amounts to choosing a sufficiently large value of N and calculating $s(N)$ and $S(N)$, with the assurance that the true value $A(D)$ lies between them. In Exercise 7, we have asked you to do this for a disc, thereby finding an estimate for the number π. (If you have access to a high-speed computer, it is an interesting task to write a program that will do the counting for a relatively large value of N, say 50; you may do it just for a quarter of the disc and use the coordinates of the upper right-hand corner of a small square to test whether the square is entirely inside the bounding circle.)

We shall say little about volume. The pattern we have outlined for the concept of area generalizes immediately to that of volume, and all the analogues are true. The volume of a region in space is defined in terms of inner and outer approximations, and these are obtained in the same way, using small rectangular blocks instead of squares or rectangles.

Exercises

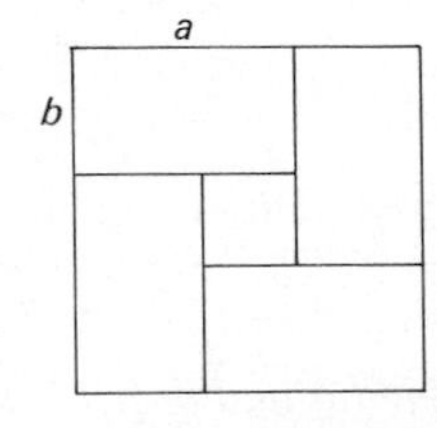

Figure 4–17

1 Assume only that the area of any square D whose side has length ℓ is given by $A(D) = \ell^2$. Prove that for any rectangle R of sides a and b, the area is given by $A(R) = ab$. (*Hint:* See Figure 4–17.)

2 (a) Show that the area of the large rectangle determined by the points A, B, C, D is equal to the sum of the areas of the small rectangles $R_1, R_2, R_3, R_4, R_5, R_6$ as displayed in Figure 4–18(a), using the formula for the area of a rectangle.

(b) Would the same argument be valid for further subdivision?

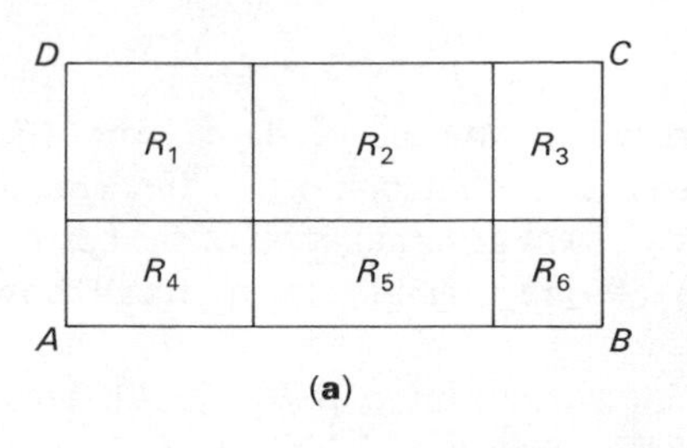

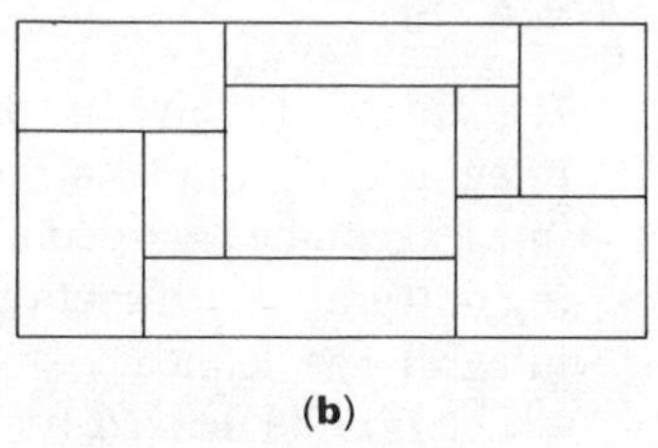

Figure 4–18

(c) Show that the sum of the areas of the small rectangles in Figure 4–18(b) is equal to the area of the large rectangle.

3 If regions D_1 and D_2, both having area, overlap,

$$A(D_1 \cup D_2) + A(D_1 \cap D_2) = ?$$

4 Assume area is defined for right-angled triangles as one half the product of the lengths of the two perpendicular sides. Assume the desired properties of the area function as discussed in this section. Complete the proof of the formula $A(D) = \frac{1}{2}bh$ for the area of an obtuse-angled triangle D (see Figure 4–19). Show that the area is also equal to $\frac{1}{2}b'h'$.

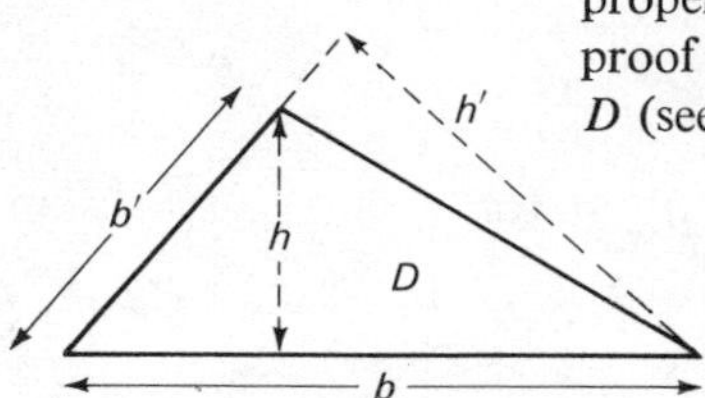

Figure 4–19

5 Use the definition of area (Definition 1) to determine the area of a right-angled triangle with perpendicular sides of lengths 1 and 2.

6 Using the definition of area, show why $A(D_1 \cup D_2) = A(D_1) + A(D_2)$ in such a case as illustrated in Figure 4–20.

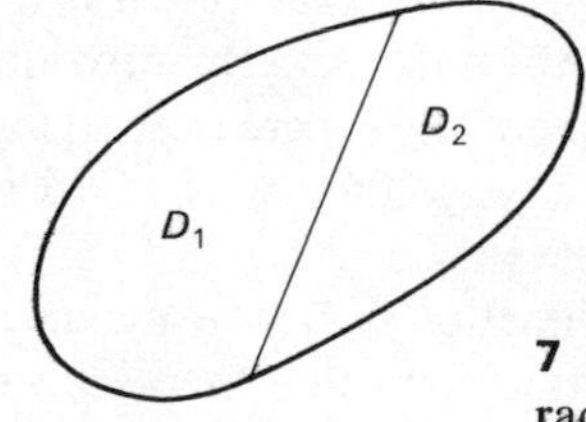

Figure 4–20

7 Use the definition of area to find the approximate area of a disc of radius 1 by drawing a quarter circle on graph or ruled paper and counting. Remember, the greater the refinement of subdivision, the better the approximation.

***8** Show that the set D composed of all the points with rational coordinates which lie in the square $0 \leqq x \leqq 1$, $0 \leqq y \leqq 1$ does not have an area.

9 Give an analogous definition of volume for a region in 3-space. Can you guess an analogous criterion for regions to have volume?

10 Assume that the area of a disc D of unit radius is π. Show that $A(S) < A(T) < A(D)$, where $S \subset D$ is an inscribed square, T is an inscribed octagon, $S \subset T \subset D$, by computing the areas of S and T using elementary geometry. (See Figure 4–21.)

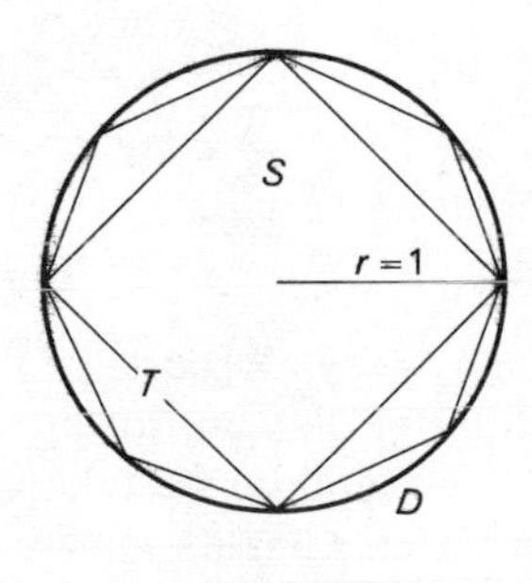

Figure 4–21

4.4

The Double Integral: Rectangular Case

We intend to give a precise definition of $\iint_R f$, the (double) integral of the function f over the rectangular set R. Since the double integral can be regarded as a direct generalization of the corresponding integral of a function of one variable, let us recall how that integral was defined.

Let $f(x)$ be defined for all x in an interval $J = [a, b]$. Then, one way to define the integral $I = \int_J f = \int_a^b f(x)\,dx$ is as follows: Choose any positive integer N, and partition the interval I into n equal subintervals by points x_i:

$$a = x_0 < x_1 < x_2 < \cdots < x_{N-1} < x_N = b.$$

Each subinterval $[x_{k-1}, x_k]$ has length $(b - a)/N$. Choose any point x_k' in the interval $[x_{k-1}, x_k]$ and form the Riemann sum

$$\sum_{k=1}^{N} f(x_k')(x_k - x_{k-1}). \tag{4.6}$$

Since $\Delta x_{k-1} = x_k - x_{k-1} = (b - a)/N$ for each k, this sum becomes

$$\frac{b-a}{N}\sum_{1}^{N} f(x_k').$$

Then, the value of the integral of f over the interval J is the number I around which all such Riemann sums cluster as N increases.

There are several other alternate ways to arrive at the number I, all of which agree for continuous functions f. For example, it is not necessary that all the subintervals have the same length, provided that the length of the largest approaches zero as N increases. Again, when f is continuous, it is sufficient to choose the point x_k' to be always an endpoint of its containing interval $[x_{k-1}, x_k]$.

Another approach to the integral of a function $f(x)$ is by means of average values. Let g be a simple function, one that is piecewise constant and whose graph looks like a bar graph (see Figure 4–22).

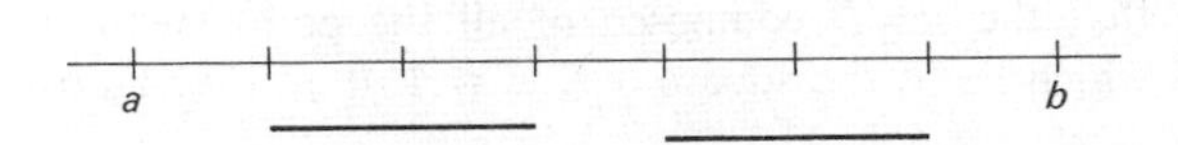

Figure 4–22

We may define $\int_J g$ and reach the definition for $\int_J f$ by a process of bracketing f above and below by simple functions g. This approach is very closely related to the first approach above. Suppose we take x_k' to be the point in the interval $[x_{k-1}, x_k]$ at which f has its minimum value m_k. Then, construct the piecewise constant function

$$g(x) = \begin{cases} m_k & \text{for all } x \text{ with } x_{k-1} \le x < x_k, \\ m_n & \text{when } x = b. \end{cases}$$

In Figure 4–23, we have sketched a function f and the function g that results from this process. Note that $g(x) \le f(x)$ for all $x \in J$, so that the graph of g lies below that of f.

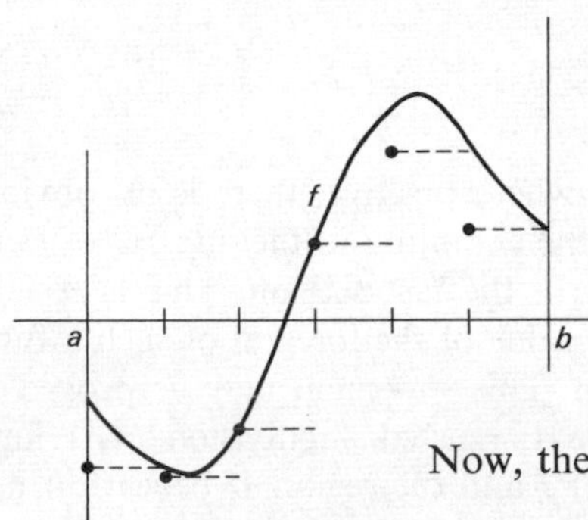

Figure 4–23

Now, the integral of g, which is taken as a lower bound for I, is

(4.7)
$$\int_J g = \sum_1^N m_k(x_k - x_{k-1}).$$

However, since $m_k = f(x_k')$, we can rewrite (4.7) as

$$\sum_1^N f(x_k')(x_k - x_{k-1}),$$

and we observe that this is itself a Riemann sum for f, as in (4.6).

In the same way, if M_k is the maximum value of f on $[x_{k-1}, x_k]$ and we define a simple piecewise constant function G by setting

$$G(x) = \begin{cases} M_k & \text{for all } x \text{ with } x_{k-1} \le x < x_k, \\ M_n & \text{when } x = b, \end{cases}$$

then $G(x) \ge f(x)$ for all $x \in [a, b]$. The graph of G lies above the graph of f, $\int_J G$ is an upper bound for I, and

(4.8)
$$\int_J G = \sum_1^N M_k(x_k - x_{k-1}) = \sum_1^N f(x_k'')(x_k - x_{k-1})$$

is another Riemann sum for f, obtained by choosing x_k'' as the point in $[x_{k-1}, x_k]$ where f takes its maximum.

This pair of Riemann sums in formulas (4.7) and (4.8) provides a standard pair called the upper and lower Riemann sums; as N increases, they will converge to I, the value of the integral of f over J, one approaching I from below and the other approaching I from above. The corresponding piecewise constant functions g and G will provide a pair that bracket f from above and below and which will do so more and more closely as N becomes larger. (See Figure 4–24.)

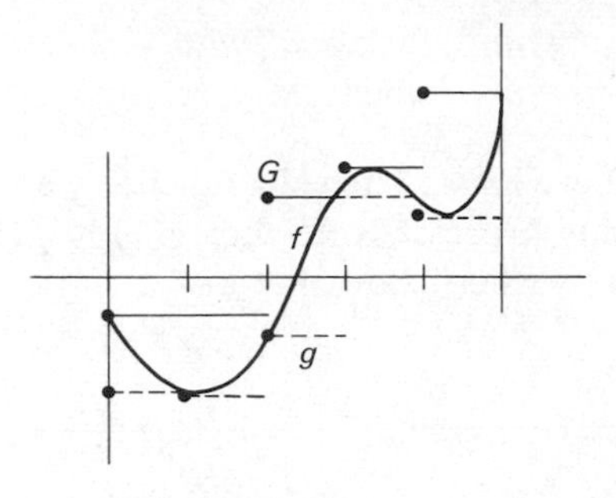

Figure 4–24

When the function f is everywhere positive, there is an obvious and direct connection between the definition of the integral of f and the theory of area as outlined in the last section. This is usually summarized by saying that the value of the integral of such a function is the area under its graph. This statement may be proved by observing that if we introduce the usual subdivision of a large rectangle enclosing the graph of f and the region D beneath it into small rectangles, then the following inequality holds:

$$s(N) \leq \int_J g \leq \int_J f \leq \int_J G \leq S(N).$$

This is illustrated in Figure 4–25, where we have interpreted the values in formulas (4.7) and (4.8) as areas.

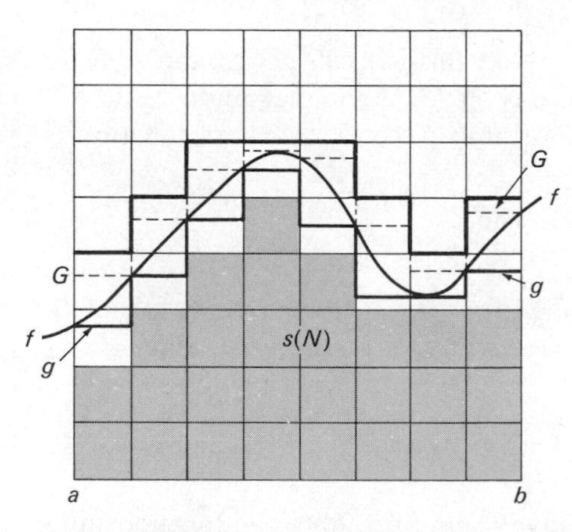

Figure 4–25

We are now ready to present a definition of integration for functions of two variables based on the preceding discussion. Let R be a rectangle in the XY-plane, the set of all points (x, y) with $x \in [a, b]$ and $y \in [c, d]$. Suppose that $f(x, y)$ is defined and continuous for all $(x, y) \in R$. Then, by analogy with the analysis of integration for functions of one variable, we arrive at the following formal definition of $\iint_R f$, which we call the definite integral of f over R, and which we indicate also by the notation $\iint_R f(x, y)\, dx\, dy$, or $\iint_R f(p)\, dA$.

Definition 2 Choose any positive integer N, and construct a partition of the rectangle R into N^2 subrectangles R_{ij} by subdividing each side of R into N equal intervals. Choose any point p_{ij} in R_{ij}, and form the Riemann sum

$$S(f, N) = \sum_{i,j=1}^{N} f(p_{ij})A(R_{ij}). \tag{4.9}$$

Then, the integral of f over R exists and is the number I if and only if $S(f, N)$ converges to I as N increases, in the sense that given any $\epsilon > 0$, there is an N_0 such that

$$|S(f, N) - I| < \epsilon$$

for all $N > N_0$, and any choice of the point p_{ij} in each R_{ij}.

Two special choices for p_{ij} are especially useful. For the first, we select p_{ij} as a point in the small rectangle R_{ij} at which the function f takes its maximum value M_{ij}. The resulting Riemann sum

$$\sum M_{ij}A(R_{ij})$$

is called the Upper Riemann Sum (URS) for the particular partition of the large rectangle R and is an upper estimate for the number I. The other arises from the choice of p_{ij} as a minimum point for f in R_{ij}, and the corresponding Riemann sum $\sum m_{ij}A(R_{ij})$ is the Lower Riemann Sum (LRS) and gives a lower estimate for I. Thus, the true value of I always lies between the upper and lower Riemann sums corresponding to any given partition of R.

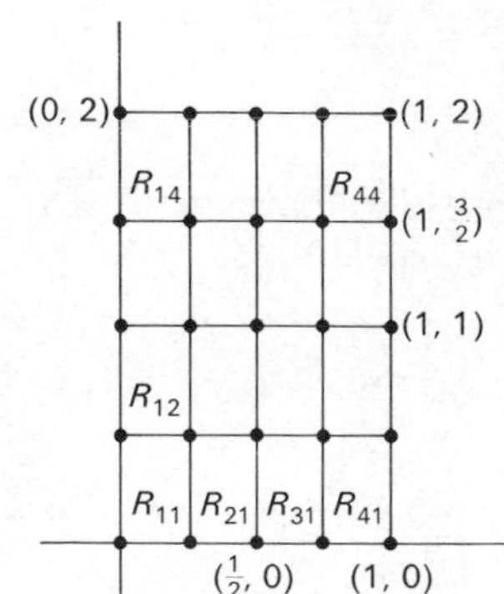

Figure 4–26

Perhaps a numerical example will make the meaning of this definition clearer. Let us take for R a rectangle in the XY-plane with vertices at (0, 0), (1, 0), (1, 2), (0, 2), and choose $N = 4$. The corresponding partition of R yields 16 smaller rectangles, as shown in Figure 4–26. For the function f, we choose $f(x, y) = 3x + y$. We notice that this function is nonnegative on R and that its value increases as (x, y) moves away from the origin in such a way that both x and y get larger. Thus, if we confine our attention to any one of the small rectangles in R, the smallest value of f will occur at the lower left corner, and the largest value at the upper right corner.

Accordingly, if we follow the instructions for calculating the Lower Riemann Sum (LRS), we have

$$\begin{aligned}\text{LRS} = {} & f(0,0)A(R_{11}) + f(\tfrac{1}{4},0)A(R_{21}) + f(\tfrac{1}{2},0)A(R_{31}) \\ & + f(\tfrac{3}{4},0)A(R_{41}) + f(0,\tfrac{1}{2})A(R_{12}) + \cdots \\ & + f(\tfrac{1}{2},\tfrac{3}{2})A(R_{34}) + f(\tfrac{3}{4},\tfrac{3}{2})A(R_{44}).\end{aligned}$$

Each of the small rectangles R_{ij} has sides $\frac{1}{4}$ by $\frac{1}{2}$, and area $\frac{1}{8}$, which is $(1/N^2)A(R)$. Using the formula for $f(x, y)$, namely $3x + y$, we find

$$\begin{aligned}\text{LRS} &= \tfrac{1}{8}\{(0+0) + (\tfrac{3}{4}+0) + (\tfrac{3}{2}+0) + (\tfrac{9}{4}+0) \\ &\qquad + (0+\tfrac{1}{2}) + \cdots + (\tfrac{9}{4}+\tfrac{3}{2})\} \\ &= \tfrac{1}{8}\{4(0+\tfrac{3}{4}+\tfrac{3}{2}+\tfrac{9}{4}) + 4(0+\tfrac{1}{2}+1+\tfrac{3}{2})\} \\ &= \tfrac{1}{2}\{3(0+\tfrac{1}{4}+\tfrac{1}{2}+\tfrac{3}{4}) + (0+\tfrac{1}{2}+1+\tfrac{3}{2})\} \\ &= \tfrac{1}{2}\{\tfrac{9}{2}+3\} \\ &= \tfrac{15}{4} \\ &= 3.75.\end{aligned}$$

In the same way, we calculate the Upper Riemann Sum (URS) by using the upper right-hand corner of the small rectangles for the p_{ij}.

$$\begin{aligned}\text{URS} &= f(\tfrac{1}{4},\tfrac{1}{2})A(R_{11}) + f(\tfrac{1}{2},\tfrac{1}{2})A(R_{21}) + \cdots \\ &\quad + f(1,\tfrac{1}{2})A(R_{41}) + f(\tfrac{1}{4},1)A(R_{12}) + \cdots \\ &\quad + f(\tfrac{3}{4},2)A(R_{34}) + f(1,2)A(R_{44}) \\ &= \tfrac{1}{8}\{(\tfrac{3}{4}+\tfrac{1}{2}) + (\tfrac{3}{2}+\tfrac{1}{2}) + \cdots + (\tfrac{9}{4}+2) + (3+2)\} \\ &= \tfrac{1}{8}\{4(\tfrac{3}{4}+\tfrac{3}{2}+\tfrac{9}{4}+3) + 4(\tfrac{1}{2}+1+\tfrac{3}{2}+2)\} \\ &= \tfrac{25}{4} \\ &= 6.25.\end{aligned}$$

This sample calculation has given for $N = 4$ the upper and lower Riemann sums URS = 6.25, LRS = 3.75. Hence, if

$$I = \iint_R (3x + y)\,dx\,dy,$$

then we know that

$$3.75 \le I \le 6.25.$$

We do not show the details, but one may calculate these sums for $N = 10$ and find better estimates:

$$\text{LRS} = 4.5 \le I \le 5.5 = \text{URS}.$$

As the number N which determines the size of the small rectangles becomes larger, the numbers URS and LRS become closer together, and squeeze the number I, which turns out to be $I = 5$. (It is interesting to notice that this value 5 is the average of the numbers URS and LRS in each of the cases we have computed; this fact is not an accident, but is true because the function f we chose to integrate is a polynomial of degree 1. However, in any case, the average of the upper and lower Riemann sums for the same choice of N is quite often a very good approximation to the exact value of the integral.)

Were we to take a more complicated function f, for example $f(x, y) = (x^2 - y^3)e^{x \sin y}$, we would be very grateful for the help of a high-speed computer to do the arithmetic for us, especially if we wanted to obtain the upper and lower Riemann sums when $N = 30$ or even $N = 100$. (How many small rectangles would there be in these cases? How many additions would we have to do?) It is very fortunate therefore that in many cases, when the function to be integrated is suitable, another method for calculating the value of $\iint_R f$ exists. This, as you may have anticipated, is the iterated integral we discussed in the last section. We state the general theorem at this point, but we postpone its proof for a while until we have more practice in applying this method.

Theorem 2 *Let R be the set of all (x, y) with $x \in [a, b]$, $y \in [c, d]$. Let $f(x, y)$ be continuous for (x, y) in R. Then,*

$$\iint_R f(x, y)\, dx\, dy = \int_a^b dx \int_c^d f(x, y)\, dy = \int_c^d dy \int_a^b f(x, y)\, dx.$$

Let us illustrate this theorem with the numerical example we have been examining, namely $I = \iint_R (3x + y)\, dx\, dy$. Since R is the rectangle described by $0 \leq x \leq 1, 0 \leq y \leq 2$, we have two iterated integrals we can use to evaluate the number I. The first is

$$I_1 = \int_0^1 dx \int_0^2 (3x + y)\, dy = \int_0^1 dx \left\{3xy + \tfrac{1}{2}y^2\right\}\Big|_{y=0}^{y=2}$$

$$= \int_0^1 (6x + 2)\, dx = (3x^2 + 2x)\Big|_{x=0}^{x=1} = 5.$$

The second is

$$I_2 = \int_0^2 dy \int_0^1 (3x + y)\, dx = \int_0^2 dy \left\{\tfrac{3}{2}x^2 + xy\right\}\Big|_{x=0}^{x=1}$$

$$= \int_0^2 (\tfrac{3}{2} + y)\, dy = (\tfrac{3}{2}y + \tfrac{1}{2}y^2)\Big|_{y=0}^{y=2} = 5.$$

Either way, we find that $I = 5$.

When the function f is everywhere positive on the rectangle R, we can produce a geometric argument that seems to show why Theorem 2 is true; it may also throw more light on the meaning of the double integral $\iint_R f$. However, it is not a valid proof of Theorem 2, and we shall point out where it fails.

We start by looking at the graph of f; by definition, this will be the set of all points (x, y, z) with $z = f(x, y)$ and with the point (x, y) in the rectangle R. If we picture R as lying in the XY-plane, we obtain a diagram such as that in Figure 4–27. In this diagram,

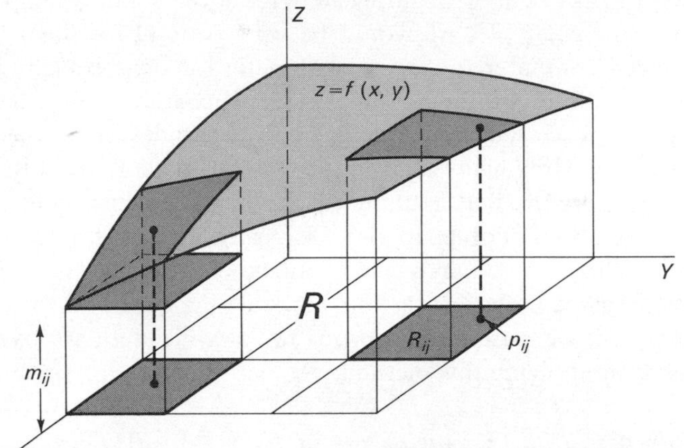

Figure 4–27

we have also shown a partition of R into smaller rectangles corresponding to some selected value of N, here $N = 3$. We can now give a geometric interpretation of a general Riemann sum of the form $\sum_{i,j=1}^{N} f(p_{ij})A(R_{ij})$. In Figure 4–27, please note that we have singled out two of the small rectangles in R and shown a solid block resting on each. The one on the right extends to the surface and has a curved top. Its base is the small rectangle R_{ij}. The dotted line starts at a point p_{ij} in R_{ij} and also extends up to the surface. Since the equation of the surface is $z = f(x, y) = f(p)$, the length of the dotted line is $f(p_{ij})$. Thus, the term $f(p_{ij})A(R_{ij})$ which occurs in the Riemann sum is the product of the length of this dotted line and the area of the base of the block and is therefore about equal to the volume of the block. Indeed, it should seem plausible to you that if we choose the right location of the point p_{ij} in R_{ij}, then this product will be exactly the volume of the block. A rigorous proof follows at once from the continuity of f and the intermediate value theorem. In general, the sum of such terms $f(p_{ij})A(R_{ij})$ will be exactly the volume of the solid whose base is the large rectangle R and whose top is the surface $z = f(x, y)$, if the points p_{ij} are chosen at the right places; in any case, the Riemann sum should be a good approximation to the exact volume of this solid, and should get better and better as we increase the number of small rectangles in the partition of the large rectangle R.

We can in fact be a little more precise about this. Look at the small block on the left side in Figure 4–27. Its height is indicated to be m_{ij}, which is clearly the minimum of the function f on the corresponding rectangle, since this is the lowest point of the surface above that small rectangle. Letting m_{ij} be the minimum of f on each of the small rectangles R_{ij}, we see at once that the special sum $\sum m_{ij}A(R_{ij})$, the Lower Riemann Sum, turns out to be exactly the volume of an inner approximation to the solid bounded above by the surface and lying above the large rectangle R.

In summary, what we have observed is that in the case of positive functions f the value of the double integral $\iint_R f$ turns out to be the volume of the solid lying under the graph of f; note that this is a direct analogue of the corresponding statement connecting single integrals with the area under a curve.

How does this analysis lead to Theorem 2? The answer lies in Cavalieri's theorem and the relationship between volumes and iterated integrals discussed in Section 4.2. If a solid S has a rectangular base and is bounded above by a surface $z = f(x, y)$, then we may calculate its volume by taking cross sections, finding the areas of these cross sections, and integrating the resulting function. Figure 4–28 shows how this process leads to

$$V(S) = \int_a^b dx \int_c^d f(x, y)\, dy.$$

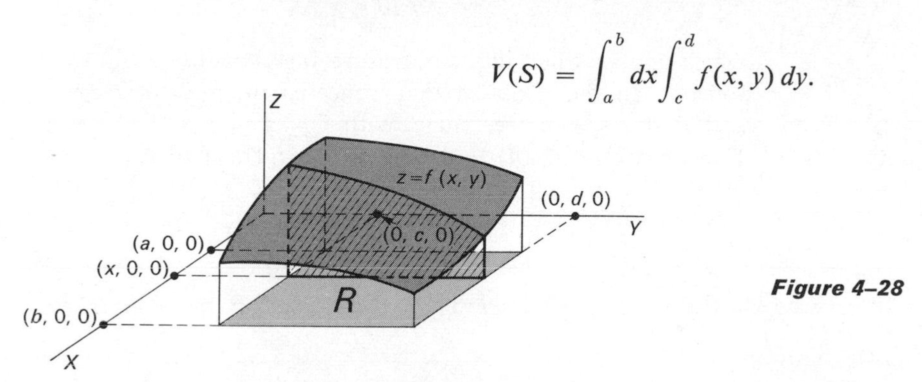

Figure 4–28

Why is this not a valid proof of Theorem 2? This is a classic example of circular reasoning, and therefore not a proof. We invoked a result attributed to Cavalieri (1598–1647), stated but certainly not proved in Section 4.2. As the dates suggest, Cavalieri did not prove it either; he stated it, supporting his statement by an intuitive geometric argument. In fact, Theorem 2 is itself the precise statement of the result we have been calling Cavalieri's theorem and must therefore be proved in quite a different way. We postpone further discussion of the proof of this basic result until Section 4.6, but we shall begin using it immediately.

Double integrals can arise in connection with problems that do not seem at all connected with the volume of solids. Suppose, for example, that we have a rectangular plate with vertices at the points $(1, -1)$, $(3, -1)$, $(1, 3)$, $(3, 3)$ and the temperature of the plate changes from point to point so that the temperature at (x, y) is given by $f(x, y) = 2xy^2 + x^2$. What is the average temperature of the plate? By average temperature we mean that temperature which if uniform across the plate would represent the same quantity of heat that is now in the plate. For example, if half the plate is now at 40° and half at 60°, we would say that the average temperature is 50.°

How does this problem lead to a double integral? Suppose the plate is made of a number of separate pieces R_{ij}, each at a uniform temperature T_{ij}; then the average temperature T of the whole plate obeys the equation

$$TA(R) = \sum T_{ij}A(R_{ij}).$$

If p_{ij} is in R_{ij}, then $T_{ij} = f(p_{ij})$, and we have

$$T = \text{avg temp} = \frac{1}{A(R)} \sum_{i,j=1}^{N} f(p_{ij})A(R_{ij}).$$

This intuitive argument leads to the following general formula:

$$T = \frac{1}{A(R)} \iint_R f(x, y)\, dx\, dy, \tag{4.10}$$

which applies when the temperature varies continuously from spot to spot. In the problem that concerns us, $f(x, y) = 2xy^2 + x^2$ and R is a specific rectangle, with $1 \le x \le 3$ and $-1 \le y \le 3$. Hence, $A(R) = (2)(4) = 8$, and invoking Theorem 2,

$$\begin{aligned}
\text{avg temp} &= \tfrac{1}{8} \iint_R (2xy^2 + x^2)\, dx\, dy \\
&= \tfrac{1}{8} \int_1^3 dx \int_{-1}^3 (2xy^2 + x^2)\, dy \\
&= \tfrac{1}{8} \int_1^3 dx\, \{\tfrac{2}{3}xy^3 + x^2y\}\Big|_{y=-1}^{y=3} \\
&= \tfrac{1}{8} \int_1^3 \{(18x + 3x^2) - (-\tfrac{2}{3}x - x^2)\}\, dx \\
&= \tfrac{1}{8} \int_1^3 \{(18 + \tfrac{2}{3})x + 4x^2\}\, dx \\
&= \tfrac{1}{8}\{(9 + \tfrac{1}{3})x^2 + \tfrac{4}{3}x^3\}\Big|_{x=1}^{x=3} \\
&= 13\tfrac{2}{3}.
\end{aligned}$$

The formula (4.10) is also used to define the average value of any function $f(x, y)$ over the rectangle R.

Other topics lead naturally to the use of double integrals; one is the relationship between pressure and force. Suppose we measure force in pounds and pressure in pounds per square inch. A force of 60 pounds exerted uniformly against a plane region of area 12 in^2 will give a uniform constant pressure of 5 lb/in^2 at each point of the region. Conversely, if the pressure against a plate of area 10 in^2 is constantly 4 lb/in^2 at each point of the plate, then the total force against the plate is 40 pounds. How do we work such problems if the pressure is not constant, but varies from point to point?

Suppose we have a rectangular plate R and the pressure at a point (x, y) in R is given by $f(x, y)$. How can we find the total force exerted against the plate? The answer is readily seen to be $\iint_R f(x, y)\,dx\,dy$. Partition R into small rectangles R_{ij} in the usual way. Let m_{ij} and M_{ij} be the smallest and largest value of f on R_{ij}. Then, as a point (x, y) moves inside R_{ij}, the pressure will vary between the numbers m_{ij} and M_{ij}. The total force on R_{ij} will lie somewhere between $m_{ij}A(R_{ij})$ and $M_{ij}A(R_{ij})$. If we add these for all of the rectangles R_{ij}, we find that the total force F exerted on the large rectangle R must be such that

$$\text{LRS} = \sum m_{ij}A(R_{ij}) \le F \le \sum M_{ij}A(R_{ij}) = \text{URS}.$$

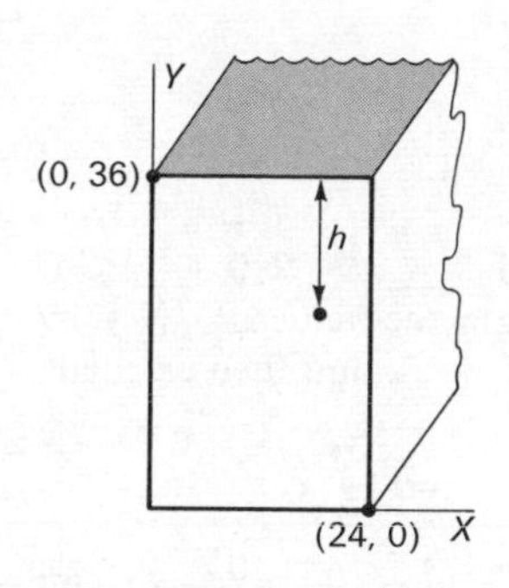

Figure 4–29

As the number of small rectangles increases and the lengths of their edges decrease, the Upper Riemann Sum and the Lower Riemann Sum both approach the value of the double integral $\iint_R f$, which must be F.

To see how this is used in a problem, let us find the total force against the side (24″ × 36″) of a rectangular water tank brimful of water. (See Figure 4–29.) We recall that water pressure is directly proportional to the depth, since the pressure at depth h can be regarded as due to the weight of a one-inch square column of water of height h. Since a cubic foot of water weighs about 62.5 pounds, we find that the pressure in lb/in.^{2} at a depth of h inches is

$$P = \frac{62.5}{12^3}h = .0361h.$$

Returning to Figure 4–29, we see that $h = 36 - y$, so that the pressure at the point (x, y) on the end of the tank is given by

$$f(x, y) = \frac{62.5}{12^3}(36 - y).$$

The total force on the end of the tank is

$$\begin{aligned} F = \iint_R f &= \int_0^{36} dy \int_0^{24} \frac{62.5}{12^3}(36 - y)\,dx \\ &= \int_0^{36} \frac{62.5}{12^3}(36 - y)(24)\,dy \\ &= \frac{62.5}{12^2}(2)\int_0^{36}(36 - y)\,dy \\ &= \frac{62.5}{12^2}(2)\{(36)^2 - \tfrac{1}{2}(36)^2\} \\ &= 62.5 \cdot 9 = 562.5 \text{ pounds.} \end{aligned}$$

The arithmetic complexity in the work above was caused by the fact that we chose to work in terms of inches rather than feet. If we choose to measure pressure in terms of pounds per square foot, then the formula for the pressure at a depth of h feet is merely

$$P = 62.5h,$$

and the corresponding integral for the total force is

$$F = \int_0^3 dy \int_0^2 62.5(3 - y)\, dx.$$

(You may wish to convince yourself that this yields the same numerical answer.)

Exercises

1 Let R be the rectangle described by (x, y) with $0 \le x \le 2, 0 \le y \le 1$. For each of the following choices of f, estimate the value of $\iint_R f$ by calculating upper and lower Riemann sums ($N = 5$), and then calculate the exact value of the integral $\iint_R f$.

(a) $f(x, y) = 2x - y$, (b) $f(x, y) = 1 + xy$.

2 For the rectangle described by (x, y) with $-1 \le x \le 1, 0 \le y \le 1$, evaluate each of the following integrals twice by using both orders of integration:

(a) $\iint_R (x^2 - y^2)\, dx\, dy$, (b) $\iint_R x^2 y\, dx\, dy$.

3 Evaluate the following over the rectangle defined by $-2 \le x \le 1$, $-1 \le y \le 0$:

(a) $\iint_R (3x^2 y - 2xy^3)\, dx\, dy$, (b) $\iint_R (x^2 e^{-y})\, dx\, dy$.

***4** Let R be a square with vertices $(0, 0)$, $(0, 1)$, $(1, 0)$, $(1, 1)$, and let f be any function continuous for $0 \le x \le 1$. Prove that

$$\iint_R f(x)f(y)\, dx\, dy \ge 0.$$

5 The average value of a function f on a rectangle R is $(1/A(R)) \iint_R f$. Find the average value of $f(x, y) = x + 3y^2 - x^2 y$ on the rectangle bounded by the lines $x = 1$, $x = 3$, $y = -1$, $y = 2$.

6 The temperature of each point of a square plate of side 4 is proportional to the square of the distance from the point to the lower left corner of the plate. The temperature at the upper right corner of the plate is 10. What is the average temperature of the plate?

7 Let R be the square with opposite vertices at $(0, 0)$ and $(1, 1)$. Estimate the value of $\iint_R (2x - y)/(1 + x + y)\, dx\, dy$ by calculating a convenient Riemann sum using $N = 5$.

8 With the same choice of R as in Exercise 7, evaluate

$$\iint_R \sqrt{x + y}\, dx\, dy \quad \text{directly.}$$

9 Calculate the exact value of $\iint_R (x + y)/(1 + xy)\, dx\, dy$ using the square R of Exercise 7. (This simple-looking integral will provide you with a review of many techniques of one-variable calculus. Don't give up!)

10 Find the average height of a rectangular room 40 ft. by 20 ft. whose roof is a paraboloid if the height (in feet) is given by

$$z = 35 - (1/20)(x^2 + y^2)$$

and $(0, 0)$ is at the center of the room.

11 Find the volume of the region bounded by $z = 0$, $x = 1$, $x = 2$, $y = 0$, $y = 1$, and $z = 1 + (y^2/x^2)$.

12 Find the pressure on a door 3 feet wide and 10 feet high at the bottom of a pool of water which is 50 feet deep.

4.5 Existence of the Integral

In the last section, we gave a formal definition of the meaning of $\iint_R f$, the double integral of the function f over the rectangle R. For the process outlined there to be valid and to assign a numerical value to the result, certain other steps must succeed. In particular, the Upper Riemann Sums and the Lower Riemann Sums must converge to the same limit as N increases. If they do not, then we say that the function f is not integrable, or that the integral of f over R does not exist.

The purpose of the present section is to outline a proof of the basic fact that a continuous function f is always integrable. This proof justifies all the calculations done in the last section, for once we know that the integral of a function f exists, we can go about finding its value either exactly or approximately. We shall also outline some of the formal properties of the double integral, and sketch proofs of the most important ones.

A complete discussion of the integral, including a detailed treatment of the conditions under which a function f is integrable, is better left to a more advanced text. For example, a bounded function f that is not continuous everywhere on R can still be integrable, provided that the set of points where f is *not* continuous is a "small" set in a special sense. We include in this section the statement of a theorem of this nature (Theorem 4) because we shall need a special result of that type in the next section, when we discuss the integration of functions over sets D that are not rectangles. We also outline a proof of the theorem used in the last section to justify evaluating a double integral as an iterated integral. Our principal result is

Theorem 3 *Let f be continuous on the rectangle R. Then the double integral $\iint_R f$ exists.*

Proof Suppose that N has been selected and that R is partitioned into N^2 small rectangles R_{ij} in the manner described in the last section. As we did there, suppose that

$$\begin{aligned} M_{ij} &= \text{maximum value of } f(p) \text{ for } p \text{ in } R_{ij}, \\ m_{ij} &= \text{minimum value of } f(p) \text{ for } p \text{ in } R_{ij}. \end{aligned}$$

Then, the Upper Riemann Sum (URS) and the Lower Riemann Sum (LRS) corresponding to this value of N are given by

$$\begin{aligned} \text{URS}_N &= \sum M_{ij}A(R_{ij}), \\ \text{LRS}_N &= \sum m_{ij}A(R_{ij}). \end{aligned}$$

A general Riemann sum is constructed by choosing a point p_{ij} in each small rectangle R_{ij} and then forming the sum

$$S(f, N) = \sum f(p_{ij})A(R_{ij}). \tag{4.11}$$

We observe that in every case

$$m_{ij} \leq f(p_{ij}) \leq M_{ij}.$$

If we multiply this inequality by the positive number $A(R_{ij})$ and then add for $i, j = 1, 2, \ldots, N$, we have

$$\sum m_{ij}A(R_{ij}) \leq \sum f(p_{ij})A(R_{ij}) \leq \sum M_{ij}A(R_{ij}).$$

This inequality is the same as

$$\text{LRS}_N \leq S(f, N) \leq \text{URS}_N. \tag{4.12}$$

If we knew that there was a number I such that $\lim_{N\to\infty} \text{LRS}_N = I$ and $\lim_{N\to\infty} \text{URS}_N = I$, then we would clearly have $\lim S(f, N) = I$, no matter how the points p_{ij} were selected, and we would have proved the theorem. As a first step in this direction, we assert

Lemma 1 *Since f is continuous on R,*

$$\lim_{N\to\infty} \{\text{URS}_N - \text{LRS}_N\} = 0.$$

Proof (of Lemma 1) We use Theorem 11 of Chapter 2, which states that any function f that is continuous on a closed bounded set D is also uniformly continuous on D. Our rectangle R is certainly closed and bounded, so that the given function f is uniformly continuous on R. Recall what this means. Given any $\epsilon > 0$, there is a δ such that $|f(p) - f(q)| < \epsilon$ whenever p and q are points in R whose distance apart is less than δ, i.e., $|p - q| < \delta$.

As the number N that determines the partition of the rectangle R increases, the size of the small rectangles R_{ij} decreases. We can find a number N_0 such that if $N > N_0$, then the diameter of any of the R_{ij} is less than δ. Accordingly, any two points in a single R_{ij} must be closer together than δ. In particular, this must be true for the point in R_{ij} at which f takes its maximum and the point at which f takes its minimum. If these points are p and q, so that $|p - q| < \delta$, $f(p) = M_{ij}$, and $f(q) = m_{ij}$, then by uniform continuity we must have $M_{ij} - m_{ij} = |f(p) - f(q)| < \epsilon$. (This number, the difference between the maximum value and the minimum value of a function f on a set, is called the **oscillation** of the function on the set. What we have shown is that if N is sufficiently large, then the oscillation of f on each of the small rectangles R_{ij} is less than ϵ.)

We next observe that

$$\begin{aligned} \mathrm{URS}_N - \mathrm{LRS}_N &= \sum M_{ij}A(R_{ij}) - \sum m_{ij}A(R_{ij}) \\ &= \sum (M_{ij} - m_{ij})A(R_{ij}) \\ &< \sum \epsilon A(R_{ij}) \\ &= \epsilon \sum A(R_{ij}) \\ &= \epsilon A(R). \end{aligned}$$

Restating what we have shown, $0 \le \mathrm{URS}_N - \mathrm{LRS}_N < \epsilon A(R)$ for all $N > N_0$. Since ϵ is arbitrarily small, this shows that $\lim_{N\to\infty} \{\mathrm{URS}_N - \mathrm{LRS}_N\} = 0$, and proves the lemma.

Return to the inequality (4.12). We have shown that the difference between the left side and the right side tends to 0 as N increases. This may seem sufficient to ensure that both LRS_N and URS_N converge to the same limit. It *would* be sufficient if we knew that either of these sequences converges at all, but we do not know this; indeed, to show this is the most difficult step in our proof.

Suppose that we know that there is a number α such that

$$\mathrm{LRS}_N \le \alpha \le \mathrm{URS}_N \tag{4.13}$$

for $N = 1, 2, 3, \ldots$. Then by using Lemma 1, we can conclude at once that both $\{\mathrm{LRS}_N\}$ and $\{\mathrm{URS}_N\}$ converge to α, which makes the proof of Theorem 3 complete. The proof that there is such a number α is contained in the next several paragraphs. The reasoning is subtle; therefore the presentation has been arranged so that this proof can be omitted by those who are willing to take the result on faith.

Let M and m be the maximum and minimum of f on the large rectangle R. Then $m \leq m_{ij} \leq M_{ij} \leq M$ for all i, j. Hence, $mA(R) = \sum mA(R_{ij}) \leq \sum m_{ij}A(R_{ij}) = \text{LRS}_N$, and $\text{URS}_N = \sum M_{ij}A(R_{ij}) \leq \sum MA(R_{ij}) = MA(R)$. Thus, $\text{LRS}_N \leq MA(R)$ for all N, and $mA(R) \leq \text{URS}_N$ for all N. The set of all numbers LRS_N for $N = 1, 2, 3, \ldots$ is bounded from above, and the set of numbers URS_N is bounded from below.

Let α be the least upper bound of all the numbers LRS_N, for all $N = 1, 2, 3, 4, \ldots$, and let β be the greatest lower bound of all the numbers URS_N. Our objective is to prove that $\alpha \leq \beta$, for then we will have

$$\text{LRS}_N \leq \alpha \leq \beta \leq \text{URS}_N$$

for all N, and we will have found a number α that always lies between the upper and lower Riemann sums.

The key is to find a way to compare a lower sum LRS_{n_1} and an upper sum URS_{n_2}, where the division numbers n_1 and n_2 are different. Let $N = n_1 n_2$. Then, n_1 and n_2 each divide N. If we now examine Riemann sums corresponding to n and N, where n divides N, we discover a special relationship.

Lemma 2 *Let n and N be whole numbers, with n dividing N. Then,*

$$\text{LRS}_n \leq \text{LRS}_N \leq \text{URS}_N \leq \text{URS}_n. \tag{4.14}$$

Proof (of Lemma 2) The essential idea of the proof can be seen from a special case. Suppose that $n = 3$ and $N = 6$. In Figure 4–30 we show the partition of the rectangle R into $n^2 = 9$ small rectangles R_{ij}, which is augmented by the dotted lines to give the partition into $N^2 = 36$ smaller rectangles R^*_{ij}. Let us look at the lower left-hand rectangle R_{11} shaded in the diagram. This is divided into four smaller rectangles $R^*_{11}, R^*_{12}, R^*_{21}, R^*_{22}$. The minimum of the function

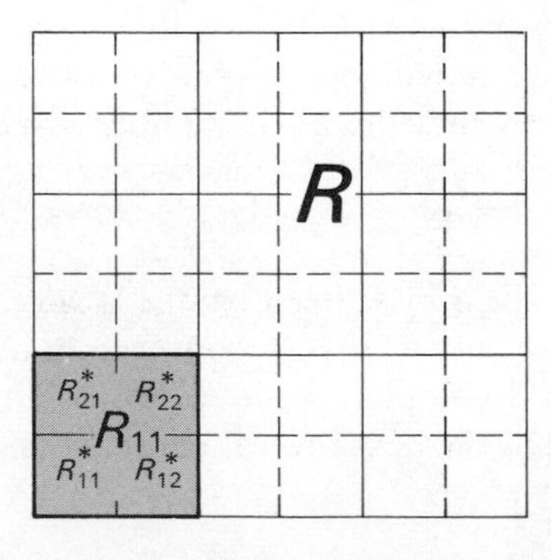

Figure 4–30

f in the rectangle R_{11} is m_{11}. Let the minimum of f in a rectangle R_{ij}^* be m_{ij}^*. Since $R_{ij}^* \subset R_{11}$ for $i, j = 1, 2$, we see that each of m_{11}^*, m_{12}^*, m_{21}^*, m_{22}^* is greater than or equal to m_{11}. We therefore have

$$\begin{aligned} m_{11}A(R_{11}) &= m_{11}A(R_{11}^*) + m_{11}A(R_{12}^*) + m_{11}A(R_{21}^*) + m_{11}A(R_{22}^*) \\ &\leq m_{11}^*A(R_{11}^*) + m_{12}^*A(R_{12}^*) + m_{21}^*A(R_{21}^*) + m_{22}^*A(R_{22}^*). \end{aligned}$$

In this inequality, the term on the left is one of the terms that occurs in the Lower Riemann Sum LRS_n. The last sum on the right takes in four of the terms in the Lower Riemann Sum LRS_N, associated with the smaller rectangles that make up the rectangle R_{11}. Thus in going from LRS_n to LRS_N we replace each term $m_{ij}A(R_{ij})$ by the sum of four terms of LRS_N; the result is either the same or larger.

A similar thing happens in going from upper Riemann sums for n divisions to upper Riemann sums for N divisions except that now the inequality signs are reversed. When we compare the maximum value M_{ij}^* of f on each of the smaller rectangles R_{ij}^* that make up the rectangle R_{11} with the maximum value M_{11} of f on R_{11}, we see that $M_{ij}^* \leq M_{11}$, so that

$$\begin{aligned} M_{11}A(R_{11}) &= M_{11}A(R_{11}^*) + M_{11}A(R_{12}^*) + M_{11}A(R_{21}^*) + M_{11}A(R_{22}^*) \\ &\geq M_{11}^*A(R_{11}^*) + M_{12}^*A(R_{12}^*) + M_{21}^*A(R_{21}^*) + M_{22}^*A(R_{22}^*). \end{aligned}$$

Thus, to replace the upper Riemann sum URS_n with the upper Riemann sum URS_N, we substitute for each term of the form $M_{ij}A(R_{ij})$ the sum of four terms of URS_N, which is either equal to URS_n or less than it. Hence (4.14) holds, completing the proof of Lemma 2.

Now, we apply this to find a relationship between upper and lower Riemann sums corresponding to arbitrary division numbers. Let n_1 and n_2 be any integers, and let $N = n_1 n_2$. (For example, we might have $n_1 = 3$ and $n_2 = 5$, with $N = 15$.) Since n_1 and n_2 each divide N, the result of Lemma 2 tells us that

$$\text{LRS}_{n_1} \leq \text{LRS}_N \leq \text{URS}_N \leq \text{URS}_{n_1}$$

and

$$\text{LRS}_{n_2} \leq \text{LRS}_N \leq \text{URS}_N \leq \text{URS}_{n_2}.$$

Noting that the two numbers in the center are the same in each of these inequalities, we conclude that for *any* choice of the integers n_1 and n_2,

$$\text{LRS}_{n_1} \leq \text{URS}_{n_2}. \tag{4.15}$$

Every lower Riemann sum is smaller than any upper Riemann sum, regardless of the number of division points used in each case.

The rest of the proof is now easy. Fix n_2 (for example, $n_2 = 47$). Then, (4.15) states that URS_{n_2} is an upper bound for *all* lower Riemann sums. Since the number α was defined to be the *least* upper bound of the lower sums, we know that $\alpha \leq \text{URS}_{n_2}$. But, this inequality says that the number α is a *lower* bound for an arbitrary upper Riemann sum; every URS_n must be larger than or equal to α. However, β is the *greatest* lower bound of all the upper sums, so that β must be larger than α. This proves that $\alpha \leq \beta$, and the proof of (4.13) and therefore of Theorem 3, the main result, is finished. The process used to define the integral of a function f over a rectangle always succeeds when the function f is continuous.

At the end of any proof, the "mathematician's question" always is, Did we need all the hypotheses, or could we prove a better theorem? In the present case, there is a better theorem, one which replaces the hypothesis that f is continuous by a weaker assumption. The result is one which will be immediately useful to us; but at this stage, it again seems better not to include a proof of the result, but instead to refer the interested reader to more advanced texts.[1] Here is the statement:

[1] *E.g., R. C. Buck,* Advanced Calculus, *2nd ed. (New York: McGraw-Hill, 1965), p. 102–3.*

Theorem 4 *Let R be a rectangle and let f be a function that is defined and bounded on R and continuous on R except on a set of points $E \subset R$ which has zero area. Then, the double integral $\iint_R f$ exists.*

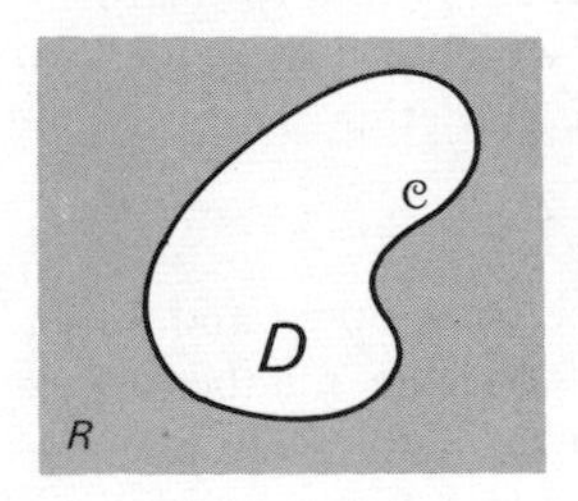

Figure 4–31

We can give an immediate and, we hope, enlightening illustration of this theorem. Let D be a region contained in a rectangle R (see Figure 4–31), and define a function g_D on R by

$$g_D(x, y) = \begin{cases} 1 & \text{if } (x, y) \text{ is in } D, \\ 0 & \text{if } (x, y) \text{ is not in } D. \end{cases}$$

(We shall again use this special function, called the characteristic function for the set D, in the next section.) The function g_D defined on the rectangle R is bounded and continuous at all points of R except on the curve that is the boundary of D. Thus, the set E of Theorem 4 is exactly this curve $\mathcal{C}$. As we pointed out in Section 4.3, the boundary of a region D always has zero area when the region itself is nonpathological. In fact, we observed (Theorem 1 in Section 4.3) that the condition that D *have* an area is precisely that the boundary of D have zero area. Thus, whenever the region D is a nice region, bounded by one or more smooth curves or line segments, then the characteristic function g_D satisfies the hypotheses of Theorem 4 and therefore has a double integral $\iint_R g_D$. A moment's thought will suggest that the value of this integral ought to be the area of D, $A(D)$. (You can check this directly by returning to the definition of the double integral. Applying it to the special function g_D, you are at once using the usual process for computing the area of D by the basic definition explained in Section 4.3 (Exercise 2 of this section).)

Many of the rules for working with double integrals are the same as those for the single integral.

Theorem 5 *If f and g are functions that can be integrated over the rectangle R, and b and c are constants, then*

$$\iint_R (bf + cg) = b\iint_R f + c\iint_R g.$$

Theorem 6 *If f can be integrated over R and $m \le f(p) \le M$ for all points $p \in R$, then*

$$mA(R) \le \iint_R f \le MA(R).$$

Each of these theorems can be proved by returning to the definition of the integral by means of Riemann sums. We see that the same relationships hold for the Riemann sums corresponding to any particular choice of N, and therefore must hold for their limit as N increases. For example, a Riemann sum for the integrand $bf + cg$ would be

$$\sum_{i,j=1}^{N} (bf(p_{ij}) + cg(p_{ij}))A(R_{ij}),$$

which can be rewritten as

$$b\sum f(p_{ij})A(R_{ij}) + c\sum g(p_{ij})A(R_{ij}),$$

yielding Theorem 5. (This property is also described by saying that the integral is a linear operation, preserving sums and scalar multiples; if the collection of all continuous functions defined on R is regarded as a vector space V, then the double integral over R is a numerical-valued linear function on V.)

Finally, as promised, we present an outline of the proof of Theorem 2 of the previous section, which asserted that a double integral over a rectangle can be evaluated as an iterated integral:

Theorem 2 *Let R be the rectangle described by $a \le x \le b$, $c \le y \le d$, and let f be continuous on R. Then,*

$$\iint_R f = \int_a^b dx \int_c^d f(x, y)\, dy = \int_c^d dy \int_a^b f(x, y)\, dx.$$

Proof Put $F(x) = \int_c^d f(x, y)\,dy$. What must be shown is that $\int_a^b F(x)\,dx = \iint_R f(x, y)\,dx\,dy$. We do this by showing that a typical Riemann sum for the single integral of F has exactly the same value as a specially chosen two-dimensional Riemann sum for the double integral of f over R.

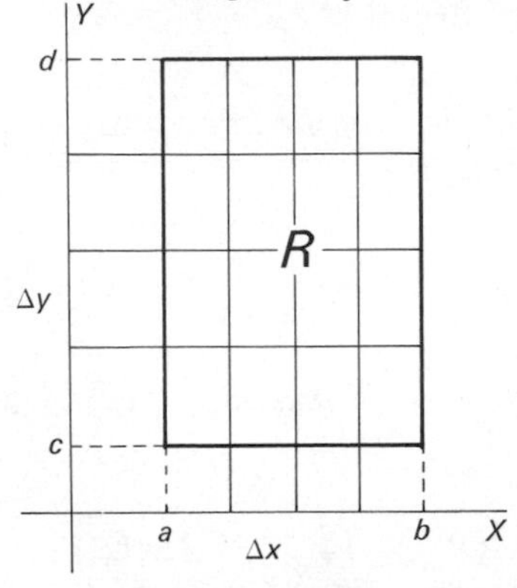

Figure 4–32

Suppose we partition R by a grid in the usual way. Note that this also divides the X-interval $[a, b]$ into subintervals all of length Δx and the Y-interval $[c, d]$ into subintervals of length Δy. (See Figure 4–32.) A typical Riemann sum for $\int_a^b F(x)\,dx$ is then

$$\sigma = \sum_{i=1}^{N} F(x_i)\,\Delta x = F(x_1)\,\Delta x + F(x_2)\,\Delta x + \cdots + F(x_N)\,\Delta x. \tag{4.16}$$

Now,

$$\begin{aligned} F(x_i) &= \int_c^d f(x_i, y)\,dy = \int_{y_0}^{y_N} f(x_i, y)\,dy \\ &= \int_{y_0}^{y_1} f(x_i, y)\,dy + \int_{y_1}^{y_2} f(x_i, y)\,dy + \cdots + \int_{y_{N-1}}^{y_N} f(x_i, y)\,dy. \end{aligned} \tag{4.17}$$

We next apply the Mean Value Theorem for integrals to each of these final integrals to find

$$\begin{aligned} \int_{y_{j-1}}^{y_j} f(x_i, y)\,dy &= f(x_i, y_j^*)(y_j - y_{j-1}) \\ &= f(p_{ij})\,\Delta y, \end{aligned}$$

where y_j^* is a point on the Y-axis between y_{j-1} and y_j. Making this replacement in (4.17), we have

$$\begin{aligned} F(x_i) &= f(p_{i1})\,\Delta y + f(p_{i2})\,\Delta y + \cdots + f(p_{iN})\,\Delta y \\ &= \sum_{j=1}^{N} f(p_{ij})\,\Delta y. \end{aligned}$$

Making the same substitution in (4.16), we then have

$$\sigma = \sum_{i=1}^{N} F(x_i)\,\Delta x = \sum_{i=1}^{N}\sum_{j=1}^{N} f(p_{ij})\,\Delta y\,\Delta x.$$

Since $\Delta x\,\Delta y$ is the area of one of the small rectangles in the partition of R, we realize that σ has the same value as a particular Riemann sum for the double integral of f, the sum corresponding to the special choice of the points $p_{ij} = (x_i, y_j^*)$.

Exercises

1 (a) Let g_D denote the characteristic function of the region D, so that $g_D(p) = 1$ if $p \in D$ and $g_D(p) = 0$ if $p \notin D$. If D_1, D_2 are regions, show that the characteristic function of the set $D_1 \cap D_2$ is $(g_{D_1})(g_{D_2})$.

(b) Express the characteristic function of $D_1 \cup D_2$ in terms of g_{D_1} and g_{D_2}.

2 Complete the details of the argument to show that $\iint_R g_D = A(D)$, where D has area and $D \subset R$.

3 Prove Theorem 6.

4 Prove that if $f(p) \geq 0$ for all $p \in R$, then $\iint_R f \geq 0$.

5 Prove that if $f \leq g$ on R, then $\iint_R f \leq \iint_R g$.

6 Prove that if f is continuous on the whole plane and if $\iint_R f = 0$ for every rectangle R, then $f \equiv 0$.

4.6 The Double Integral: General Case

Until now, we have studied only the integral of a function over a rectangle R. Our present aim is to replace R by a closed bounded set D, such as the region enclosed by a smooth curve or a polygon. It is quite possible to develop a definition and the corresponding theory for a general double integral $\iint_D F(x, y)\,dx\,dy$ in a way similar to that used for a rectangle R. We could study partitions $\mathcal{P}$ of D into subregions D_{ij} and work with Riemann sums of the form

$$S(F, \mathcal{P}) = \sum F(p_{ij})A(D_{ij}),$$

where the point p_{ij} lies in D_{ij} and $A(D_{ij})$ is the area of D_{ij}. (See Figure 4–33.) The value of the integral $\iint_D F$ would be the "limit," in a special sense, of $S(F, \mathcal{P})$ as the partitions $\mathcal{P}$ become finer and finer. Although this approach can be used and the needed theorems proved on this basis, we find it easier to reduce the general

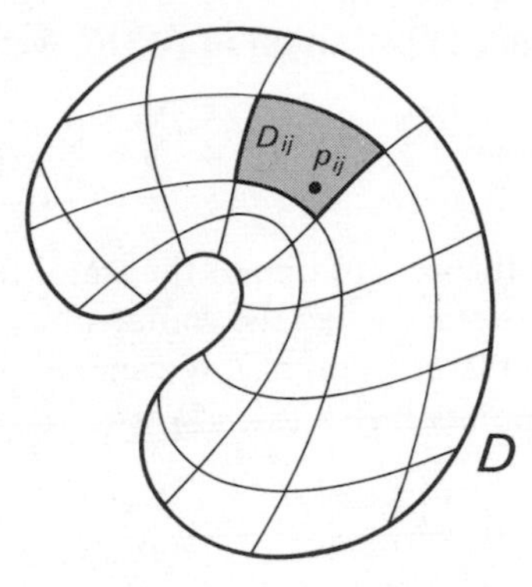

Figure 4–33

case to the rectangular case by a simple trick, and thus make use of all the results in the preceding section. We therefore adopt the following definition for the double integral of a function F over a region D.

Definition 3 Let D be a closed bounded region in the plane, and let F be a function that is defined and continuous on D. Let R be any rectangle that contains D as a subset. Let f be the function that is defined on R by

$$f(p) = \begin{cases} F(p) & \text{when } p \in D, \\ 0 & \text{when } p \notin D. \end{cases}$$

Then, the double integral of F over D is defined by

$$\iint_D F = \iint_R f.$$

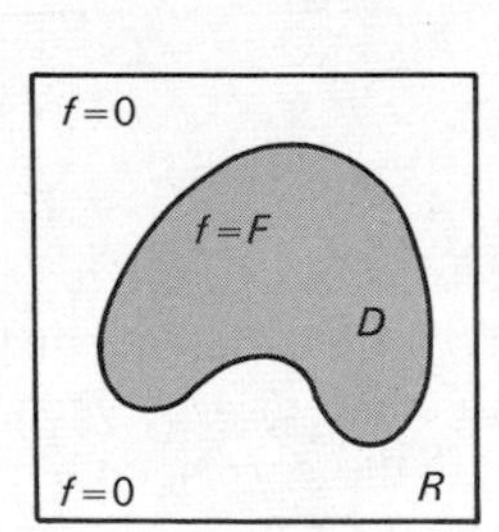

Figure 4–34

The rationale behind this definition is easily understood. If you consider any typical Riemann sum for the integral $\iint_R f$, say $\sum f(p_{ij})A(R_{ij})$, then $f(p_{ij})$ will be zero whenever p_{ij} is outside D. Thus, the total contribution to the sum arising from the part of R outside D will be zero, and only the part coming from D will count. Since $f = F$ on D, we should expect to obtain in the limit a number which ought to be the "integral of F over D." See the illustration (Figure 4–34).

In practice, we shall use a simple device to arrive at the function f, given the function F. Suppose that the function F is in fact defined over a much larger region than D; for example, we might have $F(x, y) = x^2 + y^2$, which is defined in the entire plane. Let g_D be the characteristic function for D, i.e., the function defined on the plane by

$$g_D(p) = \begin{cases} 1 & \text{if } p \in D, \\ 0 & \text{if } p \notin D. \end{cases}$$

Then, it is immediate that we can write $f = Fg_D$, for

$$F(p)g_D(p) = \begin{cases} F(p)(1) = F(p) & \text{when } p \in D, \\ F(p)(0) = \quad 0 & \text{when } p \notin D. \end{cases}$$

Thus, our definition of the general integral becomes simply

$$\iint_D F = \iint_R Fg_D.$$

Of course, the function F might have been defined only on the set D. (For example, $F(x, y) = \sqrt{1 - x^2 - y^2}$ is defined only for points $p = (x, y)$ in the unit disc, $|p| \leq 1$.) However, nothing prevents us from extending the domain of the function by defining the function F in any way we choose off D. (For example, we might extend our illustrative function by arbitrarily saying

$$F(x, y) = 13 \quad \text{for all} \quad p = (x, y) \text{ with } |p| > 1.)$$

Since it is always possible to do this, we can always assume that any function F that is the integrand in an integral of the form $\iint_D F$ has already been defined in some way at all points in the plane. In this way, we can always reduce our consideration of the general integral to the formula

$$\iint_D F = \iint_R Fg_D,$$

recalling that the key idea is that shown in Figure 4–34: We integrate a new function over R, but that function on D is the same as F, and off D is simply zero.

When we choose Definition 3 to explain the meaning of $\iint_D F$, several questions should arise at once. Does the numerical answer depend in any way upon the choice of the containing rectangle R? Is the new function $f = Fg_D$ one which *can* be integrated over D? How do we carry out this integration?

We leave the first question to Exercise 8. The second is answered by Theorem 4 of the previous section. For, the function F is assumed to be continuous at all points in D, and the constant function 0 is continuous at all points outside D. Thus, the new function f is continuous everywhere in the rectangle R except possibly on the curve or curves that form the boundary of D, when f shifts abruptly from F to 0. Thus for a region D which has an area and whose boundary is a curve $\mathcal{C}$ with zero area, the new function f is continuous everywhere on the rectangle R except possibly for the set $\mathcal{C}$ of zero area. By Theorem 4, the integral $\iint_R f$ exists.

The answer to the third question, "How do we evaluate $\iint_D F$?" is the main topic of this section. The general method can be described very briefly. Since $\iint_D F$ is equal to $\iint_R F g_D$, and since integrals over rectangles can be evaluated as iterated integrals, we calculate $\iint_D F$ as an iterated integral of $F g_D$. Perhaps the best way to explain this procedure is to carry out an example.

Let D be the triangular region with vertices (0, 0), (1, 0), (1, 1), and let $F(x, y) = x^2 + 2y$. For the rectangle R, we choose a square of side 1. (See Figure 4–35.) Then, as explained above, we have

$$\iint_D (x^2 + 2y)\,dx\,dy = \iint_R (x^2 + 2y) g_D(x, y)\,dx\,dy$$
$$= \int_0^1 dx \int_0^1 (x^2 + 2y) g_D(x, y)\,dy.$$

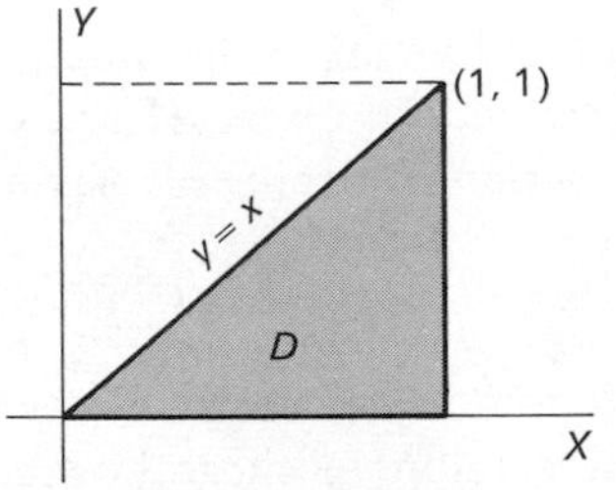

Figure 4–35

Looking at Figure 4–35, we see that

$$(x^2 + 2y) g_D(x, y) = \begin{cases} x^2 + 2y & \text{when } 0 \le y \le x, \\ 0 & \text{when } x < y \le 1. \end{cases}$$

(For, the top edge of the triangle has the equation $y = x$, and $g_D(x, y) = 0$ whenever (x, y) is outside the triangle.)

Using this formula, we can evaluate the inner iterated integral:

$$\int_0^1 (x^2 + 2y) g_D(x, y)\,dy = \int_0^x (x^2 + 2y)\,dy + 0$$
$$= (x^2 y + y^2)\Big|_{y=0}^{y=x} = x^3 + x^2.$$

Then we have

$$\iint_D (x^2 + 2y)\,dx\,dy = \int_0^1 (x^3 + x^2)\,dx = \tfrac{1}{4} + \tfrac{1}{3}.$$

Since there are always two different ways to evaluate a double integral over a rectangle in terms of iterated integrals, there is a second method that we could have used in this example. Starting as before, we write

$$\iint_D (x^2 + 2y)\,dx\,dy = \iint_R (x^2 + 2y) g_D(x, y)\,dx\,dy$$
$$= \int_0^1 dy \int_0^1 (x^2 + 2y) g_D(x, y)\,dx.$$

This time, since we will first be integrating with respect to x and therefore want to describe the integrand in terms of x, we look at Figure 4–35 again and write

$$(x^2 + 2y)g_D(x, y) = \begin{cases} 0 & \text{when } 0 \le x < y, \\ x^2 + 2y & \text{when } y \le x \le 1. \end{cases}$$

This time, the inner iterated integral becomes

$$\begin{aligned} \int_0^1 (x^2 + 2y)g_D(x, y)\,dx &= 0 + \int_y^1 (x^2 + 2y)\,dx \\ &= (\tfrac{1}{3}x^3 + 2yx)\Big|_{x=y}^{x=1} \\ &= (\tfrac{1}{3} + 2y) - (\tfrac{1}{3}y^3 + 2y^2), \end{aligned}$$

and

$$\begin{aligned} \iint_D (x^2 + 2y)\,dx\,dy &= \int_0^1 \{\tfrac{1}{3} + 2y - \tfrac{1}{3}y^3 - 2y^2\}\,dy \\ &= \tfrac{1}{3} + 1 - (\tfrac{1}{3})(\tfrac{1}{4}) - \tfrac{2}{3} = \tfrac{7}{12}. \end{aligned}$$

The general method may be described as follows. Suppose that the double integral of F over D has been reduced to an iterated integral over a rectangle of the form

$$\int_a^b dx \int_c^d F(x, y)g_D(x, y)\,dy.$$

When we evaluate the inner integral, we are thinking of x as being fixed and we are looking at $F(x, y)g_D(x, y)$ as a function of y alone. As such, the value of this integral will be zero whenever the point (x, y) is outside D, for then $g_D(x, y) = 0$. However, when this happens depends upon the location of x in the interval $[a, b]$ and upon the exact shape of the region D. Thus, if the region D is of the general shape shown in Figure 4–36, we have the general formula

$$\iint_D F(x, y)\,dx\,dy = \int_a^b dx \int_{\varphi(x)}^{\psi(x)} F(x, y)\,dy. \tag{4.18}$$

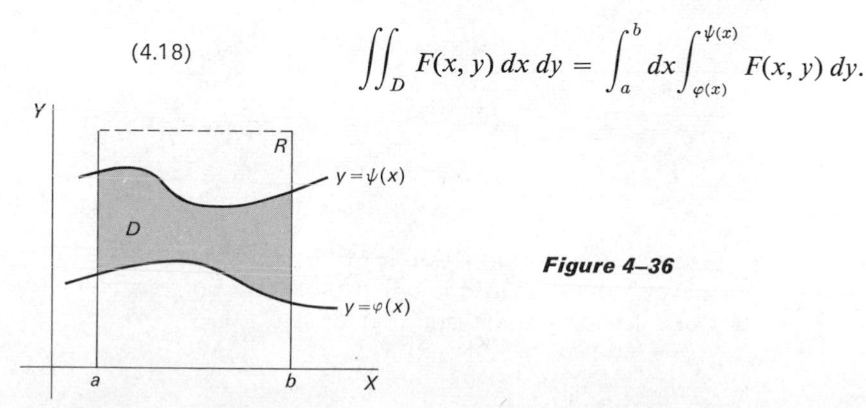

Figure 4–36

Again, if the region D is like that shown in Figure 4–37, then one would prefer to adopt the reversed order of iteration, and the correct formula is

$$\iint_D F(x, y)\,dx\,dy = \int_c^d dy \int_{\alpha(y)}^{\beta(y)} F(x, y)\,dx. \tag{4.19}$$

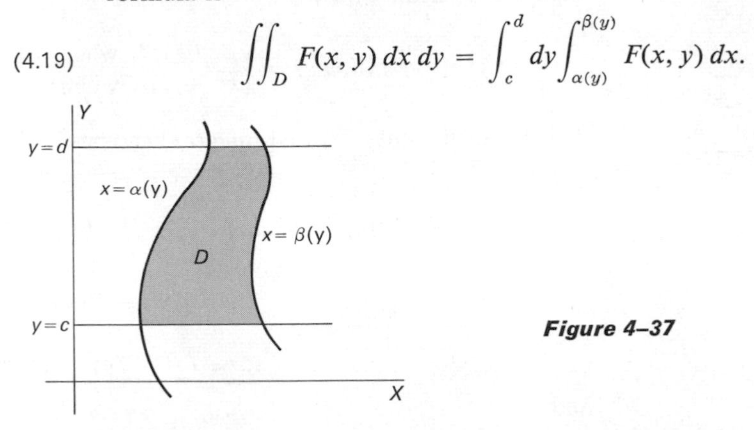

Figure 4–37

The choice depends upon the shape of D and upon the nature of its description. For example, the region in Figure 4–37 can be described as the set of all (x, y) with $c \le y \le d$ and $\alpha(y) \le x \le \beta(y)$. Note that precisely these limits are used for the limits of integration in the iterated integral in (4.19).

Some regions, such as convex sets, can be described in both ways, and for them one may calculate integrals in either order. Other regions may not be of either form, and in order to calculate integrals over them, one may have to cut them up into several parts of the above shapes. An example is the annulus A shown in Figure 4–38,

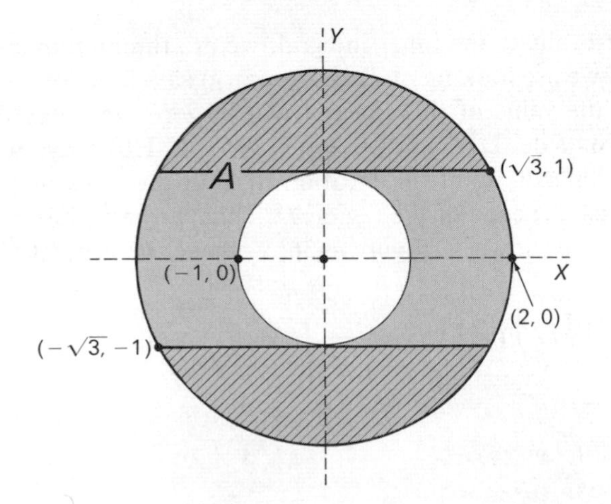

Figure 4–38

the region between two circles $x^2 + y^2 = 1$ and $x^2 + y^2 = 4$. We have shown A divided into four parts. The corresponding algorithm for calculating an integral over A is

$$\iint_A F(x, y)\, dx\, dy = \int_{-\sqrt{3}}^{\sqrt{3}} dx \int_1^{\sqrt{4-x^2}} F(x, y)\, dy + \int_{-\sqrt{3}}^{\sqrt{3}} dx \int_{-\sqrt{4-x^2}}^{-1} F(x, y)\, dy$$
$$+ \int_{-1}^{1} dy \int_{-\sqrt{4-y^2}}^{-\sqrt{1-y^2}} F(x, y)\, dx + \int_{-1}^{1} dy \int_{\sqrt{1-y^2}}^{\sqrt{4-y^2}} F(x, y)\, dx.$$

In this treatment of integration, there is one point at which there is a definite gap in logic, which may have been spotted by a careful reader. To make it more visible, let us repeat the general outline. We have defined the general double integral $\iint_D F(x, y)\, dx\, dy$ by $\iint_R Fg_D$, where R is a rectangle containing D. We observed that the new integrand Fg_D is continuous in R and bounded, except on the boundary of D, which is a set of zero area. We then used Theorem 2, proved in the last section, to replace this double integral over R by an iterated integral. However, if one checks back, one finds that in Theorem 2 it is assumed that the integrand is continuous on all of R. This may seem like a small point, and it is, for all the examples that are likely to be met in an elementary course in calculus. A proof of a stronger form of Theorem 2 can be given which permits the integrand to be discontinuous on a small set of the right sort and which in turn justifies the method we have been using for evaluating double integrals.[2]

[2] *Ibid.*, p. 114.

However, there are cases that cannot be treated in this fashion at all for which a different theory of integration must be used. This approach, discovered by Henri Lebesgue (1912), is based on a different treatment of the concepts of area and volume and is usually taken up at the graduate level. For all functions and sets that one ordinarily meets, the Riemann integral and the Lebesgue integral exist and have the same value.

Exercises

1 As seen in Exercise 2, Section 4.5, the area of a region D is given by

$$A(D) = \iint_D 1\, dx\, dy.$$

Using this formula, find the area of

(a) the triangle with vertices $(0, 0)$, $(a, 0)$, (a, b),

(b) a disc of radius R.

2 Calculate $\iint_D xy\, dx\, dy$, where D is

(a) the triangle with vertices $(0, 0)$, $(2, 0)$, $(0, 1)$,

(b) the quarter disc of radius 1 in the positive quadrant, center at the origin,

(c) the region lying between the parabola $y = x^2$ and the line $y = x$.

3 Evaluate $\iint_D (2x + 8y)\,dx\,dy$, where D is the region bounded by $y = 0$, $y = x^2$, and $x - 2y + 1 = 0$. (See Figure 4–39.)

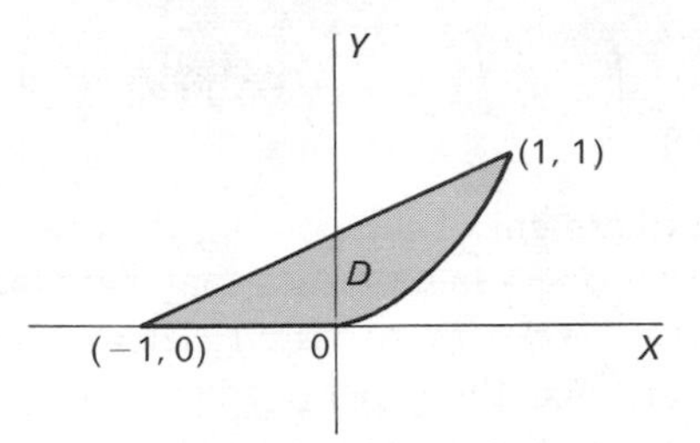

Figure 4–39

4 Evaluate $\iint_D (2x - 2xy)\,dx\,dy$, where D is the quadrilateral with vertices (0, 0), (1, 0), (2, 2), (0, 1).

5 Prove that if D_1 and D_2 are disjoint bounded regions and F is continuous on $D_1 \cup D_2$, then

$$\iint_{D_1} F + \iint_{D_2} F = \iint_{(D_1 \cup D_2)} F.$$

6 (a) Prove that if $F \geq 0$ on D, then $\iint_D F \geq 0$.
(b) Prove that if M and m are the maximum and minimum respectively of F on D, then

$$mA(D) \leq \iint_D F \leq MA(D).$$

7 Evaluate $\iint_D xy\,dx\,dy$, where D is the annular region in the positive quadrant between the circles of radius 1 and 2 centered at the origin.

8 Prove that the value of $\iint_D F$ according to Definition 3 does not depend on the choice of the rectangle R containing D.

9 The moment of inertia about the origin of a flat region D is defined to be $\iint_D (x^2 + y^2)\,dx\,dy = I$. Find the moment of inertia about the origin of each of the following regions:
(a) D is the square of side $2L$ centered at the origin,
(b) D is the triangle with vertices (0, 0), (1, 0), (1, 1),
(c) D is the disc of radius R centered at the origin.

10 Sketch the plane regions described by each of the following:
(a) $1 \leq x \leq 2, x \leq y \leq 2 + 2x - x^2$.
(b) $-1 \leq y \leq 2, y - 1 \leq x \leq 3 - y$.

11 (a) For a function f positive-valued everywhere on the region D, give an interpretation of the value of the double integral $\iint_D f$ as the volume of a certain solid.
(b) Sketch the solid in question when $f(x, y) = 1 + x^2 + y^2$ and D is the region defined by $1 \leq x^2 + y^2 \leq 2$.

(c) Set up an integral expression for the volume of the particular solid described in (b). (*Hint:* Can you make use of any symmetry?)

12 Given $\iint_D F$, where D is the plane region described in Exercise 10(b):
(a) Write the iterated forms for the evaluation of the double integral using both orders of integration.
(b) Check by taking in particular $F(x, y) \equiv 1$ and then also by finding the area of D by means of elementary geometry.

13 Given

$$\int_{-2}^{4} dx \int_{(x^2-4)/2}^{x+2} f\,dy$$

(a) Write the integral expression for integration to be performed in the opposite order.
(b) Evaluate both expressions for the case $f \equiv 1$.

14 Find the volume of the solid bounded above by $z = xy$ whose base is the region D in the XY-plane bounded by $x = 0$, $y = x^2$, $y = 8 - x^2$.

15 Find the volume of one wedge cut from the cylinder $x^2 + y^2 = 4$ by the planes $z = 0$ and $z = 2y$.

16 Find the average temperature of a parabolic plate bounded by $y^2 = 4x$ and $x = 4$ if the temperature at a point is twice the square of the distance from the vertex.

17 (a) Find the total force on a semicircular dam of radius 25 feet if the water level is at the top of the dam.
(b) Show that the force on the lower 18 feet of the dam is 88 percent of the total force.

18 The moment of inertia I_ℓ of a flat region D about a line ℓ is defined to be $\iint_D \{d_\ell\}^2$, where d_ℓ is the distance from (x, y) in D to the line ℓ. (Exercise 9 above is the special case in which line ℓ is the line perpendicular to the plane of D through the origin. A more complete discussion is found at the end of Section 4.7.) Find the moment of inertia I_y about the Y-axis of the region D of Exercise 9(a).

19 Find the moment of inertia about the line $x = 1$ for the region D of Exercise 9(b).

20 Find the moment of inertia I_x about the X-axis for the parabolic region bounded by $y = 0$ and $y = 4 - x^2$.

4.7

Triple Integrals

Except for the additional complexity created by one more spatial dimension and the corresponding problems of visualization, there is no essentially new concept in going from the double integral to the triple integral, or in fact to an n-fold multiple integral. If R is a rectangular solid in space and f is a bounded function defined on R, then the triple integral $\iiint_R f$ of f over R is defined as the unique limit, when it exists, of general Riemann sums of the form

$$S(f, N) = \sum f(p_{ijk})V(R_{ijk}),$$

where $p_{ijk} \in R_{ijk}$, the blocks R_{ijk} result from a partition of the large block R into N^3 smaller blocks, and $V(R_{ijk})$ is the volume of R_{ijk}. As before, it can be shown that these Riemann sums converge to a well-defined number when the function f is continuous on R, or in fact when f is bounded on R and continuous everywhere except for a subset $E \subset R$ which has zero volume.

The motivation for triple integrals as for double integrals can be based on physical considerations. For example, let R be a block of metal whose density (measured in grams/cm^3) varies from spot to spot. Suppose that the density at point p is given by $f(p)$, and suppose we wish to find the total mass of the block. If the density is constant, so that $f(p) = \kappa$ for all $p \in R$, then the total mass of the block is simply $\kappa V(R)$ grams. If we have chosen a partition of R into smaller blocks R_{ijk} and if the blocks are small enough so that the density is virtually constant across any one block, then clearly $f(p_{ijk})V(R_{ijk})$ represents an almost exact calculation of the total mass of the small block R_{ijk} for any choice of the point p_{ijk} in R_{ijk}, and the Riemann sum $\sum f(p_{ijk})V(R_{ijk})$ is an estimate for the total mass of R. Intuitively, this estimate should be better and better as the partition becomes finer and finer, and we are led to what can be regarded as a definition of mass:

$$\text{mass}(R) = \iiint_R f(x, y, z)\, dx\, dy\, dz.$$

Recall that the value of a single integral can be regarded as the area under the graph of the function which is the integrand and that the value of the double integral $\iint_D f(x, y)\, dx\, dy$ can be thought of as giving the volume of the solid which is bounded above by the surface $z = f(x, y)$ and whose base is the region D, at least when f is a positive-valued function. It would be tempting to hope for a comparable geometric interpretation of the triple integral. The only possibility is not very helpful, for we would be led to assert that the value $\iiint_R f(x, y, z)\, dx\, dy\, dz$ is the hypervolume (i.e., the four-dimensional volume) of the "solid" in 4-space which is bounded "above" by the surface in 4-space whose equation is $w = f(x, y, z)$ and whose base is the three-dimensional set R!

For this reason, we do not relate the meaning of the triple integral to any simple geometric concepts, but treat it as a mathematical concept, derived by analogy from the idea of the double integral.

The theorem which relates the triple integral and three-fold iterated integrals is then basic:

Theorem 7 *If R is the rectangular solid described by the inequalities $a_1 \le x \le a_2$, $b_1 \le y \le b_2$, $c_1 \le z \le c_2$ and f is continuous on R, then*

$$\iiint_R f(x, y, z)\, dx\, dy\, dz = \int_{a_1}^{a_2} dx \int_{b_1}^{b_2} dy \int_{c_1}^{c_2} f(x, y, z)\, dz$$

$$= \int_{b_1}^{b_2} dy \int_{c_1}^{c_2} dz \int_{a_1}^{a_2} f(x, y, z)\, dx$$

$= \cdots$ *(there are four other forms, corresponding to the other possible choices of order of integration).*

Integration of a function F over a more general set D is done as in the last section, by integrating the function $f = Fg_D$, where g_D is the characteristic function of the set D. Again, when $\iiint_R f = \iiint_R Fg_D$ is transformed into an iterated integral and the shape of D is used to determine the most efficient limits of integration (using the fact that the function f is zero off the set D), the result is an iterated integral with limits of integration that are apt to be functions themselves. In simple cases, the resulting integrals can be evaluated and the final result is the value of the original triple integral.

Let us examine a simple illustration. Choose D to be the solid pyramid with vertices at $(0, 0, 0)$, $(1, 0, 0)$, $(0, 1, 0)$, $(0, 0, 1)$, and take $f(x, y, z) = xy + z$ (see Figure 4–40). As indicated above, we have

$$\iiint_D F = \iiint_R Fg_D$$

$$= \int_0^1 dx \int_0^1 dy \int_0^1 F(x, y, z) g_D(x, y, z)\, dz.$$

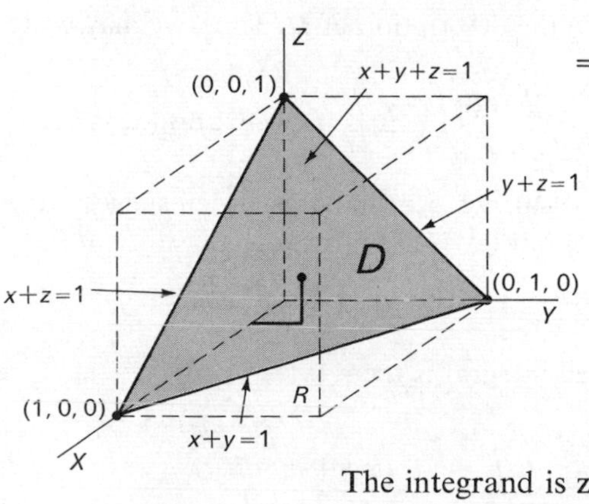

Figure 4–40

The integrand is zero off D. We therefore want to shrink the limits of integration so that (x, y, z) covers D alone, instead of the entire rectangular block R. It is convenient to do this sequentially. First, looking at Figure 4–40, we see that for any choice of x in $[0, 1]$, y is confined to the range $0 \leq y \leq 1 - x$. Then, having chosen an admissible x and y, we see that z is restricted to the interval $0 \leq z \leq 1 - x - y$, since the point (x, y, z) must lie beneath the plane surface $x + y + z = 1$. Putting these inequalities together, we specify the solid D: D = all (x, y, z) where

$$0 \leq x \leq 1, \quad 0 \leq y \leq 1 - x, \quad 0 \leq z \leq 1 - x - y.$$

The original triple integral is then computed thus:

$$\iiint_D (xy + z)\, dx\, dy\, dz = \int_0^1 dx \int_0^{1-x} dy \int_0^{1-x-y} (xy + z)\, dz$$

$$= \int_0^1 dx \int_0^{1-x} dy \{xyz + \tfrac{1}{2}z^2\}\Big|_{z=0}^{z=1-x-y}$$

$$= \int_0^1 dx \int_0^{1-x} dy \{(xy)(1 - x - y) + \tfrac{1}{2}(1 - x - y)^2\}$$

$$= \int_0^1 dx \int_0^{1-x} \{(\tfrac{1}{2} - x + \tfrac{1}{2}x^2) + (\tfrac{1}{2} - x)y^2 + (-1 + 2x - x^2)y\}\, dy$$

$$= \int_0^1 dx \{(\tfrac{1}{2} - x + \tfrac{1}{2}x^2)y + \tfrac{1}{3}(\tfrac{1}{2} - x)y^3 + \tfrac{1}{2}(-1 + 2x - x^2)y^2\}\Big|_{y=0}^{y=1-x}$$

$$= \int_0^1 dx \{(\tfrac{1}{2} - x + \tfrac{1}{2}x^2)(1 - x) + \tfrac{1}{3}(\tfrac{1}{2} - x)(1 - x)^3 + \tfrac{1}{2}(-1 + 2x - x^2)(1 - x)^2\}$$

$$= \int_0^1 \{\tfrac{1}{6} - \tfrac{1}{3}x + \tfrac{1}{3}x^3 - \tfrac{1}{6}x^4\}\, dx = \tfrac{1}{6} - (\tfrac{1}{3})(\tfrac{1}{2}) + (\tfrac{1}{3})(\tfrac{1}{4}) - \tfrac{1}{6}(\tfrac{1}{5}) = \tfrac{1}{20}.$$

If we choose to use a different order for the iterated integrals, then the limits of integration will change in accordance with the corresponding modification in the description of D. Thus, we may write

$$\iiint_D (xy + z)\, dx\, dy\, dz = \int_0^1 dy \int_0^1 dz \int_0^1 (xy + z) g_D(x, y, z)\, dx.$$

Then, examining Figure 4–40, we see that this order suggests the following description of D: all (x, y, z) with

$$0 \leq y \leq 1, \quad 0 \leq z \leq 1 - y, \quad 0 \leq x \leq 1 - y - z.$$

The corresponding iterated integral is then

$$\int_0^1 dy \int_0^{1-y} dz \int_0^{1-y-z} (xy + z)\, dx. \tag{4.20}$$

It is important to point out that the limits of integration have nothing to do with the integrand. To evaluate $\iiint_D x^2yz^3\, dx\, dy\, dz$, we can use the same iterated integral as that in (4.20), but with $xy + z$ replaced by x^2yz^3. It is sometimes true that the labor of calculation is greater with one order of integration than it is with another.

As another example, we take for D the solid tetrahedron with vertices (0, 0, 0), (1, 1, 0), (0, 1, 0), (1, 1, 1) whose faces lie on the planes $x = z$, $x = y$, $y = 1$, and $z = 0$. (See Figure 4–41.) The triple integral $\iiint_D f(x, y, z)\, dx\, dy\, dz$ can be converted into six different iterated integrals, since there are six different possible orders of integration. Each will have different limits of integration.

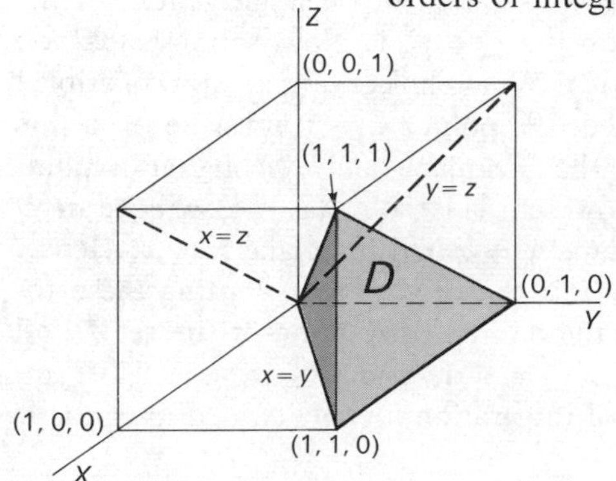

Figure 4–41

These limits may be obtained by examining Figure 4–41 to find a description of the set D. D is the region which a general point (x, y, z) is forced to cover by successive restraints in the coordinates, imposed in the order opposite to that in which the iterated integration will be carried out.

We will go through this process in detail for one of the six cases and list the results for the remaining five. It is important that the reader check these for himself, since this useful technique can be learned only by practice.

$$\begin{aligned}
\iiint_D F &= \int_0^1 dx \int_x^1 dy \int_0^x F(x, y, z)\, dz \\
&= \int_0^1 dy \int_0^y dz \int_z^y F(x, y, z)\, dx \\
&= \int_0^1 dz \int_z^1 dx \int_x^1 F(x, y, z)\, dy \\
&= \int_0^1 dx \int_0^x dz \int_x^1 F(x, y, z)\, dy \\
&= \int_0^1 dy \int_0^y dx \int_0^x F(x, y, z)\, dz \\
&= \int_0^1 dz \int_z^1 dy \int_z^y F(x, y, z)\, dx.
\end{aligned}$$

Of these, we elect to analyze the last one. Our problem is to examine Figure 4–41 and find a set of instructions which will limit the point (x, y, z) precisely to the pyramid D, imposing the instructions first on z, then on y, and finally on x. The first step is easy. The only restriction that can hold for z is that $0 \le z \le 1$, since the point (x, y, z) can be anywhere within D. Suppose now that a value of z has been selected. What is the set now covered by the coordinate y as (x, y, z) moves through D? In Figure 4–41, we have indicated the shadow thrown by D on the YZ-plane. If (x, y, z) lies in D, then the point $(0, y, z)$ must lie within this shaded triangle. Thus, for the given value of z, y must obey $z \le y \le 1$. Now, suppose that we have selected such a value of y. What choices of x are now possible? Since y and z are both fixed, the point (x, y, z) must lie on a line segment perpendicular to the YZ-plane and starting at a point within the triangular shadow set. This point (x, y, z) will lie in D when it lies between the plane $x = z$ and the plane $x = y$. Hence, the required restriction on x is that $z \le x \le y$. Putting these together, we have arrived at the desired description of the set D: all (x, y, z) with $0 \le z \le 1$, $z \le y \le 1$, and $z \le x \le y$. This immediately yields the limits of integration for the iterated integral, as listed above.

What situations provide natural sources for problems dealing with triple integrals? As we have indicated, the mass of a solid is expressed as a triple integral of the density function. If we specialize by taking the density function to be identically 1, as in the case of sets in the plane (see Exercise 2, Section 4.5), then integration over a region D yields its volume: $V(D) = \iiint_D 1\, dx\, dy\, dz$.

Two additional examples, both arising from the study of mechanics, lead to triple integrals: the calculation of the moment of inertia of a solid about an axis of rotation and the determination of the center of gravity of a solid. To define the moment of inertia, let ℓ be a line in space and let d_ℓ be the function defined by

$$d_\ell(x, y, z) = \text{distance from } (x, y, z) \text{ to } \ell.$$

Let D be the region occupied by a solid object whose density function is $\kappa(x, y, z)$. Then, the **moment of inertia** of the solid about the axis ℓ is defined to be (see Figure 4–42)

$$I_\ell(D) = \iiint_D \{d_\ell(x, y, z)\}^2 \kappa(x, y, z)\, dx\, dy\, dz.$$

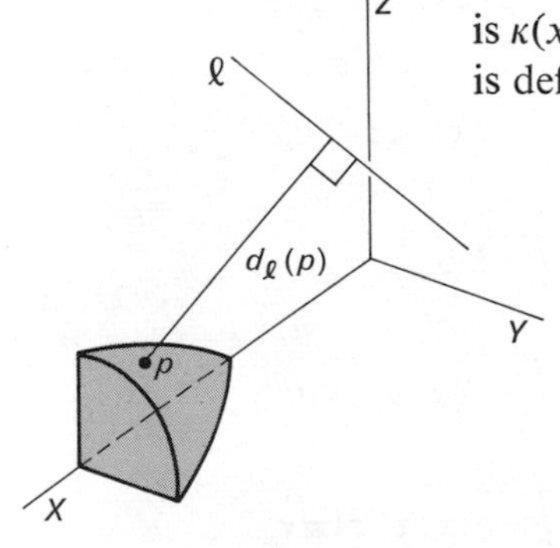

Figure 4–42

The physical interpretation of this number is that it measures the resistance to rotational motion of a solid in the same way that mass measures the resistance to straight-line motion.

When the axis ℓ is one of the coordinate axes, then the formula for d_ℓ becomes simply (see Figure 4–43)

$$
\begin{aligned}
d_X(x, y, z) &= \sqrt{y_2 + z^2}\,, \\
d_Y(x, y, z) &= \sqrt{x^2 + z^2}\,, \\
d_Z(x, y, z) &= \sqrt{x^2 + y^2}\,.
\end{aligned}
$$

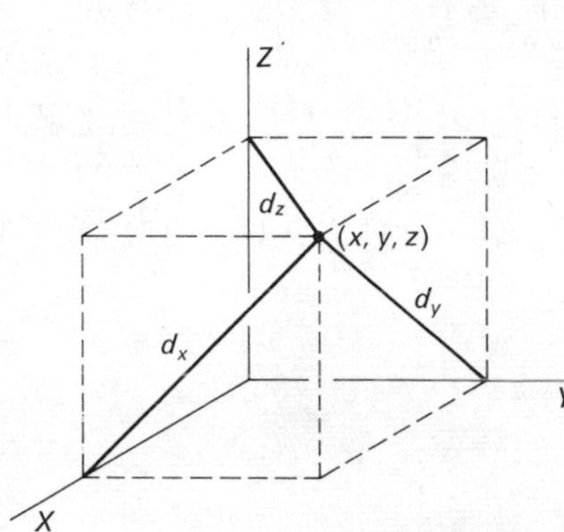

Figure 4–43

Accordingly, the moment of inertia of a solid about each of the coordinate axes is given by

$$
\begin{aligned}
I_X &= \iiint_D (y^2 + z^2)\kappa(x, y, z)\, dx\, dy\, dz, \\
I_Y &= \iiint_D (x^2 + z^2)\kappa(x, y, z)\, dx\, dy\, dz, \qquad (4.21) \\
I_Z &= \iiint_D (x^2 + y^2)\kappa(x, y, z)\, dx\, dy\, dz.
\end{aligned}
$$

Parenthetically, we observe that the last of these equations explains why the formula in Exercise 9, Section 4.6, for the moment of inertia of a *plane* region D about the origin was given as $\iint_D (x^2 + y^2)\, dx\, dy$. We were in fact computing the moment of inertia about the Z-axis of a flat two-dimensional object D of unit density. Note also that the discussion in Exercise 18 of the same set of exercises deals with the two-dimensional version of the general case.

As an example of the treatment for solids, let us find I_Y for a wedge-shaped solid with density function $\kappa(x, y, z) = Cx$ which is bounded by the planes $z = 0$, $x = 0$, $z = y$, and the cylinder $x^2 + y^2 = 1$ (see Figure 4–44). We have

$$
I_Y = \iiint_D C(x^2 + z^2)x\, dx\, dy\, dz.
$$

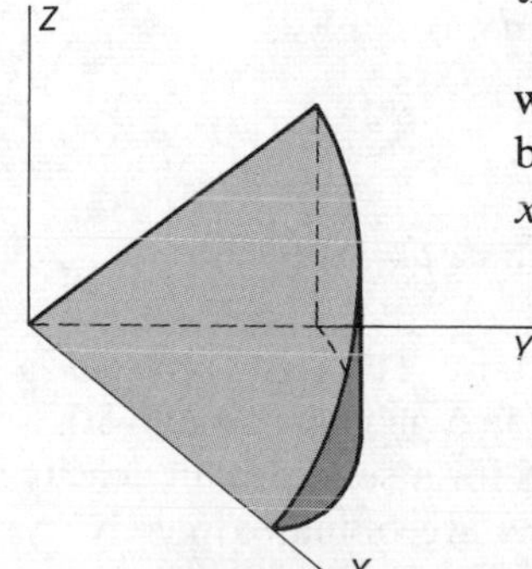

Figure 4–44

Examining the figure, we see that if we start from a point $(0, y, z)$ located in the triangle in the YZ-plane which forms one face of the solid, then x may vary from 0 to $\sqrt{1 - y^2}$, and (x, y, z) will remain in D. Thus, we can choose limits for an appropriate iterated integral

$$
\begin{aligned}
I_Y &= \int_0^1 dy \int_0^y dz \int_0^{\sqrt{1-y^2}} Cx(x^2 + z^2)\, dx \\
&= C\int_0^1 dy \int_0^y dz \left\{\frac{(1 - y^2)^2}{4} + \frac{(1 - y^2)z^2}{2}\right\} \\
&= C\int_0^1 dy \left\{\frac{(1 - y^2)^2 y}{4} + \frac{(1 - y^2)y^3}{2 \cdot 3}\right\} \\
&= C\left\{\frac{(1 - y^2)^3}{-24} + \frac{y^4}{24} - \frac{y^6}{36}\right\}\Bigg|_{y=0}^{y=1} \\
&= \frac{C}{18}.
\end{aligned}
$$

The **center of mass** (or center of gravity) of a solid is a point associated with the solid having the physical property that for many purposes the solid may be regarded as a single particle with the same mass located at this point. If the object were a two-dimensional flat region, then, for example, the object would balance on the point of a needle located at the center of mass. In general, the characteristic property of the center of mass of a solid object D is that the total moment of D about any plane through the center of mass is zero.

When the object occupies a region D and the density function is $\kappa(x, y, z)$, then the center of mass is a point $\bar{p} = (\bar{x}, \bar{y}, \bar{z})$ whose coordinates are given by the formulas

$$
\begin{aligned}
\bar{x} &= \frac{1}{M}\iiint_D x\kappa(x, y, z)\, dx\, dy\, dz, \\
\bar{y} &= \frac{1}{M}\iiint_D y\kappa(x, y, z)\, dx\, dy\, dz, \qquad (4.22) \\
\bar{z} &= \frac{1}{M}\iiint_D z\kappa(x, y, z)\, dx\, dy\, dz,
\end{aligned}
$$

where M is the mass of the object:

$$
M = \iiint_D \kappa(x, y, z)\, dx\, dy\, dz.
$$

Several simplified cases are also important. The term **centroid** is often used to refer to the point associated in a geometric way with a region. It is the same as the center of mass for a solid of unit density in the shape of the region; its coordinates are obtained merely by applying the formula (4.22) above with $\kappa(x, y, z) = 1$. One also talks about the centroid or center of gravity of a flat two-dimensional object, such as a triangle or a semicircle, and the correct formulas are merely the double integral analogues of the formulas above. For example, one would have

$$\bar{x} = \frac{1}{A}\iint_D x\kappa(x, y)\,dx\,dy,$$

where A is $\iint_D \kappa(x, y)\,dx\,dy$. A is the area of D if κ is constantly 1.

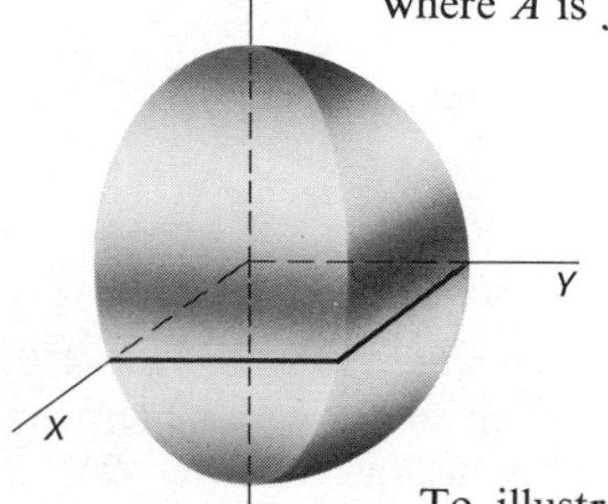

Figure 4–45

To illustrate the application of these formulas, let us find the centroid of the region shown in Figure 4–45, which is the portion of the solid formed by the intersection of two solid cylinders $x^2 + z^2 \leq a^2$ and $y^2 + z^2 \leq a^2$ which lies in the positive quadrant $x \geq 0$, $y \geq 0$ in space. By the vertical symmetry of the object, $\bar{z} = 0$. Since the region is also symmetric about the plane $x = y$, we conclude that the center of gravity must be in this plane, so that $\bar{x} = \bar{y}$. We need to calculate only $\bar{x}$. We take $\kappa = 1$, and using (4.22), we have

$$\bar{x} = \frac{1}{M}\iiint_D x\,dx\,dy\,dz,$$

where M is in fact the volume $\iiint_D dx\,dy\,dz$ of the region. We note that if $(x, 0, z)$ is in the semicircle in the XZ-plane indicated in the figure, then we can vary y from 0 to $\sqrt{a^2 - z^2}$ and (x, y, z) will remain in the region D. This at once gives us limits of integration for each of the triple integrals, and we have

$$\begin{aligned} M &= \int_{-a}^{a} dz \int_0^{\sqrt{a^2-z^2}} dx \int_0^{\sqrt{a^2-z^2}} dy \\ &= \int_{-a}^{a} (a^2 - z^2)\,dz = \tfrac{4}{3}a^3. \end{aligned}$$

Likewise,

$$\begin{aligned} \bar{x} &= \frac{1}{\frac{4}{3}a^3}\int_{-a}^{a} dz \int_0^{\sqrt{a^2-z^2}} dx \int_0^{\sqrt{a^2-z^2}} x\,dy \\ &= \frac{1}{\frac{4}{3}a^3}\int_{-a}^{a} \frac{(a^2 - z^2)^{3/2}}{2}\,dz. \end{aligned}$$

In this integral, set $z = a \sin \theta$ and make use of some elementary trigonometric identities to obtain

$$\begin{aligned}\bar{x} &= \frac{1}{\frac{4}{3}a^3} \int_{-\pi/2}^{\pi/2} \frac{a^4 (\cos \theta)^4}{2} \, d\theta \\ &= \frac{1}{\frac{4}{3}a^3} \int_{-\pi/2}^{\pi/2} \frac{a^4}{8} \{1 + 2\cos 2\theta + \tfrac{1}{2} + \tfrac{1}{2}\cos 4\theta\} \, d\theta \\ &= \frac{3a}{32} \{\tfrac{3}{2}\theta + \sin 2\theta + \tfrac{1}{8}\sin 4\theta\} \Big|_{-\pi/2}^{\pi/2} \\ &= \frac{9\pi}{64} a.\end{aligned}$$

Thus, the centroid of the solid is the point $(9\pi a/64, 9\pi a/64, 0)$.

It is an easy exercise in integration to show that the point $\bar{p} = (\bar{x}, \bar{y}, \bar{z})$ given by the formulas above has the characteristic "balancing property." First, if $\mathcal{P}$ is a plane in space and $\delta_{\mathcal{P}}$ is the function defined by

$$\delta_{\mathcal{P}}(x, y, z) = \text{algebraic distance from } (x, y, z) \text{ to the plane } \mathcal{P},$$

the distance measured positively on one side of the plane and negatively on the other, then the moment of a solid D about the plane $\mathcal{P}$ is defined to be

$$m(D, \mathcal{P}) = \iiint_D \delta_{\mathcal{P}}(x, y, z)\kappa(x, y, z) \, dx \, dy \, dz.$$

(This definition is suggested by thinking of the solid as a collection of individual particles and using the basic definition of moment: $m = $ (mass)(distance).)

Then the basic property we wish to prove is that the solid D and a single particle of mass M located at the point $\bar{p}$ have exactly the same moments about *any* plane $\mathcal{P}$, where as before M is the total mass of the solid D. The property is an immediate deduction from the fact that if $\mathcal{P}$ has the equation $Ax + By + Cz + D = 0$, then

$$\delta_{\mathcal{P}}(x, y, z) = \frac{Ax + By + Cz + D}{\sqrt{A^2 + B^2 + C^2}}.$$

For with this formula for $\delta_{\mathcal{P}}$, it is only necessary to verify the identity

$$\delta_{\mathcal{P}}(\bar{x}, \bar{y}, \bar{z}) \iiint_D \kappa(x, y, z) \, dx \, dy \, dz = \iiint_D \delta_{\mathcal{P}}(x, y, z)\kappa(x, y, z) \, dx \, dy \, dz. \tag{4.23}$$

Exercises

1 With D the region in Figure 4–41, evaluate

$$\iiint_D (x + y)\, dx\, dy\, dz,$$

using any two orders of integration.

2 Let D be the tetrahedron with vertices $(1, 0, 1)$, $(0, 0, 0)$, $(0, 0, 2)$, $(0, 1, 0)$.

(a) Show that the planes that contain the faces of D are $y = 0$, $x = 0$, $x = z$, $x + 2y + z = 2$.

(b) Fill in the missing pieces of information in the formula:

$$\iiint_D F(x, y, z)\, dx\, dy\, dz = \int_0^1 dx \int_0^{1-x} dy \int_x^{?} F(x, y, z)\, dz$$

$$= \int_0^1 dx \int_{?}^{2-x} dz \int_0^{?} F(x, y, z)\, dy.$$

3 Set up but do not evaluate a triple integral for the mass of a sphere of radius R whose density at a point is proportional to the distance from the point to the center of the sphere.

4 Explain why an appropriate formula for the moment of inertia of a homogeneous solid D of density κ revolving about the Z-axis is

$$I = \iiint_D \kappa(x^2 + y^2)\, dx\, dy\, dz.$$

5 (a) Sketch the region having the following description:

$$0 \le x \le 1,\ x \le y \le 1,\ 0 \le z \le 1 - \tfrac{1}{3}x - \tfrac{1}{2}y.$$

(b) Find the moment of inertia of a homogeneous solid of constant density κ filling the region of part (a) when it is revolved about the Z-axis.

6 Find the mass of a wedge in the first octant bounded by $x = 0$, $y = 0$, $z = 0$, $y + x = 1$, $z + x = 1$ whose density at (x, y, z) is the product of the distances to the coordinate planes, that is, $\kappa(x, y, z) = xyz$.

7 Find the mass of the object in the first octant which is bounded by the cylinders $z^2 = y$, $x^2 = y$, and the planes $z = 0$, $x = 0$, $y = 1$ if the density $\kappa(x, y, z) = z$.

8 Find the volume of the region in the first octant bounded by $y^2 + x = 1$ and $z^2 + x = 1$.

9 Find the volume of the region bounded by the paraboloids $z = x^2 + y^2$ and $z = 27 - 2x^2 - 2y^2$.

10 Find the volume of the region interior to the cylinders $x^2 + z^2 = a^2$ and $y^2 + z^2 = a^2$. (Make use of symmetry; part of the region is shown in Figure 4–45.)

11 Find the mass of the solid occupying the region of Exercise 10 if the density κ is equal to the square of the distance from the origin.

12 (a) Find the moment of inertia about the Z-axis for a solid bounded by $x = y = z = 0$, $y + z = 1$, and $x + z = 1$ if the density $\kappa = x$.

(b) Write the integral for the moment of inertia about the X-axis for the same object. (You need not evaluate.)

***13** Find the moment of inertia about the X-axis of a solid of constant density κ which is bounded by $y = 1$, $y = 0$, and

$$z^2 = y^2(1 - x^2).$$

14 Find the moment of inertia about the Y-axis of a solid figure in the first octant whose density is given by $\kappa = Cx$ and which is bounded by $z = 1$, $x^2 = z$, and $y^2 = z$.

15 Find the centroid of the plane region bounded by the semicircle $y = \sqrt{a^2 - x^2}$ and $y = 0$.

16 Find the centroid of the plane region bounded by the curve $y^2 + x = 0$ and the line $y = x + 2$.

17 Find the centroid of the solid region bounded by the cone $z^2 = x^2 + y^2$ and the plane $z = 1$. (You may make use of the geometric formula for the volume of a cone.)

18 Find the center of gravity for the object in Exercise 7.

19 Verify the identity (4.23) at the end of this section.

4.8

Other Coordinate Systems

There are times when the choice of a different coordinate system will make it considerably simpler to calculate the value of a multiple integral, either because the region becomes much easier to describe in the new system, or because the integrand becomes easier to integrate. For example, if the expression $x^2 + y^2$ occurs in the formulas, one might be tempted to change to polar coordinates, since this expression would become merely r^2. To make this change, we would replace x by $r \cos \theta$ and y by $r \sin \theta$. How does this change affect the double integral?

There are two ways to approach this question. The simplest, for an elementary course, is to return to the definition of a double integral and see how it is modified by this change. The second, which we shall describe only briefly in the Postlude, ends by creating a separate branch of mathematical analysis called the theory of differential forms and gives a method for making arbitrary changes of variable in multiple integrals analogous to substitutions in single integrals.

If we are using the polar coordinate system to locate points in the plane, what set is the natural one to replace a rectangle? Clearly, the analogue is the set of all points whose polar coordinates r, θ obey inequalities such as $R_1 \leq r \leq R_2$, $\alpha \leq \theta \leq \beta$. Such a region D is shown in Figure 4–46. Suppose that f is a function defined on D,

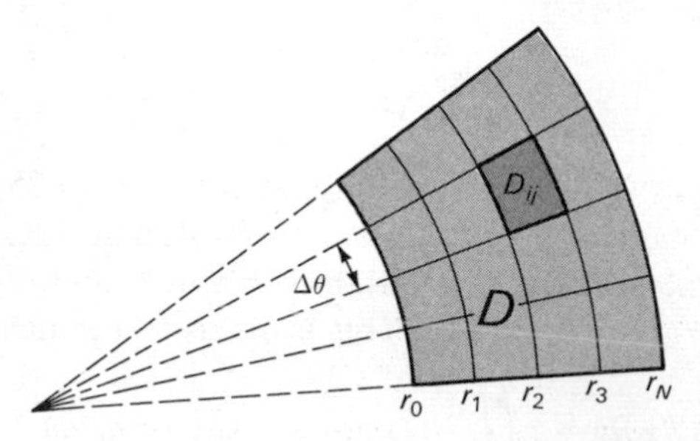

Figure 4–46

and suppose that we wish to calculate the integral $\iint_D f$. A natural way to partition D is to subdivide the r-interval $[R_1, R_2]$ and the θ-interval $[\alpha, \beta]$ each into N subintervals and thus to subdivide the set D into N^2 subsets D_{ij} in the manner shown in Figure 4–46. If we again choose a point p_{ij} in each D_{ij} and form the sum

(4.24) $$S(N) = \sum_{i,j=1}^{N} f(p_{ij})A(D_{ij}),$$

then the resulting Riemann sum will converge to the value of $\iint_D f$ as the number N increases, provided that the function f is continuous on D, or at least bounded and continuous almost everywhere on D (i.e., except on a set of zero area). If the partitions of the r-interval and the θ-interval are of the form

$$R_1 = r_0 < r_1 < r_2 < \cdots < r_N = R_2,$$
$$\alpha = \theta_0 < \theta_1 < \theta_2 < \cdots < \theta_N = \beta,$$

then we may take the point p_{ij} to be $r = r_i$, $\theta = \theta_j$ (in polar coordinates). We can also calculate the areas of the region D_{ij} exactly. Recall that the area of a sector of a circle of radius R and angular opening φ is $R(R\varphi)/2 = (\frac{1}{2})R^2\varphi$. Then, the area of a typical region D_{ij} as shown in Figure 4–46 is

$$\begin{aligned} A(D_{ij}) &= \tfrac{1}{2}(r_{i+1})^2(\theta_{j+1} - \theta_j) - \tfrac{1}{2}(r_i)^2(\theta_{j+1} - \theta_j) \\ &= \tfrac{1}{2}\{(r_i + \Delta r)^2 - (r_i)^2\}\,\Delta\theta \\ &= r_i\,\Delta r\,\Delta\theta + \tfrac{1}{2}(\Delta r)^2(\Delta\theta), \end{aligned}$$

where $\quad \Delta r = r_{i+1} - r_i \quad$ and $\quad \Delta\theta = \theta_{j+1} - \theta_j.$

Substituting this expression for $A(D_{ij})$ in (4.24), we obtain

(4.25) $$S(N) = \sum_{i,j=1}^{N} f(p_{ij})r_i(\Delta r)(\Delta\theta) + \tfrac{1}{2}\sum_{i,j=1}^{N} f(p_{ij})(\Delta r)^2(\Delta\theta).$$

However, $\Delta r = (R_2 - R_1)/N$, $\Delta\theta = (\beta - \alpha)/N$, and f is a bounded function, so the second sum in (4.25) can be estimated by taking $|f(p_{ij})| \leq M$ and adding its N^2 terms:

$$\left| \sum_{i,j=1}^{N} f(p_{ij})(\Delta r)^2(\Delta\theta) \right| \leq M(R_2 - R_1)^2(\beta - \alpha) \sum_{i,j=1}^{N} \frac{1}{N^3}$$
$$\leq CN^2 \frac{1}{N^3} = \frac{C}{N},$$

which approaches 0 as N increases. Hence, in formula (4.25), only the first sum is important, and it has the form of a standard Riemann sum for a double integral for the variables r and θ and the new function $g(r, \theta) = rf(r \cos\theta, r \sin\theta)$. This leads us to the identity

$$\iint_D f(x, y)\, dx\, dy = \iint_{D^*} f(r\cos\theta, r\sin\theta) r\, dr\, d\theta, \tag{4.26}$$

where D^* is the rectangle described by the inequalities $R_1 \leq r \leq R_2$, $\alpha \leq \theta \leq \beta$. Replacing the right side of (4.26) in the usual way by iterated integrals, we arrive at the final formula

$$\iint_D f = \int_{R_1}^{R_2} r\, dr \int_\alpha^\beta f(r\cos\theta, r\sin\theta)\, d\theta$$
$$= \int_\alpha^\beta d\theta \int_{R_1}^{R_2} f(r\cos\theta, r\sin\theta).$$

This rule is usually summarized by saying that when you change to polar coordinates, you "replace $dx\, dy$ by $r\, dr\, d\theta$." This injunction is incomplete, since it does not describe what else one must do to make the conversion, such as replace x and y by $r\cos\theta$ and $r\sin\theta$, and—most important—change the limits of integration in the appropriate way according to the description of the region D in terms of r and θ.

Several examples may illustrate the entire process better than further discussion.

Let D be the region in the XY-plane shown in Figure 4-47. With $f(x, y) = x + y$, let us calculate $\iint_D f$. The region is described very neatly in polar coordinates, namely by the inequalities $1 \leq r \leq 2$, $0 \leq \theta \leq \pi/2$. At the same time, $f(x, y)$ becomes $r\cos\theta + r\sin\theta$. The new integral becomes

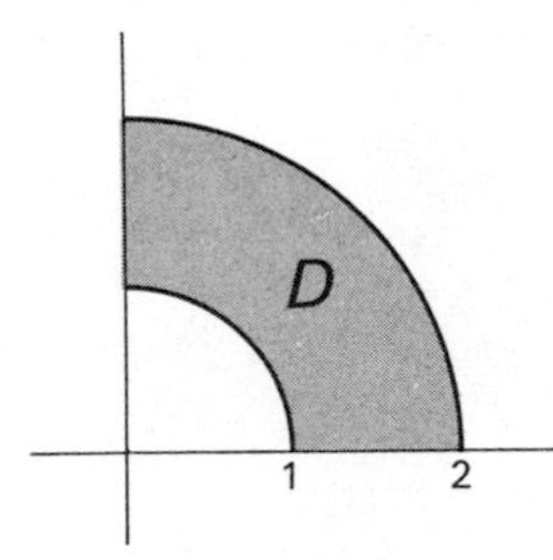

Figure 4–47

$$I = \int_1^2 dr \int_0^{\pi/2} (r\cos\theta + r\sin\theta) r\, d\theta$$
$$= \int_1^2 r^2\, dr \int_0^{\pi/2} (\cos\theta + \sin\theta)\, d\theta$$
$$= \int_1^2 2r^2\, dr = \tfrac{2}{3}r^3 \Big|_1^2 = \tfrac{14}{3}.$$

The same technique can be used with other regions, if they admit of simple descriptions in polar coordinates. For example, the special curve known as an equiangular spiral has the horrible equation $\tan\left[\frac{1}{2}\log(x^2 + y^2)\right] = y/x$ in rectangular coordinates but the simple equation $r = e^{\theta}$ in polar coordinates. Suppose we wish to evaluate the double integral $\iint_D (x^2 + y^2)\,dx\,dy$, where D is the shaded region in Figure 4–48, bounded in part by a piece of this

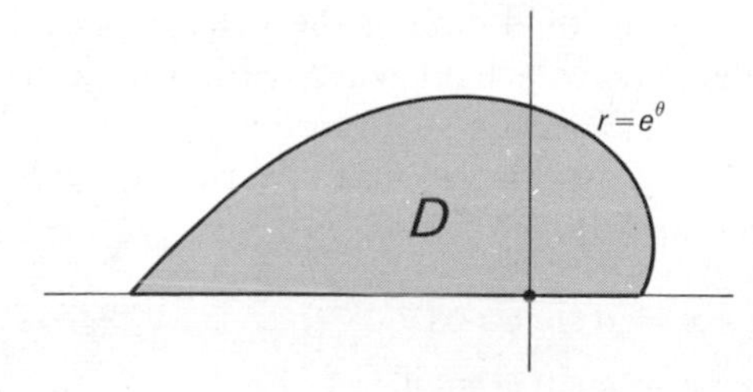

Figure 4–48

curve. In polar coordinates, this region can be described by the inequalities $0 \le \theta \le \pi$, $0 \le r \le e^{\theta}$, and the integrand becomes merely $x^2 + y^2 = r^2$. Replacing $dx\,dy$ by $r\,dr\,d\theta$, as directed by the algorithm, we obtain

$$
\begin{aligned}
I = \iint_D (x^2 + y^2)\,dx\,dy &= \int_0^{\pi} d\theta \int_0^{e\theta} (r^2)r\,dr \\
&= \int_0^{\pi} d\theta \{\tfrac{1}{4}r^4\}\Big|_0^{e\theta} = \int_0^{\pi} \tfrac{1}{4}e^{4\theta}\,d\theta \\
&= \tfrac{1}{16}\{e^{4\pi} - 1\}.
\end{aligned}
$$

A number of additional examples, chosen to show some of the cases in which this technique is especially useful, are given in the exercises. Although other changes of variable are also useful at times, the change to polar coordinates is the most common together with its three-dimensional analogues, cylindrical coordinates and spherical coordinates. Of the latter, cylindrical coordinates are easier, since they are obtained by adjoining a vertical Z-axis to a polar coordinate plane. The corresponding algorithm for change of variable calls upon one to replace "$dx\,dy\,dz$" by "$r\,dr\,d\theta\,dz$."

As an illustration, let us calculate the moment of inertia about its central axis of a solid cylinder of height h, radius R, and unit density. Calling this solid D, we must evaluate the triple integral $\iiint_D (x^2 + y^2)\,dx\,dy\,dz$. In cylindrical coordinates, the cylinder (see Figure 4–49) can be described by the inequalities

$$
\begin{aligned}
0 &\le r \le R, \\
0 &\le \theta \le 2\pi, \\
0 &\le z \le h.
\end{aligned}
$$

Figure 4–49

The corresponding iterated integral will be

$$I = \int_0^h dz \int_0^{2\pi} d\theta \int_0^R (r^2) r\, dr$$
$$= (2\pi)h(R^4/4) = \frac{\pi}{2} hR^4.$$

(This answer shows how much greater is the effect of doubling the radius of such a cylinder than of doubling the height! It explains why it is so much harder to spin a light wheel with a large radius than it is to spin a heavy wheel with a small radius.)

The change of coordinates associated with spherical coordinates (see formula (1.17), Section 1.6) is

$$x = \rho \sin \phi \cos \theta,$$
$$y = \rho \sin \phi \sin \theta,$$
$$z = \rho \cos \phi.$$

An analysis similar to that given above for polar coordinates leads to the conclusion that $dx\, dy\, dz$ should be replaced by $\rho^2 \sin \phi\, d\rho\, d\phi\, d\theta$. This seems reasonable after an examination of Figure 4–50, which suggests a way to estimate the volume of a typical portion obtained by subdividing intervals of ρ values, of ϕ values, and of θ values.

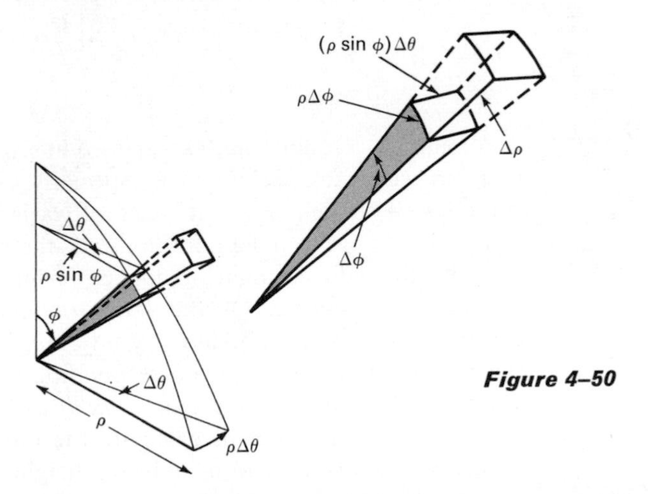

Figure 4–50

To see how spherical coordinates are used, let us return to an earlier exercise and find the mass of a sphere of radius R whose density varies so that at a point it is proportional to the distance from that point to the center of the sphere. The description of a solid sphere in spherical coordinates is very simple: $0 \leq \rho \leq R$, $0 \leq \phi \leq \pi$, $0 \leq \theta \leq 2\pi$. The density function, which is the integrand, is merely $\kappa\rho$. Hence, the desired mass is given by

$$M = \int_0^{2\pi} d\theta \int_0^{\pi} d\phi \int_0^{R} (\kappa\rho)\rho^2 \sin\phi \, d\rho$$

$$= \frac{2\pi\kappa R^4}{4} \int_0^{\pi} \sin\phi \, d\phi = \kappa\pi R^4.$$

Exercises

1 (a) Show that the disc D with radius 1 and center $(1, 0)$ can be described in polar coordinates by

$$-\pi/2 \le \theta \le \pi/2, \; 0 \le r \le 2\cos\theta.$$

(b) Using polar coordinates, evaluate $\iint_D x \, dx \, dy$ where D is the disc of part (a).

2 The equation of a cardioid is $r = 1 + \cos\theta$, $-\pi \le \theta \le \pi$.

(a) Find the area of the region D enclosed by the cardioid.

(b) Evaluate $\iint_D (x^2 + y^2) \, dx \, dy$ for the same region.

3 Use cylindrical coordinates to find the moment of inertia about its axis of a solid cone of radius R and height h.

4 By integrating the function with constant value 1 over a sphere of radius R using spherical coordinates, check the formula $V = \frac{4}{3}\pi R^3$ for the volume of a sphere.

5 Using polar coordinates, integrate $f(x, y) = xy$ over the quarter annulus in the positive quadrant between the circles of radius 1 and 2, centered at the origin. (See Exercise 7, Section 4.6.)

6 What is the average value of the function $x + y + z$ over the cylindrical region $0 \le z \le 1$, $0 \le r \le 2$, $0 \le \theta \le 2\pi$?

7 Find the mass of a cylindrical tube of thickness 1 inch, inside radius 3 inches, length 10 inches if the density at a point is equal to its distance from the top of the tube.

8 Find the volume of the solid bounded by the paraboloid $x^2 + y^2 = 4z$, the cylinder $x^2 + y^2 = 4y$, and the plane $z = 0$. (It is easier to use cylindrical coordinates.)

9 (a) Set up the necessary integral formulations in all three systems of coordinates—rectangular, cylindrical, and spherical—for the moment of inertia with respect to a diameter of a sphere of radius R.

(b) Evaluate, using the simplest formulation.

10 Find the mass of a spherical shell with inside radius a and outside radius b, if the density is proportional to the distance from the center.

11 Find the volume cut from the cone $x^2 + y^2 \leqq z^2$ by the sphere $x^2 + y^2 + z^2 = 2az$. (*Hint:* Use spherical coordinates.)

12 Find the center of gravity of the upper half of the sphere of radius R if the density is proportional to the distance from the center of the sphere.

13 Find the center of gravity of a solid sector of a sphere $\rho \leq a$ which is bounded below by the cone $\phi = \pi/6$ if the density is proportional to the distance from the vertex of the cone.

***14** Find the moment of inertia about the Z-axis of the solid with density function $\kappa = \sqrt{x^2 + y^2}$ which is bounded by $z^2 = x^2 + y^2$ and $(x^2 + y^2)^2 = a^2(x^2 - y^2)$. (*Hint:* Use cylindrical coordinates.)

15 Find the center of gravity of a homogeneous solid bounded by $x^2 + y^2 + z^2 = 2$ and $z = x^2 + y^2$. (Which of the nonrectangular systems of coordinates simplifies this problem most?)

---

differential equations

chapter 5

5.1

Models and Reality

Most people study mathematics because they are interested in using mathematics as a tool in some other discipline. Indeed, one of the rewards for many research mathematicians is seeing how a newly discovered mathematical theory makes it possible to better understand and deal with a portion of the world that was formerly a mystery—for example, the changing structure of the economy, or the strange nature of the electromagnetic winds that sweep across the solar system, or the mixture of human decisions that lies behind the play of a poker game.

In this respect, mathematics seems to have a unique versatility, and many scientists with a philosophical turn of mind have tried to understand why mathematics has proved to be so useful. Without attempting to answer this question, we can describe in very general terms the way a scientist uses mathematics. The key idea is given in the title of this section. One may think of the totality of mathematics thus far discovered or invented as contained in a large warehouse. Each mathematical concept or structure has a definite pattern and a definite collection of associated rules and techniques. A scientist who is interested in studying some aspect of reality comes to the warehouse to find a mathematical system which he thinks will be a model of that aspect of reality. He uses mathematical techniques to explore the behavior of the model and to answer questions about the model which reflect questions about nature.

Perhaps the diagram in Figure 5–1 will help to describe this process. We start from reality (whatever that may be), pass to an idealized situation, and then construct (or choose) a mathematical model. Both of these steps can be of great difficulty, and in many

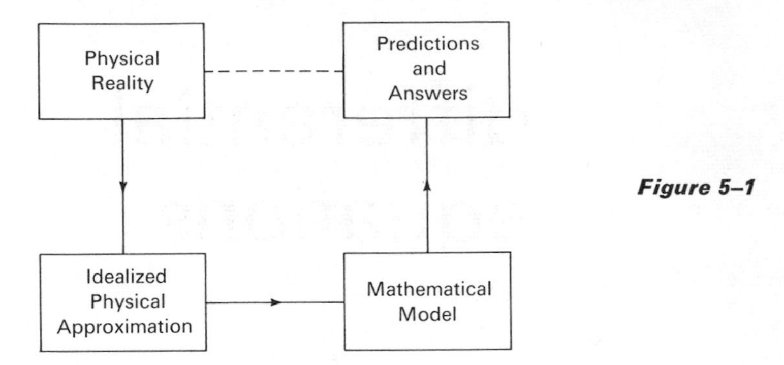

Figure 5–1

cases represent major scientific advances. The model itself can be as simple as a set of algebraic equations or so complex that an entire book is required to describe it. It should be possible to translate each question concerning reality into a parallel question about the model. The next step is to use specialized mathematical techniques to answer the questions about the model. Sometimes the questions are so difficult that no known techniques will answer them, and in this case, new techniques must be found or a new model chosen. When the questions can be answered, one is then ready to translate the answers back into the language of reality and if possible test them by experiment.

At this point, we suggest that the reader review Section 3.1, "Models for Motion," in which exactly such a treatment was given to the study of the phenomenon we experience which we call "motion."

Almost every branch of mathematics has served as a source for models of some aspect of reality. However, the most common models that have led to the most striking advances in science have been those associated with the calculus, and in particular, with differential equations. Differential equations will be the main topic of the present chapter. You have already looked at some special cases and have studied some of the models that correspond to special problems. Starting with the next section, we shall present a systematic treatment of some of the most useful techniques for analyzing a simple ordinary differential equation. Thus, in terms of Figure 5–1, we shall concentrate upon the right side of the diagram, the passage from the mathematical model to the box called "predictions and answers." Before doing so, however, we shall examine a number of simple illustrations of the modeling process. We shall not be able to carry through to the final step until we have obtained the needed mathematical techniques, but we hope to be able to show how such models arise and how some of the translations are made.

(a) The Pendulum Problem Suppose that "physical reality" is a pendulum consisting of a spherical bob mounted at the end of a rod and swung from a pivot. We say that we "understand" the physics of this pendulum if we can predict the behavior of this pendulum. Can we answer such questions as the following: (1) How does the nature of the motion depend upon the mass of the spherical weight? Or on the length of the rod or the material of which it is made? (2) If I hold the rod out to form an angle θ_0 and release it, what will be the speed at the bottom of the swing and how long will it take the pendulum to reach the bottom? (3) Will the pendulum eventually stop swinging, and how long will that take?

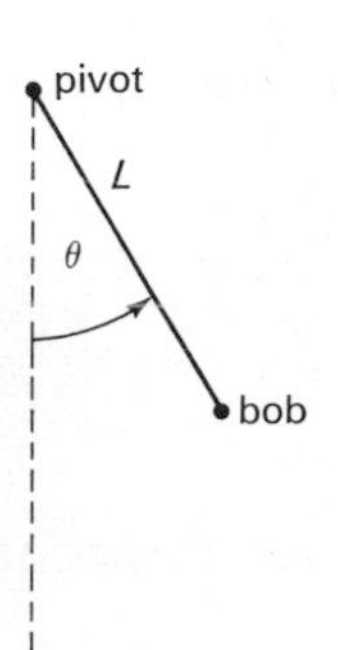

Figure 5–2

Experience and experimentation suggest that we had better start with a simplified situation. (Example: Try swinging a weight on the end of a rubber band!) We restrict ourselves to the following physical approximation: We assume that the rod is completely rigid and of zero mass, that the bob is a point of mass M, that the pivot has no friction, that the force of gravity is constant, that there is no air resistance, and that Newton's laws are sufficient to describe the situation. (Note that the last assumption is equivalent to saying that we are willing to buy the Newtonian model for motion.) With these assumptions, we can now move to the mathematical model. (See Figure 5–2.) The position of the pendulum swinging in a plane can be given at any moment of time by the angle θ. Thus, the motion of the pendulum can be translated into a function $\varphi(t)$ defined for all $t \geq 0$ with $\theta = \varphi(t)$ giving us the value of the angle at time t. We regard $t = 0$ as the instant at which we released the pendulum, so that we start with $\varphi(0) = \theta_0$. Since we released the pendulum from rest, so that its velocity was zero at $t = 0$, we also have $\varphi'(0) = 0$.

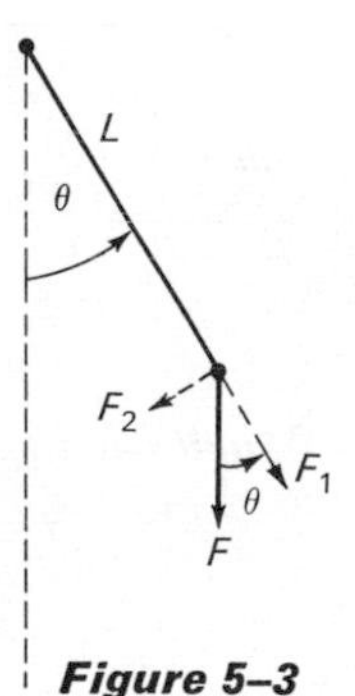

Figure 5–3

If we use Newton's laws to further restrict the nature of the function φ, we can derive a differential equation that must hold. In Figure 5–3, we show the force of gravity F acting on the bob of the pendulum in a vertical direction. The magnitude of F is Mg, where g is the acceleration of gravity and M is the mass of the bob. F can be resolved into two components, F_1 and F_2, as shown. Since the rod is absolutely rigid, F_1 has no effect. The motion of the pendulum bob is on a circle of radius L. Newton's Second Law in the form

$$\begin{aligned} \text{force} &= \text{mass times acceleration} \\ &= \text{rate of change of momentum} \\ &= \text{rate of change of (mass times velocity)} \end{aligned}$$

becomes

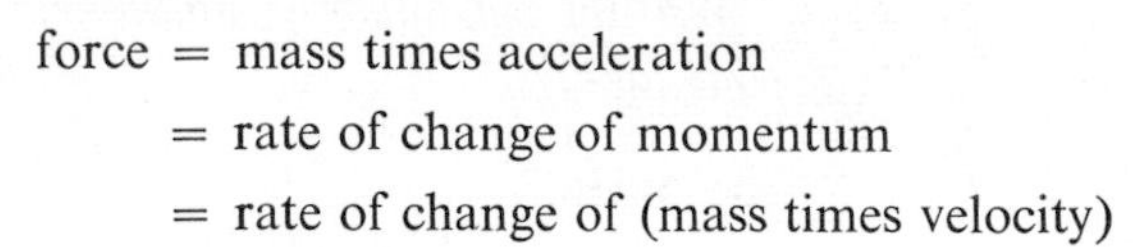

$$F_2 = \frac{d}{dt}\left\{(M)\left(L\frac{d\theta}{dt}\right)\right\} = ML\frac{d^2\theta}{dt^2}.$$

Since by geometry $F_2 = -(Mg)\sin\theta$, we obtain

$$\frac{d^2\theta}{dt^2} = -\frac{g}{L}\sin\theta. \tag{5.1}$$

Restating in terms of the function φ, we obtain the following:

A mathematical model for the motion of this pendulum is a function φ, defined and continuous on the interval $t \geq 0$, which has a continuous second derivative and is such that

$$\varphi''(t) = -\frac{g}{L}\sin\varphi(t) \quad \text{for all } t \geq 0, \tag{5.2}$$

$$\begin{aligned} \varphi(0) &= \theta_0, \\ \varphi'(0) &= 0. \end{aligned} \tag{5.3}$$

How do the questions we asked above translate into questions about this model? The first, which was

"*How does the nature of the motion depend on the mass of the bob or the length of the rod?*"

becomes

"*How does the graph of the function φ depend on the number M or the number L?*"

Since the number M does not appear in (5.2), we see that the motion is independent of M. We are not ready yet to explore the nature of the dependence upon L.

The second question about reality,

"*What will be the speed at the bottom of the swing and how long will it take the pendulum to reach the bottom?*"

becomes

(5.4) "*For what value of t, say t_0, is it true that $\varphi(t_0) = 0$, and what is the value of $L\varphi'(t_0)$?*"

Finally, the last question, which was

"*Will the pendulum eventually stop swinging, and if so, how long will this take?*"

becomes

"*Is there a number T such that*

$$\varphi'(t) = \varphi(t) = 0 \tag{5.5}$$

for all $t > T$?"

We have not yet presented the mathematical techniques needed to answer the questions about the function φ. However, here are several of the results that one can obtain.

(5.6) *There is a number $t_0 > 0$ such that $0 < \varphi(t)$ for all t, $0 < t < t_0$, and $\varphi(t_0) = 0$. Moreover,*

$$t_0 = \sqrt{\frac{L}{2g}} \int_0^{\theta_0} \frac{d\theta}{\sqrt{\cos\theta - \cos\theta_0}} \tag{5.7}$$

and the value of $L\varphi'$ at t_0 is given by

$$V_0 = 2\sqrt{gL}\,\sin\left(\tfrac{1}{2}\theta_0\right). \tag{5.8}$$

Furthermore, for all $t > 0$,

$$\varphi(t + 4t_0) = \varphi(t) \tag{5.9}$$

and $\lim_{t\to\infty} \varphi(t)$ does not exist.

These statements are next translated back into statements about reality, or rather into statements about the idealized physical pendulum from which our problem arose. (In each case that follows, you should check back to see exactly how the mathematical statement answers one of the questions asked about the mathematical model, and how it in turn relates back to the original physical reality.)

Thus, from (5.6) we see that the length of time it takes the pendulum bob to reach the bottom of its first swing is t_0, the number given in formula (5.7). Moreover, the velocity which it has reached at that moment is the number V_0 given in formula (5.8). Finally, the motion of the pendulum is periodic, repeating exactly after the time interval $4t_0$. It never stops moving nor does it ever slow down.

These are the results which the mathematical model predicts about the behavior of the real pendulum. The next step is to compare them with the results of experimentation with real pendulums. For example, we can test (5.6) and (5.7) by measuring the length of time it takes the pendulum bob to descend to the bottom of its swing and comparing this with the predicted value t_0 from (5.7) for various choices of the initial angle θ_0 and the pendulum length L. We could also do the same for formula (5.8) for the velocity V_0. The degree of agreement is some sort of measure of the degree to which the model is a true reflection of reality and of the confidence one could place in its predictions in areas where its agreement with reality has not been tested by experiment.

Regarding the latter, one must never lose sight of the fact that a model is not the same as reality. Although a model may behave in many ways like the real situation which gave rise to it, it will also predict behavior which does not occur. For example, (5.9) says that once set moving, a pendulum will never stop, and any experiment will show that this is false. Furthermore, the values of t_0 and V_0 given above in formulas (5.7) and (5.8) will not match exactly the values obtained from measurement. A partial answer is to say that our model was constructed for an "ideal" pendulum, and that requirements such as a weightless perfectly rigid rod and a frictionless pivot cannot be satisfied in practice. It is, of course, possible to start from a less ideal physical picture, permitting the rod to have mass and to be slightly elastic, the bob to have nonzero radius, and the pivot to have frictional forces, etc. When all this is done, the resulting mathematical model will be immensely complicated, undoubtedly beyond the limits of presently understood techniques of mathematical solution. Nevertheless, it would still be our contention that this last model would predict kinds of behavior not exhibited by real pendulums and that in no sense can we identify such a model with reality, whatever indeed that is.

As indicated above, it is often useful to consider several different models for the same physical situation. Perhaps the reader has noticed that the formula (5.7) for the quarter period t_0 of a pendulum depends in a complicated way upon the initial angle θ_0. If you recall the frequently quoted statement that the period of a pendulum does not depend upon the amplitude (which is determined by the angle θ_0), then you may wonder where the error lies. The facts are easily checked, and a simple experiment will show you that the length of the period of a pendulum is visibly different for $\theta_0 = 10°$ and for $\theta_0 = 80°$.

The explanation is that in many discussions of the motion of a pendulum a much cruder mathematical model is used than that set forth in (5.2), (5.3). In the cruder model, the differential equation (5.2) is replaced by the equation

$$\varphi''(t) = -\frac{g}{L}\varphi(t), \tag{5.10}$$

or in more familiar notation,

$$\frac{d^2\theta}{dt^2} = -\frac{g}{L}\theta.$$

The predictions which result from this model are different from those of the previous model, but are more easily obtained. We have

$$t_0 = \frac{\pi}{2}\sqrt{L/g}, \tag{5.11}$$

$$V_0 = \sqrt{gL}\,\theta_0, \tag{5.12}$$

and it is evident that the value of t_0 does not involve θ_0 at all.

The connection between these two models for the motion of a pendulum is a mathematical one and not a physical one, although it can be given a physical interpretation. When θ is small, $\sin\theta$ is very nearly the same number as θ. Thus, if the motion of the pendulum is confined to a small neighborhood of the bottom of its arc, then the term sin $\varphi(t)$ which occurs in (5.2) is nearly equal to $\varphi(t)$; this replacement turns (5.2) into (5.10). For the same reason, one should expect that the predicted values of t_0 and V_0 given in formulas (5.7) and (5.8) turn into those in (5.11) and (5.12) as θ_0 tends toward zero.

Another simple illustration may help here. Physicists speak of light sometimes as a wave and sometimes as a particle. Which is it? The answer, of course, is that light is neither. Both words describe specific mathematical models for light, but neither *is* light; one must not confuse a model with reality itself. Both of these models successfully predict ("explain") some of the observed phenomena of light; both predict behavior that light does not exhibit. A model is most useful when it imitates nature closely, but there will always be aspects of reality it does not reproduce, and it will always predict events that do not in fact occur. The skill of a scientist lies in knowing how far and in which contexts to rely on a particular mathematical model. The final moral is clear; there is never a unique *correct* mathematical model for an aspect of reality. Those "laws of nature" that are not merely tautologies, akin to the statement that there are three feet to every yard, are models, and as such subject to revision and replacement by others. None is "correct" or "true" in some eternal sense. (At this time, we have moved far onto the quicksands of metaphysics; those interested in further speculation are encouraged to read H. Poincaré, *Science and Hypothesis* (New York, Dover Publications, Inc., 1952).

(b) The Rocket Flight Problem The force of gravity is not constant, but depends upon the distance between objects, and in the case of the Earth, it also depends upon local rock densities and other factors. In dealing with objects at large distances from the Earth, it is sufficient to regard the Earth as a sphere, and to assume that the force on an object of mass M far above the atmosphere is given by

$$F = \frac{kM}{x^2}, \tag{5.13}$$

where k is a negative constant and x is the distance from the object to the center of the Earth.

During the initial powered flight period, the motion of a rocket obeys rather complicated laws, since the force of the exhaust and the mass of the rocket are both changing. After burnout, things are much simpler; since the mass is then constant, the rocket is coasting, and its velocity is slowly being decreased by the effects of the Earth's gravity.

If we assume that the rocket is moving directly away from the Earth and we know its height and velocity at burnout, what will its

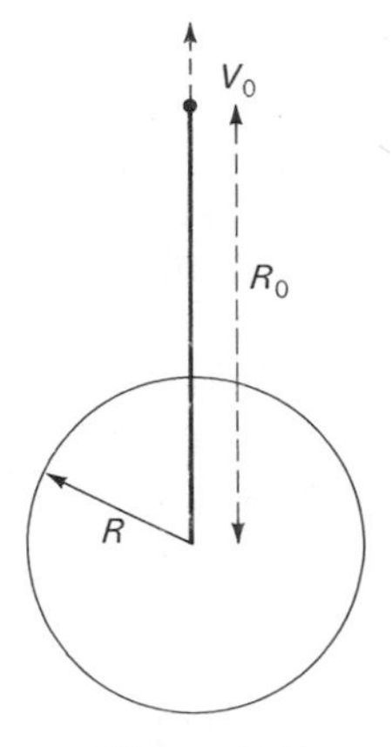

Figure 5–4

future motion be? Will it continue moving upward and never stop, or will it reach a maximum height and then fall back?

Under the conditions we have described, an appropriate mathematical model for the motion of the rocket (now considered to be a particle of mass M) will be a function f such that $f(t) = x$ is the distance from the center of the Earth to the rocket at time t. Let $t = 0$ correspond to the instant of burnout. Then, the distance from the Earth at this moment is $R_0 = f(0)$, and the vertical velocity of the rocket at this moment is $V_0 = f'(0)$. We know both R_0 and V_0, and we want to predict the shape of the graph of f for $t > 0$. The final step in constructing the model is to produce a differential equation which the function f must satisfy. Again, this comes by the use of Newton's Second Law. From Figure 5–4 and formula (5.13), we obtain first

$$M\frac{d^2x}{dt^2} = \frac{kM}{x^2}.$$

However, we know that when $x = R$ (the radius of the Earth), then d^2x/dt^2 must be $-g$, the acceleration of gravity on the surface of the Earth. This gives us the final form of the desired equation:

$$\frac{d^2x}{dt^2} = -\frac{gR^2}{x^2}. \tag{5.14}$$

We combine all the above to formulate the mathematical model for our rocket flight problem:

The function f is defined and of class C^2 for all $t \geq 0$, and obeys the differential equation

$$f''(t) = -\frac{gR^2}{\{f(t)\}^2} \quad \text{for all } t > 0.$$

In addition, $f(0) = R_0$ and $f'(0) = V_0$.

The questions which we raised about the motion of the rocket are now translated into the following questions about the function f in our model.

"Is the function f monotonic increasing for all $t > 0$ and unbounded, or does f have a maximum value; and if so, what is it, and for what value of t is it achieved? More generally, what is the shape of the graph of f?"

Without entering into the methods used to answer these questions, it may be proved that the graph of f has one of two typical shapes. The first, shown in Figure 5–5, occurs whenever $V_0 \geq R\sqrt{2g/R_0}$, and the second, shown in Figure 5–6, occurs when $V_0 < R\sqrt{2g/R_0}$.

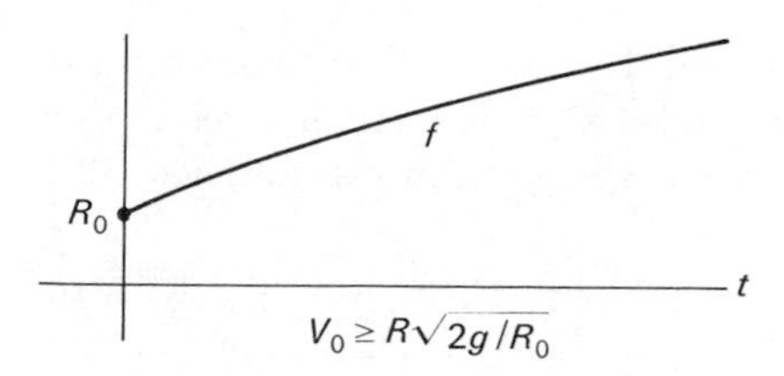

Figure 5–5

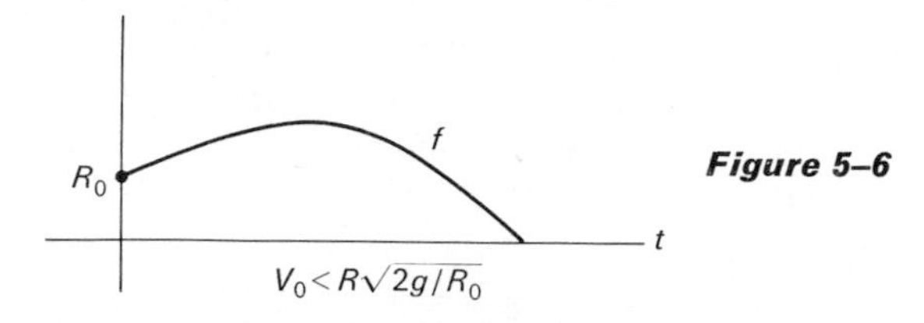

Figure 5–6

These facts about the function f can be translated back to yield useful statements about the expected behavior of the rocket. Thus, if the distance and velocity of the rocket at burnout are such that $V_0 \geq R\sqrt{2g/R_0}$, then the model predicts that the rocket will continue to move upward forever; it will recede indefinitely from the Earth, eventually reaching any assigned distance. On the other hand, if $V_0 < R\sqrt{2g/R_0}$, then the rocket will reach a maximum height and then fall back toward the Earth, reaching it after a specific length of time. The most frequently examined case occurs when R_0, the distance at burnout, is approximately the same as R, the radius of the Earth. In this case, the condition for nonreturn becomes $V_0 \geq \sqrt{2gR}$. Taking $g = 32$ ft/sec^2 and $R = 4{,}000$ miles, computation gives the critical escape velocity as $V_0 = 40/\sqrt{33} = 6.9$ mi/sec. Thus, a rocket that at burnout is traveling vertically at about 7 miles per second will (if the model is successful) never fall back to the Earth.

(c) The Two-Tank Problem Suppose that tanks I and II are placed as shown in Figure 5–7 and provided with input and output pipes as indicated. Each tank contains 100 gallons of liquid, and the flow capacity of each pipe is as shown on the diagram. The topmost pipe

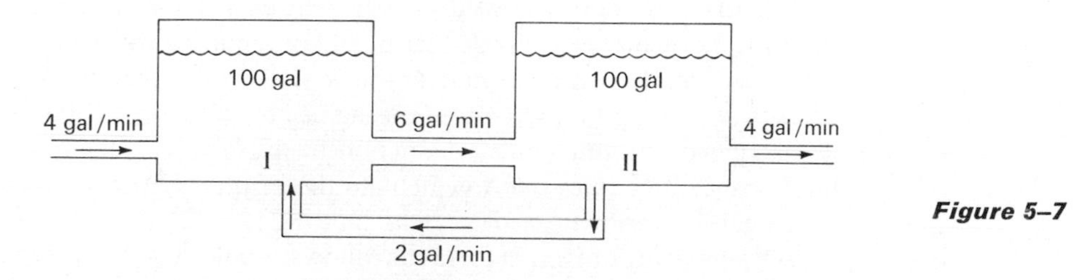

Figure 5–7

admits pure water to tank I. Suppose that at the moment that all pipes are opened, tank II contains 10 grams of a toxic substance, and tank I contains only pure water. How long will it take until the liquid flowing out of the lowest output pipe is virtually free of the substance (e.g., until the liquid reaches concentration level .001 gm/gal.)?

As a starting point in building a model for this situation, let x be the amount of the toxic substance in tank I at time t and y the amount in tank II at time t. For example, we know that at $t = 0$, when the system starts, $x = 0$ and $y = 10$. The rates of change of x and y are governed by a pair of related differential equations which we shall derive shortly:

$$\frac{dx}{dt} = \frac{2}{100}y - \frac{6}{100}x, \qquad \frac{dy}{dt} = \frac{6}{100}x - \frac{6}{100}y. \tag{5.15}$$

Thus a model for the two-tank situation is provided by a pair of continuous functions $\varphi(t)$ and $\psi(t)$ such that $x = \varphi(t)$ and $y = \psi(t)$ satisfy the equations (5.15), and in addition, are such that $\varphi(0) = 0$, $\psi(0) = 10$. The basic question which we must ask of the model is

"*For what value of t is it true that $\psi(t) < .1$?*"

Other questions about the model are suggested in the exercises. Again, we have not yet discussed the techniques needed to obtain an answer to questions about this model, and we therefore postpone further discussion until later.

The differential equations (5.15) can be obtained in the following manner. It should first be noted that the process of going from reality to the mathematical model is not a logical exercise like the proof of a theorem in geometry. The concept of deduction that is involved is of a different sort. We must present an argument that suggests that equations (5.15) are an appropriate mathematical image of the process of mixing that takes place in the pair of tanks. The procedure is often a combination of precise mathematical reasoning and arguments based on physical intuition. In the creation of a model, there is room for wild guesses and even luck. The role of rigorous mathematics is the deduction of consequences of the model, and the ultimate test of a model is how well its predictions match what is seen in reality. Quantum mechanics is not accepted because it is a rigorous mathematical deduction from observed phenomena, but because it enables one to calculate the results of measurements which can be verified.

We can arrive at formula (5.15) as follows. Looking at the diagram in Figure 5–7, we suppose that liquid is flowing as indicated. At a particular moment of time, we suppose that x and y are the amounts of the toxic substance in tanks I and II in grams. Its concentration in each tank in gm/gal is $x/100$ and $y/100$. During a short interval of time Δt, what change will take place? Looking at each pipe in turn, we find

(i) $4(\Delta t)$ gallons of pure water enters tank I,
(ii) $2(\Delta t)$ gallons enters tank I carrying the toxic substance at concentration $y/100$,
(iii) $6(\Delta t)$ gallons leaves tank I at concentration $x/100$,
(iv) $6(\Delta t)$ gallons enters tank II at concentration $x/100$,
(v) $2(\Delta t)$ gallons leaves tank II at concentration $y/100$,
(vi) $4(\Delta t)$ gallons leaves tank II at concentration $y/100$.

From the above, we can calculate the total change in the amount of the toxic substance in each tank at the end of the Δt-long interval:

$$\Delta x = 0 + (2)(\Delta t)(y/100) - (6)(\Delta t)(x/100),$$
$$\Delta y = (6)(\Delta t)(x/100) - (2 + 4)(\Delta t)(y/100).$$

Dividing by Δt, we are immediately led to impose the relations given in (5.15) upon the functions that govern the values of x and y.

This is an example of one of the methods used in building a model. It may be noted that in building a model, we have assumed that mixing is slow enough to allow us to use $x/100$ as the concentration of the liquid leaving tank I while we are adding new toxic substance at the top, but fast enough so that we need not worry about how long it takes for the new material to diffuse throughout the tank for the next step. Perhaps these assumptions seem inconsistent. One might therefore desire a different model which takes account of the true speed of diffusion and the unequal concentration in different parts of either tank. Such a model, although a better reflection of reality, is likely to be so difficult to work with that we may not be able to learn much from it.

In the exercises that follow, we ask you to examine a number of models and to attempt to build others. There are few guidelines that can be given; remember also that there is no uniquely correct model of any portion of nature, although some are more useful than others.

Exercises

1 Verify analytically that for values of θ_0 near 0, formula (5.8) is approximately the same as formula (5.12).

2 According to either of the models for a simple pendulum, what is the effect of lengthening the rod? What happens if you make it twice or four times as long? How could you test your answer?

3 When $\theta_0 = \pi/2$, calculate the predicted value of V_0 by each model. Can you design an experiment that would tell you which better represents reality? Is this easier to do if $\theta_0 = \pi$?

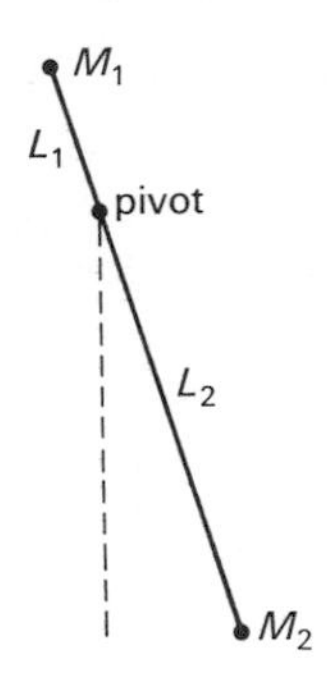

Figure 5–8

4 Construct simple pendulums in which the rod is replaced by (a) a string, (b) a long rubber band, (c) a heavy rod whose mass is comparable to the bob. Can you decide if either of the models in the text represent any of these real pendulums?

5 (If you have enough physics experience) Can you devise a mathematical model for the pendulum shown in Figure 5–8 similar to either of those in the text?

6 A company which makes radios has two factories and five warehouses, all of which are in different cities. The various costs of shipping a radio from either factory to any of the warehouses are all known. We are given the available storage room at each warehouse, and the maximum production capacity of each factory. Devise a model which could enable you to answer a question such as "What plan for producing and shipping radios will be the cheapest?" (*Note:* The simplest model will be an algebraic one, not involving functions or differential equations at all.)

7 A lake is being polluted constantly by the addition of a fixed amount of a certain chemical per day. At one end of the lake, pure water is entering at a constant rate, and at the other end, water is flowing out at the same rate. Can you construct a model for this lake which will enable you (after you have learned to solve differential equations) to discuss the future behavior of the lake and to predict when the pollution level will exceed a certain threshold?

8 How would you modify the model for Exercise 7 if the amount of chemical being added per day varies in a known way? If the amount of water flowing in and out per day is not a constant, but also changes with time?

9 It is observed that a quantity of material A left by itself emits radiation and after a while is found to be a mixture of substance A and substance B. It is conjectured that the atoms of substance A are "unstable," and that they randomly turn into atoms of substance B, emitting radiation as they do so. If the quantity of material is large, so that there are many, many atoms involved, it is conjectured that the rate at which substance A turns into substance B can be regarded as proportional to the amount of A still present. Construct a mathematical model which would enable you to answer questions such as the following: "If it takes 10^5 years for 1 gram of A to turn into $\frac{1}{2}$ gram A and $\frac{1}{2}$ gram B, then how long will it take 1 gram of A to become .8 gram A and .2 gram B?" What is the connection here with carbon dating of human artifacts?

10 Construct a model for human population growth. (Since no answer to this question can be either simple or unique, it is an ideal topic for class discussion; an optimal answer may be a list of the factors that should be taken into account in building such a model and some of the functions which behave in the required way.)

11 (a) Use your experience and physical intuition to guess the shape of the functions φ and ψ in the Two-Tank Problem.

(b) How do you think these graphs would change if we had started with 10 grams of the toxic substance in tank II and 50 grams in tank I? (Everything else stays the same.)

(c) Under the circumstances holding in part (b), what would be the effect of changing the flow in the two pipes connecting tank I and tank II from 2 gal/min and 6 gal/min to 12 gal/min and 16 gal/min?

12 Can you devise a model for a system consisting of three tanks and a number of interconnecting pipes, and one or more pollutants?

5.2

First Order Equations

Having seen how physical problems can give rise to differential equations, we shall begin a systematic study of these equations. From the mathematical point of view, a differential equation is one in which the unknown is a function, rather than a number as in the case of an algebraic equation; the unknown function is to be determined from the identities and relations which it and its derivatives are known to satisfy. When we are dealing with functions of one (independent) variable, the equations are called **ordinary differential equations**; when we are dealing with functions of several variables and the identities involve partial differentiation, then they are called **partial differential equations.** As an example, the study of heat distribution leads to the partial differential equation

$$\frac{\partial^2 U}{\partial x^2} + \frac{\partial^2 U}{\partial y^2} + \frac{\partial^2 U}{\partial z^2} = 0.$$

Functions U that satisfy this identity are called potential or harmonic functions. In this chapter, we shall be concerned only with ordinary differential equations.

Differential equations are classified in various ways. The term "order" is used to refer to the highest order derivative of the unknown function that appears in the identity to be solved. Thus, a first order equation may involve the unknown function and its first derivative, but will not involve its second derivative or higher derivatives. Since time is so often the independent variable in physical applications, we choose to label points in the plane as (t, y) rather than (x, y). Then, the normal form of a first order differential equation is

$$\frac{dy}{dt} = f(t, y), \tag{5.16}$$

where f is a function of two variables defined on a region D in the TY-plane.

To solve (5.16) means to find an interval I and a function φ such that (see Figure 5–9)

(i) $\varphi'(t)$ exists for all $t \in I$.
(ii) The graph of φ lies in the region D.
(iii) For each $t \in I$, $y = \varphi(t)$ satisfies the equation (5.16), meaning that

$$\varphi'(t) = f(t, \varphi(t)) \quad \text{for all } t \in I.$$

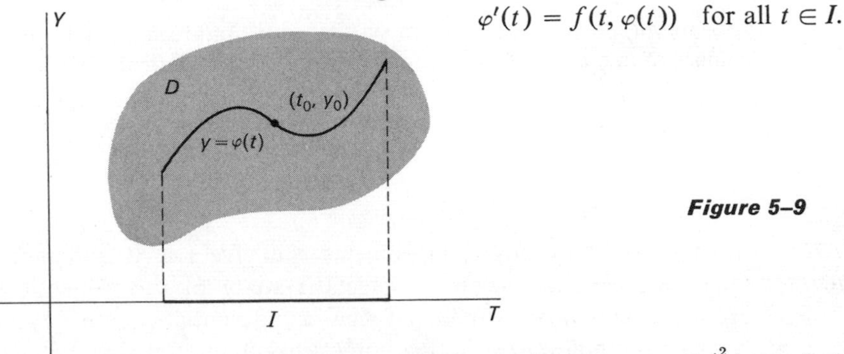

Figure 5–9

As an example, the function $y = 3e^{t^2}$ is a solution of the equation $dy/dt = 2ty$ on the interval $-5 < t < 5$ because for every such t,

$$\frac{d}{dt}(3e^{t^2}) = (3e^{t^2})(2t) = 6te^{t^2},$$

and

$$2ty = (2t)(3e^{t^2}) = 6te^{t^2}.$$

In this case, there is nothing special about the choice of the interval I, and in fact $3e^{t^2}$ is a solution of the given equation on the whole T-axis, $-\infty < t < \infty$.

A more typical example is provided by the equation

$$\frac{dy}{dt} = -\frac{t}{y}. \tag{5.17}$$

Here, the function f is defined for any choice of t and y, with $y \neq 0$. Thus, D could be taken to be any region in the upper half plane. The function given by $y = \sqrt{1 - t^2}$ is defined only on the interval I: $-1 < t < 1$; but on that interval, it is a solution of the equation, for

$$\frac{dy}{dt} + \frac{t}{y} = (1/2)(1 - t^2)^{-1/2}(-2t) + \frac{t}{\sqrt{1 - t^2}}$$
$$= 0.$$

It is also easily checked that $y = \sqrt{2 - t^2}$ is also a solution of this same equation—this time on the larger interval $(-\sqrt{2}, \sqrt{2})$.

There is a simple geometric relationship between a first order differential equation $dy/dt = f(t, y)$ and the functions that are its solutions. Moreover, this relationship is the basis for a simple and effective method for finding out something about the solutions of an equation, and also lies at the heart of the procedures by which high-speed computers obtain approximate solutions of differential equations.

Since dy/dt is associated with the concept of slope, we can use the given differential equation to construct a direction field or tangent field in the region D where f is defined. We proceed as follows: At each point $p = (t, y)$ in D, we evaluate f and use this value $f(p) = f(t, y)$ as the slope of a short line segment which we draw centered on p. The totality of such line segments for all points p in D is called the direction field or tangent field for the differential equation. In Figure 5–10, we have drawn a portion of the tangent

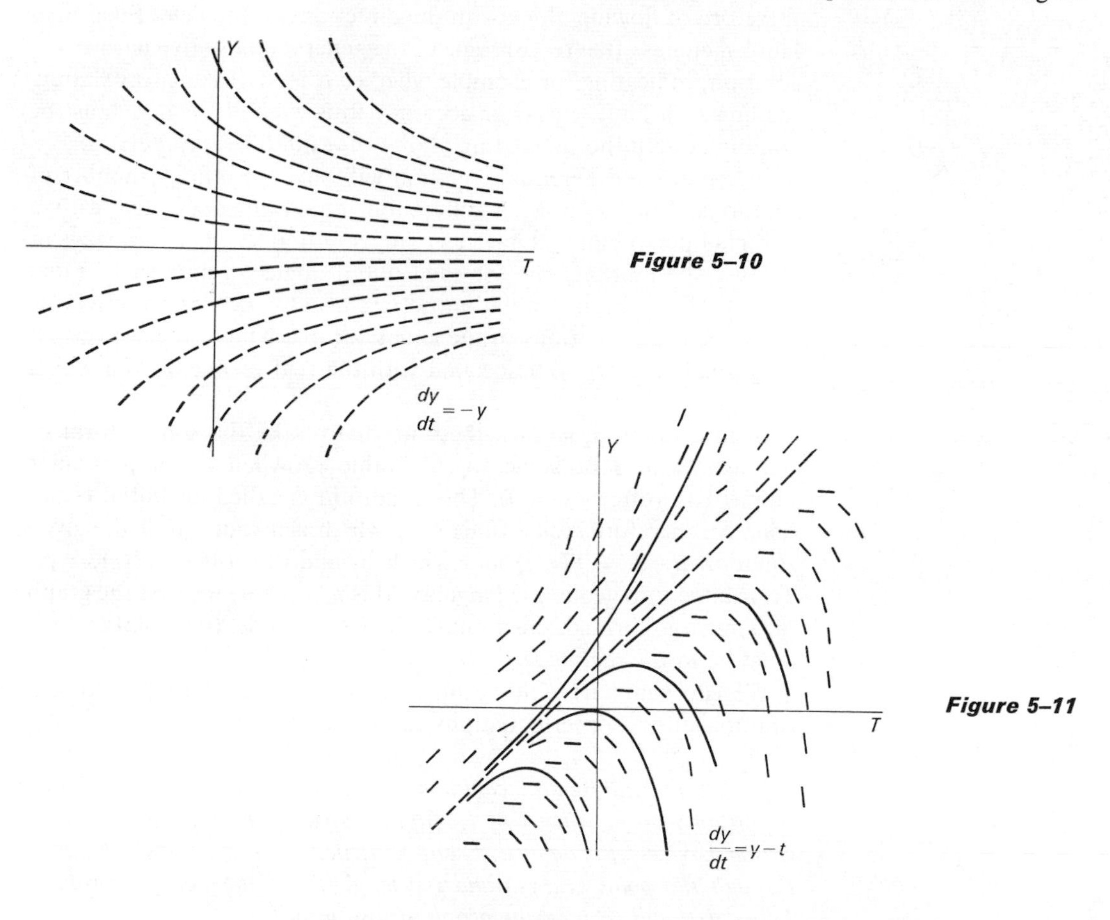

Figure 5–10

Figure 5–11

field defined by the equation $dy/dt = -y$, and in Figure 5–11, the field defined by $dy/dt = y - t$. (A special trick which makes it easier to construct a tangent field, called the method of isoclines, is given in the exercises (Exercise 4).)

What does the direction field for an equation have to do with solutions of the equation? The fact that $y = \varphi(t)$ satisfies the equation $dy/dt = f(t, y)$ means that at any point p_0 on the curve that is the graph of φ the slope of the curve must coincide with the slope of the line segment in the direction field passing through p_0. In other words, the curve must be tangent to each of the line segments whose midpoints it passes through. In Figure 5–11, we have drawn a number of the solution curves to show this tangency property.

It is now evident that this procedure can be used in reverse to find out something about the solutions of a given differential equation. In fact, given a carefully drawn tangent field, a skilled person can often produce reasonably accurate sketches of many of the solutions by drawing curves that have this property of being tangent to and therefore of flowing along with the directions of the field. Such freehand sketches often reveal some of the general qualitative nature of a solution, indicating for example whether it is increasing, oscillating, tending to a limit, etc. (For accuracy, this eyeball method must be combined with the calculating ability of a digital computer.)

In general, a differential equation will have an infinite number of solutions. For example, a differential equation associated with a swinging pendulum will have as many solutions as there are ways of starting the motion, one for each initial angle θ_0 and each initial velocity. In a physical situation, we are usually looking for only one of the possible solutions of the equation, one which satisfies certain additional restrictions associated with the real problem from which the model came.

In first order equations, these normally take the simple form of requiring that y be some specific value y_0 when t is a particular moment t_0 (often $t_0 = 0$). This condition is called an **initial condition**. We therefore seek a function φ which is a solution of the given equation $dy/dt = f(t, y)$ and which in addition obeys $\varphi(t_0) = y_0$. Translated into geometric language, this merely means that the graph of φ must go through the point (t_0, y_0). Of course, this point (t_0, y_0) must be in the region D.

We can summarize this geometric approach to the approximate solution of first order equations as follows:

To solve the differential equation $dy/dt = f(t, y)$ with given initial condition $y = y_0$ when $t = t_0$, first construct the tangent field in the region D where f is defined. Then, construct a curve in D which passes through the point (t_0, y_0) and is "tangent" to the field everywhere. This curve will then be the graph of a solution.

In most cases, and certainly in all the most useful ones, there will be one and only one solution curve passing through an interior point of the given region D. However, as we shall see, this depends upon the nature of the defining function f and may fail if the function f isn't sufficiently smooth.

A physical problem may involve two or more interrelated quantities that depend upon time. In this case, a standard model for the situation may be a system of differential equations. This was the case, for example, in the Two-Tank Problem in the last section (see formula (5.15)). Systems of equations are also classified by order, and the standard form for a system of two first order differential equations would be

$$\frac{dx}{dt} = f(t, x, y), \qquad \frac{dy}{dt} = g(t, x, y). \tag{5.18}$$

The general approach is quite similar to that of a single equation. We assume that both f and g are defined in a region D in TXY-space. A solution of the system (5.18) is a pair of functions φ and ψ, each defined on the same T-interval I, such that φ' and ψ' are also defined there. Furthermore, when $t \in I$, the points $(t, \varphi(t), \psi(t))$ must all lie in D and $x = \varphi(t)$, $y = \psi(t)$ must satisfy the equations as an identity, i.e.,

$$\varphi'(t) = f(t, \varphi(t), \psi(t)),$$
$$\psi'(t) = g(t, \varphi(t), \psi(t))$$

for all $t \in I$.

It is possible to give a geometric interpretation here. The pair of functions f and g can be used to create a direction field on the region D, and the set of points $(t, \varphi(t), \psi(t))$ trace a curve in 3-space which must be tangent to the direction field in order for $x = \varphi(t)$ and $y = \psi(t)$ to be a solution. To go into this further would take us beyond the intended limits of the text.

Exercises

1 Verify that for $C > 0$, each of the functions $y = \sqrt{C - t^2}$ is a solution of the differential equation in (5.17). What about the functions $y = -\sqrt{C - t^2}$? Sketch some of these solutions for several choices of C.

2 Show that the functions $y = 1/(t - C)$ are solutions of the equation $dy/dt = -y^2$ on certain intervals which depend upon the choice of C. Sketch these solutions for $C = 0, \pm 1, \pm 2$.

3 Verify that for each C, the function $y = (1 + Ce^t)/(1 - Ce^t)$ is a solution on an appropriate interval of the equation $dy/dt = (y^2 - 1)/2$. Sketch the graphs of these solutions for $C = 0$, $C = 1$, and $C = -1$.

4 The isocline method for constructing the direction field for the equation $dy/dt = f(t, y)$ is to draw line segments of slope m at all the points on the level curve of f whose equation is $m = f(t, y)$. Apply this method to produce the direction field for the equation $dy/dt = 2y$, and sketch several of the solution curves.

5 Sketch the direction field by the isocline method for the equation $dy/dt = t + y$ and draw several of the solutions.

6 Do the same for the equation $dy/dt = t/(t - y)$.

7 Do the same for the equation $dy/dt = t^2 + y^2$. Do you think that the solution passing through the point $(0, 1)$ can be extended to arbitrarily large values of t?

8 Among the solutions discussed in Exercise 1 above, can you find one which satisfies the condition $y = 2$ when $t = 1$? for which $y = -1$ when $t = 3$?

9 Among the solutions discussed in Exercise 3 above, is there one which passes through the point $t = 1$, $y = 5$? Is there one such that $y = 2$ when $t = 0$?

10 A class of functions is described by the formula $y = Ct/\sqrt{C^2 - t^2}$ for each real number C. Is there a function in this family such that $y = 2$ when $t = 1$? such that $y = 1$ when $t = 2$?

11 (a) Verify that the functions given by $x = e^{4t} + e^{-6t}$, $y = e^{4t} - e^{-6t}$ are solutions of the system of equations

$$dx/dt = 5y - x, \quad dy/dt = 5x - y.$$

(b) Verify that all functions of the form $x = Ae^{4t} + Be^{-6t}$ and $y = Ae^{4t} - Be^{-6t}$ are solutions of the same system.

12 (a) Find a solution of the system in Exercise 11 which obeys the initial condition $t = 0$, $x = 1$, $y = 3$.

(b) Do the same for the initial condition $t = 1$, $x = 1$, $y = 3$.

13 Show that $x = -t^3/3$ and $y = 3/t$ are solutions of the system of first order equations $dx/dt = xy$, $dy/dt = 3/(xy)$. Do you think there are others?

14 Show that for every choice of A, the functions $x = t + At^2$, $y = 1 + 2At - At^2$ are solutions of the system $dx/dt = x + y - t$, $dy/dt = 2y/t - 2x/t^2$. Do you think there are other solutions?

5.3

Higher Order Equations

A differential equation may involve higher order derivatives, as in the case of the first two illustrations of mathematical models given in Section 5.1. One obvious reason is that many models involve Newton's Laws of Motion and therefore forces and accelerations which involve second derivatives with respect to time.

Let f be a function defined in a region D in 3-space. Then, with f we can associate a general second order differential equation whose normal form is

$$\frac{d^2y}{dt^2} = f(t, y, y'). \tag{5.19}$$

(Here, we have written y' for dy/dt. Another common notation is $\dot{y}$.)

A function φ is a solution of the equation (5.19) on an interval I of the T-axis if the following conditions hold:

(i) $\varphi(t)$, $\varphi'(t)$, and $\varphi''(t)$ are defined for all $t \in I$.
(ii) For each $t \in I$, the point $(t, \varphi(t), \varphi'(t))$ lies in the region D.
(iii) $y = \varphi(t)$ satisfies the equation (5.19) on I, meaning that the identity

$$\varphi''(t) = f(t, \varphi(t), \varphi'(t))$$

holds for all $t \in I$.

As an example, consider the equation

$$y'' - 2y' + y = 0,$$

which is an equivalent way to write

$$\frac{d^2y}{dt^2} - 2\frac{dy}{dt} + y = 0.$$

This can be rewritten as

$$\frac{d^2y}{dt^2} = 2\frac{dy}{dt} - y,$$

which has the form (5.19) with $f(t, y, y') = 2y' - y$. The set D can be taken to be all of 3-space. Then, the function $\varphi(t) = e^t$ is easily seen to be a solution of the equation on the interval $-\infty < t < \infty$. For, $y' = e^t$ and $y'' = e^t$, so that $y'' - 2y' + y = e^t - 2e^t + e^t = 0$ for all values of t.

It is possible to attach a geometric interpretation to the relationship between an equation and one of its solutions as with first order equations, but it is far less useful either as a practical guide in finding solutions or as a theoretical tool. If we look at a particular solution function φ which satisfies the equation (5.19), then the knowledge of the coordinates $t = t_0$, $y = y_0$ of a point on the graph of φ and the knowledge of the slope y'_0 of this graph at (t_0, y_0) enables us to use f to compute the value $f(t_0, y_0, y'_0)$, which is the value of $\varphi''(t_0)$. Thus, we can determine the curvature of the graph of φ at any point where we know its slope. This does not lead to a practical scheme for obtaining freehand sketches of solutions to (5.19).

A technique to single out one among the infinite number of solutions of a second order equation is to specify initial conditions;

this means that the desired solution $y = \varphi(t)$ must have prescribed values A and B for y and $y' = dy/dt$ corresponding to $t = t_0$. In terms of the geometric interpretation given above, this means that we have asked for a solution of the given equation which passes through a specified point (t_0, A), having there the assigned slope B.

For example, the differential equation

$$\frac{d^2y}{dt^2} = -\frac{t}{1+t}y' + \frac{1}{1+t}y$$

has for solutions the functions

$$y = Ae^{-t} + Bt$$

for any choice of the numbers A and B. Suppose we want a solution such that at $t = 1$, $y = 3$ and $y' = 2$. We have

$$y' = \frac{dy}{dt} = -Ae^{-t} + B.$$

Setting $t = 1$, and $y = 3$, $y' = 2$, we obtain the algebraic equations

$$3 = Ae^{-1} + B,$$
$$2 = -Ae^{-1} + B,$$

which can be solved to find $B = \frac{5}{2}$ and $A = e/2$.

A second type of condition which picks one of the solutions of an equation from all the rest occurs in the study of higher order equations; it has no analogy in the study of first order equations. Sometimes, the nature of the problem that gave rise to a particular second order differential equation leads to the requirement that the unknown function φ must satisfy a condition like $\varphi(t_1) = a_1$, $\varphi(t_2) = a_2$, where t_1 and t_2 are different moments of time. (For example, we may ask that the moving object be at specified locations at the start and at the end of its motion.) This type of condition is called a **two-point boundary condition** and is of great significance in the theory and application of differential equations. As a statement about the graph of a solution, it merely says that the graph of φ passes through two specified points, but says nothing about the slope at either point.

For example, if we return to the solutions $Ae^{-t} + Bt$ of the previous example and ask that $y = 2$ when $t = 0$, and $y = 3$ when $t = 1$, then substitution again gives a set of algebraic equations:

$$2 = Ae^0 + B(0) = A,$$
$$3 = Ae^{-1} + B,$$

from which we find $A = 2$ and $B = 3 - 2/e$. Thus the desired solution is $y = 2e^{-t} + (3 - 2/e)t$.

Finally, we point out a mathematical fact about differential equations of higher order that is of great importance for creating a general theory. It simplifies the general viewpoint and even helps in the discussion of an individual equation. The first special case of the general principle is that any second order equation can be regarded as a system of two first order equations. In general, a single nth order equation can be regarded as a special system of n first order equations.

To see how this is possible, suppose we have the second order equation

$$y'' = \frac{d^2y}{dt^2} = f(t, y, y'). \tag{5.20}$$

Now, put $x = dy/dt$ so that $dx/dt = d^2y/dt^2$. Then, equation (5.20) is clearly equivalent to the system

$$\frac{dx}{dt} = f(t, y, x),$$

$$\frac{dy}{dt} = x.$$

As another example, the system of first order equations

$$\frac{dx}{dt} = f(t, y, z, x),$$

$$\frac{dz}{dt} = x,$$

$$\frac{dy}{dt} = z$$

is equivalent to the general third order equation

$$\frac{d^3y}{dt^3} = f(t, y, y', y''). \tag{5.21}$$

Exercises

1 (a) Verify that the functions $y = Ae^{4t} + Be^{-6t}$ are solutions of the equation $d^2y/dt^2 = 24y - 2(dy/dt)$ for every choice of A and B.

(b) Find a solution such that $y = 2$ and $y' = 3$ when $t = 0$.

2 (a) Show that the functions $y = A/(t + B)$ are solutions of the equation $d^2y/dt^2 = 2(y')^2/y$ for every A and B with $A \neq 0$. Can you guess some other solutions?

(b) Can you find a solution which satisfies the initial conditions $y = 2$, $y' = 3$ when $t = 0$? Are there any admissible initial conditions you cannot satisfy?

3 (a) The equation $y'' = (5/2t)y' - (3/2t^2)y$ has solutions of the form $y = t^\alpha$. Find two of these, φ and ψ.

(b) Show that every function $y = A\varphi + B\psi$ is also a solution, and find such a solution that satisfies the initial conditions $y = 3$, $y' = 2$ when $t = 1$.

4 Can you find a solution of the equation discussed in Exercise 1 above which satisfies the following two-point boundary condition: when $t = 0$, $y = 0$ and when $t = 1$, $y = 1$?

5 Can you find a solution of the equation discussed in Exercise 2 above which satisfies the following two-point boundary condition: when $t = 0$, $y = 1$ and when $t = 1$, $y = 3$?

6 Can you find a solution of the equation in Exercise 3 which obeys the two-point condition $t = 1$, $y = 2$; $t = 4$, $y = 1$?

7 Check the equivalence of the equation in (5.21) to the system of three first order equations given.

8 Write D for d/dt so that, for example,

$$\begin{aligned}\frac{d^2x}{dt^2} - 4\frac{dx}{dt} + 5x &= D^2x - 4Dx + 5x \\ &= (D^2 - 4D + 5)x.\end{aligned}$$

Show that the system $dx/dt = 3x + 2y$, $dy/dt = 7x - 5y$ can be written in the form

$$\begin{aligned}(D - 3)x &= 2y, \\ (D + 5)y &= 7x.\end{aligned}$$

9 Using the previous exercise, show that the functions x and y in this particular system must satisfy the identity

$$\varphi'' + 2\varphi' - 29\varphi = 0.$$

10 If x and y are functions satisfying the system $dx/dt = x + y$, $dy/dt = x - y$, then show that x and y each must satisfy the relation $\varphi'' = 2\varphi$.

11 Show that a general nth order differential equation is equivalent to a special system of n first order equations.

5.4 Theory vs. Experiment

We have seen that the task of formulating a mathematical model for a physical problem (e.g., the motion of a pendulum) can lead to a differential equation. Different ways of approaching the problem, different physical approximations, may result in different models. In the pendulum example of Section 5.1, we arrived at the equation

$$\frac{d^2\theta}{dt^2} = -\frac{g}{L}\sin\theta. \tag{5.22}$$

In order to answer some of the questions raised about the motion of the pendulum, we must find a solution of this equation satisfying the initial conditions $\theta = \theta_0$, $d\theta/dt = 0$ when $t = 0$.

It would certainly be disconcerting if there were to be no such solution! Since an actual pendulum really moves, this would mean that our model is quite useless, and we would have to construct a new model. This means that the only useful models are those which have solutions; an important area of mathematical study has to do with proving that certain very general classes of differential equations have solutions, even when we do not have any techniques for finding these solutions.

Nor is the existence of solutions the only reasonable requirement that we might wish for in constructing a differential equation that is to lead to a useful model. Suppose that we take an actual pendulum, displace it an angle θ_0, and release it from rest. Experience suggests that if we could repeat this experiment exactly we would obtain precisely the same motion. The time of descent of the bob and the speed attained would be the same each time. This is really a philosophical assumption, connected with the hypothesis of determinism. If we accept this assumption, with regard to a mathematical model for the motion of a pendulum it means that a differential equation such as (5.22) should have exactly one solution satisfying a given set of initial conditions. If φ and ψ are solutions of (5.22) for which $\varphi(0) = \psi(0) = \theta_0$ and $\varphi'(0) = \psi'(0) = 0$, then we must have $\varphi(t) = \psi(t)$ for all $t > 0$. The model must also predict that repetitions of the same experiment have the same results. The corresponding mathematical study is concerned with the uniqueness of solutions of differential equations which have assigned initial conditions or boundary condition.

Finally, experience also suggests a third requirement for models. We know that in practice we cannot repeat an experiment in exactly the same way. However, if all the initial conditions are almost the same, then we expect the outcomes to be almost the same. If the pendulum is released at an angle very nearly θ_0, then the time of descent ought to be very nearly that obtained when the angle was exactly θ_0. The solutions of our model ought to have the same behavior. Stating this in mathematical language, we say that the solutions of a differential equation for a physical situation ought to depend in a continuous way upon the numerical values imposed in the initial conditions. For brevity, we refer to this as "continuity of the solution."

We digress for a moment to observe that there are areas of modern physics where it has seemed necessary to relax some of these assumptions in order to build adequate models. Especially at the atomic and nuclear level, there seem to be phenomena that may violate both the determinism and the continuity hypotheses.

Summarizing the three mathematical requirements that seem desirable for a differential equation used to model a physical situation, we have

(i) *Existence. The equation has a solution satisfying the given set of initial conditions.*
(ii) *Uniqueness. Two solutions of the equation that satisfy the same initial conditions are identical.*
(iii) *Continuity. Solutions depend continuously on the initial conditions.*

Much of the research of mathematicians who study the theory of differential equations has dealt with the effort to show that wider and wider classes of equations satisfy these three requirements. For example, they have succeeded in showing that equations so complicated that there is no known method for finding formulas for their solutions satisfy (i), (ii), (iii) above. This accomplishment is particularly important for applications; faced with such an equation, an applied mathematician can turn to a high-speed computer and calculate an approximation to the solution, knowing by (ii) and (iii) that there is only one exact solution and that his values must be close to the true values.

The proofs of theorems dealing with existence, uniqueness, or continuity are beyond this text. However, we shall explain the meaning of certain of the basic theorems of this type without attempting to prove them. The first, dealing with first order equations, is the following.

Theorem 1 *Let D be an open convex set in the TY-plane. Let $f(t, y)$ be defined for all (t, y) in D, and suppose that f and its partial derivative f_y are continuous in D. Let $p_0 = (t_0, y_0)$ be any interior point of D. Then the differential equation $dy/dt = f(t, y)$ has a unique solution $y = \varphi(t)$ which satisfies the initial condition $\varphi(t_0) = y_0$. Its graph passes through p_0, and φ is defined on an interval I containing t_0. Moreover, if t is in I, then the value of $\varphi(t)$ is a continuous function of the initial conditions (i.e., of the point p_0.)*

To give a simple illustration of how this result can be used, consider the equation

$$\frac{dy}{dt} = t^2 + te^{-y}. \tag{5.23}$$

Here, we have $f(t, y) = t^2 + te^{-y}$, which is defined and continuous in the whole TY-plane. Since $f_y(t, y) = 0 - te^{-y}$, which is also continuous everywhere, we conclude that (5.23) can be solved uniquely to satisfy any given initial condition. For example, there is exactly one solution of (5.23) that satisfies the condition that y shall be 103 when $t = 1.71$.

When would an equation fail to satisfy the hypotheses of this theorem? The standard example is the equation

$$\frac{dy}{dt} = 3y^{2/3}. \tag{5.24}$$

Here, $f(t, y) = 3y^{2/3}$, which is continuous for all t and y. However, $f_y(t, y) = 2y^{-1/3} = 2/\sqrt[3]{y}$. This is not continuous or even defined at points (t, y) with $y = 0$. Thus, the theorem would not apply if we were trying to discuss solutions of (5.24) on a region D that contained any points of the T-axis. What makes this equation interesting is that although it has solutions for any given initial condition, the uniqueness condition breaks down completely. If the point $p_0 = (t_0, y_0)$ is not on the horizontal T-axis, then there is exactly one solution curve passing through p_0. However, if p_0 is of the form $(t_0, 0)$, then there are infinitely many solutions of the equation that pass through p_0.

For example, if $p_0 = (0, 0)$, then two solutions are easily found. Both $y = 0$ for all t and $y = t^3$ for all t are solutions, and both fulfill the requirement $y = 0$ when $t = 0$. What is not so obvious is that each of the functions defined by

$$y = \begin{cases} 0 & -\infty < t < b, \\ (t - b)^3 & b \leq t < \infty \end{cases} \tag{5.25}$$

is a continuous function having a continuous derivative satisfying the equation (5.24) and the assigned initial condition. (See Figure 5–12.)

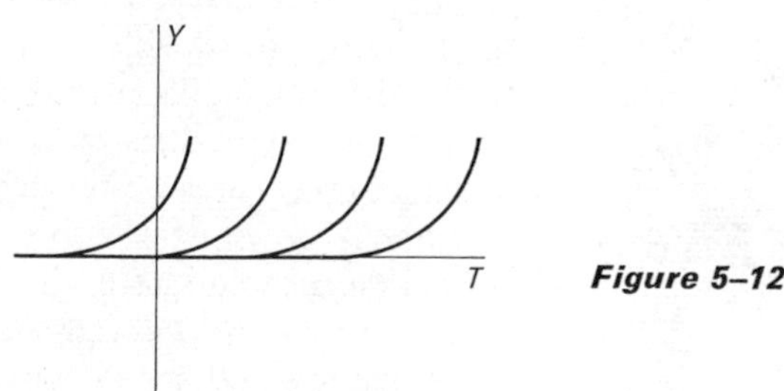

Figure 5–12

This example shows that uniqueness is not automatic, and that an adequate theory of differential equations is needed to enable us to know conditions which will ensure that solutions for assigned initial conditions are unique.

A similar result can be stated for second order equations.

Theorem 2 *Let D be a convex set in 3-space. Let f and its partial derivatives f_2 and f_3 be defined and continuous in D. Let (t_0, A, B) be an interior point of D. Then, the differential equation $d^2y/dt^2 = f(t, y, y')$ has a unique solution $y = \varphi(t)$ which satisfies the initial conditions $y = \varphi(t_0) = A$,*

$y' = \varphi'(t_0) = B$. The function φ is defined on an interval I containing t_0, and if $t \in I$, then the value of $\varphi(t)$ depends continuously on the initial values A and B.

To illustrate, consider the following equation which represents a model for a pendulum that is submerged in something like oil:

$$\frac{d^2\theta}{dt^2} = -\frac{g}{L}\sin\theta - C\left(\frac{d\theta}{dt}\right)^2. \tag{5.26}$$

When we set $\theta = y$ and $d\theta/dt = y'$, the right-hand side becomes $f(t, y, y') = -(g/L)\sin y - C(y')^2$. This is continuous for all choices of the three numbers t, y, y', so that the region D can be taken as all of 3-space. The notation f_2, f_3 signifies the partial derivatives of the function f with respect to the second and the third occurring variables. In this case, these are y and y', so that

$$f_2(t, y, y') = -(g/L)\cos y,$$
$$f_3(t, y, y') = -2Cy'.$$

Since f_2 and f_3 are again continuous everywhere, we conclude that equation (5.26) has a unique solution corresponding to any possible assignment of initial conditions.

Another development by mathematicians in the theory of differential equations lies in the determination of qualitative properties of the solutions of classes of equations, particularly in those cases where there is no known method for obtaining formulas for the solutions themselves. In many cases, it is important to know if the functions which solve an equation are bounded functions, or if they are periodic like a sine curve, or if they have any special behavior for large values of t. Also, if we cannot solve an equation $dy/dt = f(t, y)$ and we replace the defining function f by a simpler function g which is an approximation to f, then we need to know how close the solutions of $dy/dt = g(t, y)$ are to the solutions of the original equation. The study of such matters is the theory of differential equations and should be studied by any person who expects to make further use of differential equations in the construction of models.

Exercises

1 Apply Theorem 1 to the equation $dy/dt = 1 + y^2$. What is the region D in which one might expect to find solutions? Are there any points in this region which should not be used as initial points?

2 Do the same for the equation $dy/dt = \sqrt{y^2 - 1}$. (Note that $y = \varphi(t) \equiv 1$ is a solution.)

3 Check that the equations (5.25) do provide solutions of the differential equation $dy/dt = 3y^{2/3}$ and that Figure 5–12 is correct.

4 Use Theorem 2 to discuss the existence of solutions to the equation $d^2y/dt^2 = a(t) + b(t)y + c(t)(dy/dt)$.

5 Do the same for the equation

$$y'' = \frac{y}{4 - (y')^2}.$$

Are there any initial conditions that cannot be expected to be usable for solutions?

6 Do the same for the equation $y'' = (t - y)/(t - y')$.

7 Speculative question: Suppose a differential equation is a model for a certain type of chemical reaction. Could the fact that the equation does not have a solution indicate that the reaction cannot take place? Would the fact that the equation has a solution guarantee that the reaction *does* take place?

5.5

Techniques of Solution

The general existence and uniqueness theory discussed in the preceding section tells us when to expect solutions for a particular differential equation with given initial conditions. The next step is to learn some methods for finding these solutions. It must first be said that there are no general techniques that work for all equations, even for all first order equations. The best that can be expected is to discover procedures for obtaining a sequence of better and better approximations to the true solution. We would prefer to have a formula that expresses the solution in terms of polynomials or other relatively simple functions which we regard as elementary. As an analogy, we may be able to express the solution of an algebraic equation as the real number defined by a converging sequence of rational numbers (e.g., in its decimal expansion), but we often prefer to have an answer that is expressed in terms of rational numbers and certain familiar real numbers like π, e, $\sqrt{3}$, etc.

Lacking a general universal method, it is useful to have a number of special methods that have been discovered over the years but which apply to only certain equations. Note that *any* device for obtaining a solution is legitimate, provided that the function obtained satisfies the equation. The reason is that the theory tells us that the equation has only one solution for a given set of initial conditions, so that the function we have found must be this unique solution. Most of what we discuss in this section applies only to first order equations. More general methods will be treated in sections 5.6 and 5.7.

Recall the general form for a first order equation:

$$\frac{dy}{dt} = f(t, y). \tag{5.27}$$

Two special cases can be disposed of at once, namely when the right side contains only one of the variables t, y. For example, if $f(t, y) = g(t)$, then (5.27) merely tells us that the unknown function $y = \varphi(t)$ is an antiderivative of g. For example, the solutions of the equation

$$\frac{dy}{dt} = te^{-t^2}$$

consist of the functions $y = -\frac{1}{2}e^{-t^2} + A$ for any choice of the number A.

The case in which the right side of (5.27) has the form $h(y)$ is almost as simple. Make the tentative assumption that we are dealing with a point where $h(y) \neq 0$. Since this implies $dy/dt \neq 0$, we can regard the equation $y = \varphi(t)$ as one that can be solved for t, in the form $t = \psi(y)$. Recall that we then have $dt/dy = 1/(dy/dt)$. We have now converted the equation $dy/dt = h(y)$ into the equation

$$\frac{dt}{dy} = \frac{1}{h(y)},$$

and again we can solve this directly by integrating:

$$t = \int \frac{dy}{h(y)}.$$

We then regard the result as an equation defining y as a function of t.

For example, to solve the equation

$$\frac{dy}{dt} = \frac{y}{y+1},$$

we first rewrite it as

$$\frac{dt}{dy} = \frac{y+1}{y}.$$

We then have, by integration,

$$\begin{aligned} t &= \int \frac{y+1}{y}\,dy = \int (1 + y^{-1})dy. \\ &= y + \log y + C. \end{aligned}$$

This equation, for any C, is taken as a relation that implicitly defines the desired function $y = \varphi(t)$.

We inject the comment that a more usual procedure is to go from the equation

$$\frac{dy}{dt} = \frac{y}{y+1} \tag{5.28}$$

to the expression

$$\frac{y+1}{y}\,dy = dt \tag{5.29}$$

and then to the integral form

$$\int \frac{y+1}{y}\,dy = \int dt = t. \tag{5.30}$$

Although it is possible to become very technical about the meaning of the expression in (5.29), it is much easier to regard this as nothing more than a notational device which leads one conveniently from (5.28) to (5.30).

A more general case, which includes both of the above, is of wide usefulness. A differential equation is said to have **variables separable** if it has the form

$$\frac{dy}{dt} = g(t)h(y). \tag{5.31}$$

Proceeding as in the previous two cases, we may go from (5.31) to the expression

$$\frac{dy}{h(y)} = g(t)\,dt$$

(which explains the name of the method, since the variables are now separated) and then to the integrated equation

$$\int \frac{dy}{h(y)} = \int g(t)\,dt. \tag{5.32}$$

Each side yields an indefinite integral (or antiderivative)—the left a function of y, and the right a function of t. The resulting equation is again regarded as implicitly defining $y = \varphi(t)$, and we obtain a solution of (5.31). We note that each side of (5.32) will include constants of integration, since antiderivatives are not unique. However, we do not need to add a constant on *both* sides, since we can achieve the same result by adding a constant on only one side. In general, for each choice of a numerical value for this constant, we obtain a solution of the equation (5.31). It is possible that this process does not yield a full set of solutions of the original equations, so that we may not be able to satisfy an arbitrary set of initial conditions among these solutions, but must look elsewhere.

An illustration may help to clarify the last remarks. Suppose we consider the first order equation $dy/dt = 4y^2t$. This is clearly of separable type, so that we arrive at

$$\int \frac{dy}{y^2} = \int 4t\,dt,$$

or

$$-\frac{1}{y} = 2t^2 + C,$$

from which we obtain the desired family of solutions of the given equations:

$$y = \frac{-1}{2t^2 + C} \quad \text{for any } C. \tag{5.33}$$

(This can be checked by direct substitution into the equation.) Since the right side of the equation, $4y^2t$, is continuous and has continuous partial derivatives everywhere, we can be sure that the equation has a unique solution passing through any given point (t_0, y_0). If, for instance, $t_0 = 1$ and $y_0 = 3$, then we must choose C so that $3 = -1/(2 + C)$, or $C = -\frac{7}{3}$. This works for every assignment of initial condition except an assignment such as $(t_0, 0)$, for the equation $0 = -1/(2t_0^2 + C)$ has no solution. Does this mean that the given equation has no solution through the point $(1, 0)$ or $(0, 0)$? This would contradict Theorem 1, which we cannot accept. The answer lies in the simple fact that there is another solution of the equation, $y = \varphi(t) \equiv 0$, which is not described by the formula (5.33) but which fulfills all initial conditions corresponding to points $(t_0, 0)$.

How did the above method happen to overlook this missing solution? The answer lies in the first step of the solution. We divided by $h(y)$, which in this case was y^2, and thus we inadvertently limited ourselves to solutions other than $y = 0$. It is exactly the solution $y = 0$ that was omitted.

The same thing happens in the general case. If a separable equation of the form $dy/dt = g(t)h(y)$ is solved by the method described above, then the final solutions will not take into account solutions which satisfy initial conditions (t_0, y_0) where $h(y_0) = 0$. However, they will be covered by straight-line solutions of the form $y \equiv y_0$, and the combination usually gives a solution of the given equation going through the admissible point (t_0, y_0).

Another illustration will show how the method of separated variables is used in a more complicated case. Let

$$\frac{dy}{dt} = \frac{y^2 + y}{t}. \tag{5.34}$$

We first write

$$\frac{dy}{y(y+1)} = \frac{dt}{t}. \tag{5.35}$$

Noting that

$$\frac{1}{y(y+1)} = \frac{1}{y} - \frac{1}{y+1},$$

we integrate (5.35), obtaining

$$\int \left(\frac{1}{y} - \frac{1}{y+1}\right) dy = \int \frac{1}{t}\, dt,$$

or

$$\log y - \log(y+1) = \log t + C.$$

In order to make the work of simplifying easier, we replace C by $\log A$ and then combine the terms on each side, obtaining

$$\log\left(\frac{y}{y+1}\right) = \log(At),$$

$$\frac{y}{y+1} = At.$$

This can now be solved for y, yielding the desired family of solutions

$$y = \frac{At}{1 - At}. \tag{5.36}$$

Can we now satisfy any admissible initial condition by choosing one of these solutions of (5.34)? Since the right side of (5.34) is not defined or continuous when $t = 0$, the region D for which the equation is meaningful cannot contain any points on the line $t = 0$. However, there are no other restrictions, so we might wish to consider any initial point (t_0, y_0) where $t_0 \neq 0$, but y_0 is arbitrary. Putting $t = t_0$ and $y = y_0$ in (5.36), we can solve for A, obtaining the (unique as expected) solution

$$A = \frac{y_0}{(1 + y_0)t_0}.$$

However, this solution breaks down when $y_0 = -1$! This means that no number can be chosen for A to give a solution of this form which passes through a point such as $(2, -1)$ or $(1, -1)$. Return to (5.34) and observe that the right side factors as $h(y)g(t)$, $h(y) = y^2 + y$, which vanishes when $y = -1$. As explained in the general discussion preceding this example, we can conclude that we must adjoin the special solution $y = \varphi(t) \equiv -1$ to the set described by

(5.36). We then will have a complete set of solutions, one passing through every admissible initial point (t_0, y_0).

At this stage, you are prepared to solve any of Exercises 1 through 8 of this section.

Sometimes the form of the function f in a differential equation

$$\frac{dy}{dt} = f(t, y)$$

makes it possible to solve the equation directly or to change it to an equation which can be solved directly. We will illustrate two general methods for the latter.

Suppose it is possible to expand the right side in powers of y, as

$$\frac{dy}{dt} = a_0(t) + a_1(t)y + a_2(t)y^2 + \cdots. \tag{5.37}$$

An equation is said to be a linear equation of first order if only the first two terms in this expansion are present. For example,

$$\frac{dy}{dt} = t^3 - 2ty \tag{5.38}$$

is a linear equation. An equation that is not linear can sometimes be replaced by a linear equation simply by throwing away all the terms with higher powers of y. The new equation will be a usable model for the original problem as a first approximation. This process, called linearization, makes sense if we expect that the values of y will always be very small. (It is the process that led from pendulum model (5.2) to pendulum model (5.10).)

Any first order linear equation can be solved. The technique is easily learned, but in many cases the solution in its final form is not very useful because it is given in terms of integrals that can be evaluated only in special cases. We illustrate the process first for the equation (5.38), and then we discuss the general case.

Let us rewrite (5.38) as

$$y' + 2ty = t^3$$

and then multiply both sides by e^{t^2}, obtaining

$$e^{t^2}y' + 2te^{t^2}y = t^3e^{t^2}. \tag{5.39}$$

As happens with many mathematical discoveries, this technique, attributed to one of the Bernoulli brothers, was probably the result of a lucky insight. Inspiration is very difficult to motivate.

We observe that the two terms on the left arise from differentiating the expression $e^{t^2}y$. Thus, we can rewrite (5.39) as

$$\frac{d}{dt}\{e^{t^2}y\} = t^3e^{t^2}.$$

If we now integrate, we first obtain

$$e^{t^2}y = \int t^3e^{t^2}\,dt + C$$

and then the desired solution of the equation (5.38),

$$y = e^{-t^2}\int t^3e^{t^2}\,dt + Ce^{-t^2}.$$

We can carry out the integration to obtain a simpler answer. Set $u = t^2$, so that

$$\int t^3e^{t^2}\,dt = \tfrac{1}{2}\int ue^u\,du = \tfrac{1}{2}(u-1)e^u$$
$$= \tfrac{1}{2}(t^2-1)e^{t^2},$$

and we obtain the final form

$$y = \tfrac{1}{2}(t^2-1) + Ce^{-t^2}$$

for the set of all solutions to (5.38).

How does one generalize from this special case to obtain a general method? Since a linear first order equation is one in which the right side has the form $f(t, y) = a_0(t) + a_1(t)y$, we can always rewrite the equation so that it takes the form

$$y' + p(t)y = q(t). \tag{5.40}$$

The trick we used to solve (5.38) was to turn the left side into an expression that was exactly the derivative of a product of the form (some function of t)(y) by multiplying the equation by a function. We want to choose a function $G(t)$ such that

$$\frac{d}{dt}\{G(t)y\} = G(t)(y' + p(t)y).$$

Rewriting, we want to have

$$G(t)y' + G'(t)y = G(t)y' + G(t)p(t)y.$$

Thus we need

$$G'(t) = p(t)G(t),$$

which is clearly satisfied if $G(t) = e^{P(t)}$, where $p = P'$. Thus,

$$\begin{aligned}\frac{d}{dt}\{e^{P(t)}y\} &= e^{P(t)}y' + P'(t)e^{P(t)}y \\ &= e^{P(t)}(y' + P'(t)y) \\ &= e^{P(t)}(y' + p(t)y).\end{aligned}$$

We now have the key to the desired general method. To solve (5.40), we find a function P such that $P'(t) = p(t)$. Then, we multiply both sides of (5.40) by $e^{P(t)}$, and observe that the result can be rewritten

$$\frac{d}{dt}\{e^{P(t)}y\} = q(t)e^{P(t)}.$$

An integration immediately gives

$$e^{P(t)}y = \int q(t)e^{P(t)}\,dt + C, \tag{5.41}$$

from which we obtain a formula for y.

We give several illustrations of this method.

(a) Solve $$y' = \frac{1}{t^2}y + t^3.$$

We first rewrite the equation as

$$y' - \frac{1}{t^2}y = t^3. \tag{5.42}$$

Hence, $p(t) = -1/t^2$, so we can take $1/t$ as $P(t)$. Multiplying (5.42) by $e^{1/t}$, we obtain

$$\frac{d}{dt}\{e^{1/t}y\} = t^3e^{1/t},$$

and thus

$$e^{1/t}y = \int t^3e^{1/t}\,dt + C,$$

or

$$y = e^{-1/t}\int t^3e^{1/t}\,dt + Ce^{-1/t}. \tag{5.43}$$

This example illustrates the basic handicap in the method. The formula in (5.43) for the solution is of very little use; the integral cannot be evaluated in terms of the ordinary functions of analysis. So there is no simple way to calculate the value of the solution function $y = \varphi(t)$ for a prescribed value of t. However, it is sometimes possible to make some practical use of formulas such as (5.43). By the use of more subtle techniques, the qualitative behavior of some functions defined by such formulas can be determined.

(b) Solve $$y' = -\frac{3}{t}y + t^2 - t.$$

Rewrite the equation as $y' + (3/t)y = t^2 - t$. Then, $p(t) = 3/t$, so that $P(t) = 3 \log t$. Hence, $e^{P(t)} = t^3$. Thus, if we multiply both sides of the equation by t^3, obtaining

$$t^3\left(y' + \frac{3}{t}y\right) = t^3(t^2 - t),$$

or

$$t^3y' + 3t^2y = t^5 - t^4,$$

then it can be rewritten as

$$\frac{d}{dt}\{t^3y\} = t^5 - t^4;$$

and our solution is

$$\begin{aligned} t^3y &= \int (t^5 - t^4)\,dt \\ &= \tfrac{1}{6}t^5 - \tfrac{1}{5}t^5 + C, \end{aligned}$$

or

$$y = \tfrac{1}{6}t^3 - \tfrac{1}{5}t^2 + \frac{C}{t^3}.$$

This example shows that there are some cases in which the method is very successful in obtaining solutions easily.

These two classes of differential equations of first order, those that are linear and those that have separable variables, are the two main types that can be solved. There are a number of other special types, several of which are described in the exercises, that can be converted into one of these two by a particular substitution or other device. (See Exercises 14, 18.)

It is also possible to use these methods to solve some second order equations if they happen to be of the right form. For example, consider the equation

$$\frac{d^2y}{dt^2} = -2y' + 4t, \tag{5.44}$$

and suppose that we wish a solution satisfying the initial conditions $y = 1$ and $y' = 2$ when $t = 0$. This is a very special case of the general second order equation $d^2y/dt^2 = f(t, y, y')$, since y is entirely missing from the right-hand side. Let us put $u = y'$ in (5.44). Since $d^2y/dt^2 = y'' = u'$, we can rewrite (5.44) in terms of t and u, obtaining a first order equation

$$\frac{du}{dt} = -2u + 4t.$$

Moreover, this is a linear equation; we can apply the method outlined above to obtain

$$\begin{aligned} e^{2t}(u' + 2u) &= 4te^{2t}, \\ \frac{d}{dt}\{e^{2t}u\} &= 4te^{2t}, \\ e^{2t}u &= \int 4te^{2t}\,dt + C_1, \\ &= (2t - 1)e^{2t} + C_1, \end{aligned}$$

to find that $u = 2t - 1 + C_1e^{-2t}$. Now replace u by y', and obtain another first order equation

$$y' = \frac{dy}{dt} = 2t - 1 + C_1e^{-2t}. \tag{5.45}$$

Solving, we have

$$y = t^2 - t - \tfrac{1}{2}C_1e^{-2t} + C_2. \tag{5.46}$$

To apply the initial conditions, we make the substitutions $t = 0$, $y = 1$, $y' = 2$ in both (5.45) and (5.46), which lead to the algebraic linear equations

$$\begin{aligned} 2 &= 0 - 1 + C_1, \\ 1 &= 0 - 0 - \tfrac{1}{2}C_1 + C_2, \end{aligned}$$

from which we find $C_1 = 3$, $C_2 = \frac{5}{2}$. The desired solution function is then $y = t^2 - t - (\frac{3}{2})e^{-2t} + \frac{5}{2}$.

There is another class of second order equations that can be reduced to first order equations and thus sometimes solved completely. Since this class happens to include many of the equations that are used for models for important physical problems, we must discuss it and illustrate its use. The characteristic property of the class of equations is that the independent variable, usually t, is *not* present. Thus, if we use the standard form for a second order equation, $y'' = f(t, y, y')$, then we mean that in fact the equation has the form $y'' = g(y, y')$. To illustrate, we recall equations that were used for several different models of a pendulum:

$$\frac{d^2\theta}{dt^2} = -\frac{g}{L}\sin\theta, \tag{5.1}$$

$$\frac{d^2\theta}{dt^2} = -\frac{g}{L}\theta, \tag{5.10}$$

$$\frac{d^2\theta}{dt^2} = -(g/L)\sin\theta - C\left(\frac{d\theta}{dt}\right)^2, \tag{5.26}$$

as well as the fundamental equation used in the rocket flight problem,

$$\frac{d^2x}{dt^2} = -gR^2\left(\frac{1}{x^2}\right). \tag{5.14}$$

All of these equations have the special property that the independent variable t is missing. It was probably Newton who first conceived of the special trick that makes it possible to change such an equation into a first order differential equation. Let us try to recapture his reasoning, using modern notation.

Suppose we are interested in solving the equation

$$y'' = g(y, y'), \tag{5.47}$$

where as before $y' = dy/dt$ and $y'' = d^2y/dt^2$. Let us set $u = y'$. A solution of (5.47) would be a function φ such that $y = \varphi(t)$ satisfies (5.47). Note that we then have $u = \varphi'(t)$. Suppose that we can solve the equation $y = \varphi(t)$ for t, so that we have $t = \psi(y)$. Then, we can write $u = \varphi'(t) = \varphi'(\psi(y)) = F(y)$. In other words, we can regard y as an independent variable and express u in terms of y. With this speculation in mind, Newton might have written down a form of the chain rule:

$$\frac{d^2y}{dt^2} = \frac{du}{dt} = \frac{du}{dy}\frac{dy}{dt} = \frac{du}{dy}u.$$

If we make this substitution in (5.47), we obtain

$$u\frac{du}{dy} = g(y, u),$$

which is a *first order* equation in u and y. If we are lucky, this equation is one we can solve in the form $u = F(y)$. Since this equation is the same as $dy/dt = F(y)$, the latter equation too can be solved, giving us the final solution in the form $t = \psi(y)$.

Let us apply this method first to the rocket problem. Recall that the motion of the rocket after burnout is modeled by the equation

$$\frac{d^2x}{dt^2} = -gR^2\left(\frac{1}{x^2}\right) \tag{5.48}$$

and that the initial conditions are $x = R_0$, $dx/dt = V_0$ when $t = 0$ (the rocket is at distance R_0 from the center of the earth and has velocity V_0 at the moment of burnout). As in the general method, we set $u = dx/dt$ and assume that u can be taken to be a function of x. Then by the chain rule,

$$\frac{d^2x}{dt^2} = \frac{du}{dt} = \frac{du}{dx}\frac{dx}{dt} = u\frac{du}{dx}.$$

Equation (5.48) becomes

$$u\frac{du}{dx} = -gR^2\left(\frac{1}{x^2}\right).$$

This is a separable equation, so we write

$$u\,du = -gR^2\frac{dx}{x^2}.$$

Integrating, we have

$$\frac{u^2}{2} = gR^2\left(\frac{1}{x}\right) + C. \tag{5.49}$$

We can evaluate the number C by using the initial conditions. When $t = 0$, we have $x = R_0$ and $u = dx/dt = V_0$. Hence, making these substitutions,

$$\tfrac{1}{2}V_0^2 = gR^2/R_0 + C.$$

Solving this for C, we use in (5.49) the value thus obtained and derive the formula

$$\left(\frac{dx}{dt}\right)^2 = u^2 = \frac{2gR^2}{x} + V_0^2 - \frac{2gR^2}{R_0}. \tag{5.50}$$

Already we have enough information to answer one of the questions asked about the flight of an unpowered rocket. How far up will it go? As with a thrown rock, the vertical speed will be zero at the moment (if any) that it reaches its maximum height and starts to return toward the Earth. If we set $dx/dt = 0$ in (5.50) and solve for x, we obtain

$$x = \frac{2gR^2}{\dfrac{2gR^2}{R_0} - V_0^2}.$$

If the number in the denominator is positive, then this formula gives us the maximum distance from the Earth achieved by the rocket. What interpretation should we give to this formula when the denominator is zero or negative? In that case, it is clear from (5.50) that no choice of x, however large, will lead to a zero value of dx/dt. Thus, we conclude that if $V_0^2 \geq 2gR^2/R_0$, the rocket will recede indefinitely far and its velocity will constantly decrease toward the limiting value $(V_0^2 - 2gR^2/R_0)^{1/2}$. When the initial height of the rocket is small compared to the radius R of the Earth, then $R_0 \sim R$ and the condition for the rocket to escape the Earth completely becomes $V_0 \geq \sqrt{2gR}$, which is the conventional formula for the escape velocity.

It is possible to start with formula (5.50), solve for

$$u = \frac{dx}{dt} = \sqrt{\frac{2gR^2}{x} + B},$$

where $B = V_0^2 - 2gR^2/R_0$, and solve this equation as a first order differential equation, again separable. By this procedure we can derive a formula

$$F(x) = t$$

which will give the time required for the rocket to reach a given distance from the Earth. We will not carry out the work, since the integration calculation is complicated.

As our second illustration, consider the first model for a pendulum

$$\frac{d^2\theta}{dt^2} = -(g/L)\sin\theta, \tag{5.51}$$

where we have the initial conditions $t = 0$, $\theta = \theta_0$, $d\theta/dt = 0$ (we release the pendulum from rest at an angle θ_0). Following Newton's device, we set $u = d\theta/dt$, regard u as a function of θ, and use the chain rule

$$\frac{d^2\theta}{dt^2} = \frac{du}{dt} = \frac{du}{d\theta}\frac{d\theta}{dt} = u\frac{du}{d\theta}$$

to replace (5.51) by the equation in u and θ

$$u\frac{du}{d\theta} = -(g/L)\sin\theta.$$

This is separable and has the solution

$$\tfrac{1}{2}u^2 = (g/L)\cos\theta + C.$$

Checking the initial conditions, we have $d\theta/dt = 0$ when $\theta = \theta_0$, which corresponds to $u = 0$ when $\theta = \theta_0$. This gives

$$C = -(g/L)\cos\theta_0,$$

and we have a partial solution

$$\left(\frac{d\theta}{dt}\right)^2 = u^2 = \frac{2g}{L}(\cos\theta - \cos\theta_0).$$

From this equation, we obtain a formula for the angular speed of the swinging pendulum at any position:

$$\left|\frac{d\theta}{dt}\right| = \sqrt{2g/L}\sqrt{\cos\theta - \cos\theta_0}. \tag{5.52}$$

In particular, at the bottom of its swing $\theta = 0$ and the speed of the bob of the pendulum is

$$V_0 = L\left|\frac{d\theta}{dt}\right| = \sqrt{2gL}\sqrt{1 - \cos\theta_0}$$
$$= 2\sqrt{gL}\sin\left(\tfrac{1}{2}\theta_0\right),$$

as given earlier in formula (5.8).

Can we carry on beyond this point to determine the actual motion of the pendulum by solving for the function $\theta = \varphi(t)$? Let us take a part of the motion where $d\theta/dt$ has constant sign, for example during the first downward swing when θ is decreasing from θ_0 to 0. As in Section 5.1, we suppose that t_0 is the time at which the pendulum first reaches the vertical position. During the interval $0 < t < t_0$, θ is decreasing and $d\theta/dt$ is negative. Hence, $|d\theta/dt| = -d\theta/dt$, and from (5.52) we find

$$\frac{d\theta}{dt} = -\sqrt{2g/L}\sqrt{\cos\theta - \cos\theta_0}.$$

This is a separable equation, and we obtain the integrated solution

$$\int \frac{d\theta}{\sqrt{\cos\theta - \cos\theta_0}} = -\sqrt{\frac{2g}{L}}\int dt. \tag{5.53}$$

Unfortunately, it is not possible to express the indefinite integral on the left in terms of the elementary functions of analysis. (It *can* be expressed in terms of what are called elliptic functions.) However, we can answer one of the questions we have asked about the motion by finding an exact expression for the time t_0. We use the device of inserting corresponding appropriate limits in the indefinite integrals in (5.53). We know that $\theta = \theta_0$ when $t = 0$; we also know that $t = t_0$ when $\theta = 0$, since at that moment the pendulum is at the bottom of its swing. Take $\theta = \theta_0$, $t = 0$ for the bottom limits of integration and $\theta = 0$, $t = t_0$ for the upper limits. Then, (5.53) becomes

$$\int_{\theta_0}^{0} \frac{d\theta}{\sqrt{\cos\theta - \cos\theta_0}} = -\sqrt{\frac{2g}{L}}\int_0^{t_0} dt.$$

Evaluating the right side, we end with the relation

$$t_0 = \sqrt{\frac{L}{2g}}\int_0^{\theta_0} \frac{d\theta}{\sqrt{\cos\theta - \cos\theta_0}}$$

as stated in formula (5.7), Section 5.1.

Exercises

For each of Exercises 1–13, find solutions $y = \varphi(t)$ which go through the given initial points (t_0, y_0).

1 $y' = t^2y$; $(1, 2)$.

2 $y' = t^2y$; $(1, 0)$.

3 $y' = t/y$; $(1, 3)$, $(0, 1)$, $(1, -3)$.

4 $y' = (t - t^2)/(2y + 1)$; $(2, 1)$, $(1, 1)$.

5 $y' = (y^2 - y)/t$; $(1, 2)$ $(1, 1)$.

6 $y' = e^{t+y}$; $(0, 1)$.

7 $y' = t^2y - t^2$; $(1, 3)$, $(1, 2)$, $(1, 0)$, $(0, 2)$.

8 $y' = (y - y^2)/(t + t^2)$; $(1, 2)$.

9 $y' = -2y + e^t$; $(0, 1)$.

10 $y' = 2y + e^t$; $(0, 1)$.

11 $y' = y + e^t$; $(0, 1)$.

12 $y' = 2y/t + t^4$; $(1, 1)$.

13 $t^2y' = 1 - 2ty$; $(1, 1)$, $(1, 0)$.

14 A function f is said to be homogeneous (of degree 0) if $f(cu, cv) = f(u, v)$ for all c, u, and v for which the expressions are defined. For example, $f(u, v) = (u + 4v)/(2u - v)$ is homogeneous. Show that if f is homogeneous, then the differential equation $y' = f(t, y)$ can be changed to a separable equation in u and t by setting $y = ut$.

15 Apply the device of Exercise 14 to solve $y' = (t - 3y)/(3t + y)$.

16 Show that the equation $y' = a(t)y + b(t)y^2$ becomes a linear equation in t and v if you set $y = 1/v$.

17 Use the method of Exercise 16 to solve the equation

$$y' = 2ty + ty^2.$$

18 Show that the equation $dy/dt = a(t)y + b(t)y^m$ can be turned into a linear equation in v by the substitution $y = v^{1/(1-m)}$.

19 Use the device in Exercise 18 to solve the equation

$$ty' = 2y + 4t^3\sqrt{y}.$$

20 Find the complete solution of the equation

$$y'' = -2y' + e^t.$$

21 Find the complete solution of the equation

$$y'' = \frac{2}{t}y' + t^4.$$

22 Use Newton's trick to solve the equation $y'' = 2(y')^2/y$. (Also see Exercise 2, Section 5.3.)

23 Use Newton's trick to solve the equation used for the second model for the pendulum (formula (5.10), Section 5.1).

24 Combine Newton's trick with the device in Exercise 18 above to show that the solution of the equation $y'' = Ay + B(y')^2$, with A and B constants, can be reduced to a problem in integration.

***25** Combine Newton's trick and the device of Exercise 18 to show that it is possible to solve the equation in (5.26), which is used to model the motion of a pendulum with a special nonlinear resistance law.

5.6 Approximate Methods

If "solving a differential equation" means finding a formula for the unknown function, then as we have seen in the last section, only a few special classes of first order equations can be solved. In most cases, however, we do not need to have a formula if ways can be found to calculate the values of the solution function for the desired values of t, or if an accurate graph of the solution can be obtained. When the solution is expressed in terms of an integral which we cannot evaluate because we cannot find an antiderivative, then we can always resort to a technique of numerical integration (e.g., Simpson's Rule) to find the value as accurately as we wish.

But, if we are going to use a desk calculator or a high-speed digital computer to evaluate the final answer, perhaps it would be more efficient to forget all the special tricks for finding a solution and use a numerical method at the start, one that is designed specifically to solve differential equations. There are many, and the right place to make a thorough study of them is in a course either in numerical analysis or differential equations. However, it is important to know something about why such methods work and to see how one applies them and where their errors come from.

The technique we shall discuss was discovered by Euler (c. 1750) and can be used on any first order equation. It is nothing more than a careful, systematic use of the simple graphical approach described in Section 5.2. Let us consider the problem of finding a solution of the equation

$$\frac{dy}{dt} = f(t, y) \tag{5.54}$$

passing through the point (t_0, y_0). We assume that f obeys the hypotheses of Theorem 1, Section 5.4, which assures us that there exists one and only one function φ such that $\varphi(t_0) = y_0$ and such that $y = \varphi(t)$ satisfies the equation (5.54). We want to find the value of $\varphi(T)$ for a chosen value of $T > t_0$; what we in fact obtain is an approximation to this value, together with a crude estimate of the error.

Choose a large value of n, and divide the interval $t_0 \leq t \leq T$ into n subintervals by picking points t_i,

$$t_0 < t_1 < t_2 < t_3 < \cdots < t_{n-1} < t_n = T.$$

In practice, these points are usually evenly spaced, so that $h = t_{i+1} - t_i = (T - t_0)/n$, but this is not necessary. We now start at the given initial point (t_0, y_0) and attempt to follow (predict) the solution (see Figure 5–13). We know that the desired curve $y = \varphi(t)$ goes through (t_0, y_0). From the equation (5.54) we also know the slope of this curve at this point, namely $m_0 = dy/dt = f(t_0, y_0)$.

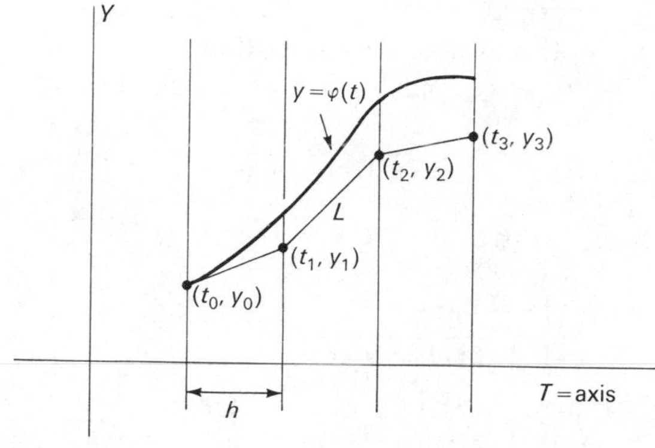

Figure 5–13

We assume that we can replace the actual solution curve by a straight line, and so we construct a line of slope m_0 through the point (t_0, y_0). We follow this line until we reach the vertical line $t = t_1$. The reasoning is that if t_1 is sufficiently near t_0, then the straight line is a very good approximation to the actual shape of the solution curve, so that the y value reached by the line is very little different from the true y value that we would find if we could follow the curve.

We calculate the equation of this line, obtaining

$$y = y_0 + m_0(t - t_0),$$

so the estimate for the true y value, $\varphi(t_1)$, is

$$y_1 = y_0 + m_0(t_1 - t_0).$$

We now pretend that this new point (t_1, y_1) is really on the solution curve $y = \varphi(t)$, and we proceed in the same way to predict the value of y for $t = t_2$ by assuming that the curve is again a straight line. If the solution curve were really to go through (t_1, y_1), then its slope there, by (5.54), would have to be $m_1 = f(t_1, y_1)$. Let L be the line through (t_1, y_1) whose slope is m_1. The equation of this new line is $y = y_1 + m_1(t - t_1)$, and if we assume that this line is the true solution curve, then we shall predict the y value when $t = t_2$ to be

$$y_2 = y_1 + m_1(t_2 - t_1).$$

We continue in this way, ending with a curve made up of segments of lines as the approximation to the true solution curve $y = \varphi(t)$. The entire process can be written out as a set of instructions for calculating a sequence of y values $y_1, y_2, \ldots,$ where the last one, y_n, is the desired estimate for the true value of y for $t = T$. The recursive formula is

$$y_{k+1} = y_k + (t_{k+1} - t_k)f(t_k, y_k). \tag{5.55}$$

Knowing the function f and the values t_i, we start from the given value of y_0 and use (5.55) to calculate each of the successive y_i.

As an illustration, suppose that we take the simple equation $y' = y$, with $t_0 = 0$, $y_0 = 1$, and take $T = 1$. With $n = 4$, so that $t_0 = 0$, $t_1 = \frac{1}{4}$, $t_2 = \frac{1}{2}$, $t_3 = \frac{3}{4}$, and $t_4 = T = 1$, the formula (5.55) becomes

$$y_{k+1} = y_k + (\tfrac{1}{4})y_k,$$

and the calculation proceeds thus:

$$\begin{aligned}
y_0 &= 1, \\
y_1 &= 1 + (\tfrac{1}{4})y_0 = \tfrac{5}{4}, \\
y_2 &= y_1 + (\tfrac{1}{4})y_1 \\
&= \tfrac{5}{4} + (\tfrac{1}{4})(\tfrac{5}{4}) \\
&= \tfrac{25}{16}, \\
y_3 &= y_2 + (\tfrac{1}{4})y_2 \\
&= (\tfrac{5}{4})y_2 = \tfrac{125}{64}, \\
\varphi(1) \sim y_4 &= y_3 + (\tfrac{1}{4})y_3 \\
&= (\tfrac{5}{4})(\tfrac{125}{64}) = \tfrac{625}{256} = 2.441.
\end{aligned}$$

The sources for error in the method are evident. Since we start by assuming that the solution curve is a straight line, the first computed value y_1 is not apt to be the true value $\varphi(t_1)$. However, we assumed that the point (t_1, y_1) *was* correct, and we used it in (5.54) to calculate the value of $\varphi'(t_1)$, which we estimate to be $f(t_1, y_1)$. Thus, we compounded the first error by using the wrong slope at the second stage. As we continue the process, the effect of the various errors accumulate, and it would be astonishing if the estimate for the true value of y when $t = T$ were accurate. In the example we have used, the equation $y' = y$ has $y = Ce^t$ for its general solution, and the solution that passes through $(0, 1)$ is $y = e^t$. The true value of $\varphi(T)$ is therefore $e = 2.71828\ldots$. (Compare with the estimate $\varphi(T) \sim 2.441$ obtained above.)

It can be shown, however, that the error diminishes when the number of steps increases and hence the length of the steps $\Delta t_k = t_{k+1} - t_k$ tends toward zero. Indeed, if all the intervals are equal and of step length $h = (T - t_0)/n$, then it can be proved that the difference between the true value $\varphi(T)$ and the estimated value y_n does not exceed a constant times h. Thus, we know that if we want to find the value of y when $t = T$ with any preassigned maximum error, we can apply Euler's method using a sufficiently small step size h. However, h small means n large, and we will gain accuracy at the expense of a great number of calculations. The table below shows the effect of decreasing h from $\frac{1}{4}$ to 0.1 in our sample problem, $y' = y$.

$k =$	0	1	2	....	8	9	10
$t_k =$	0	.1	.2	....	.8	.9	$1.0 = T$
$y_k =$	1.0	1.1	1.21	....	2.143	2.357	2.593

By doing 10 computations instead of 4, we have obtained the estimate 2.593 in place of 2.441; the true value is $e = 2.71828$. It is here that the high-speed digital computer makes a difference, since it can do this type of repetitive calculation very quickly, making it possible to use $n = 500$ or even $n = 5{,}000$. In addition, better schemes than Euler's have been discovered. For example, with $n = 10$, the method of Milne applied to our sample problem yields the estimate $\varphi(T) = 2.717$, which is quite close to the true value.

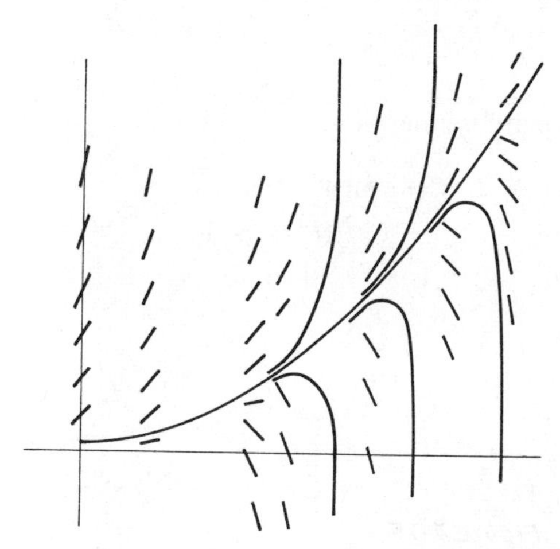

Figure 5–14

It should also be mentioned that some equations have solutions that cannot be obtained in this fashion at all, for the errors accumulate so fast that the approximation slides away from the desired solution and never gets near it again. In Figure 5–14 we show the direction field for the equation $y' = 5y - t^2$. If we were using the Euler method to try to approximate the solution shown by the heavy curve, we would be in trouble immediately, since the moment we got off that curve, we would find ourselves driven steeply upward or downward, never moving back toward the desired curve. (Other numerical methods have been discovered that compensate for this type of difficulty.)

There are other ways to find approximate solutions of a differential equation. For example, one method uses the equation itself to find additional properties of the desired solution. Suppose we want to determine something about the solution of the equation

$$y' = t^2 + y^2 \tag{5.56}$$

passing through the point $(1, -1)$. From the equation, we can immediately read off the slope of the solution curve at the point $(1, -1)$: $(1)^2 + (-1)^2 = 2$. Now, regarding (5.56) as an identity in t, as it would be if we were to replace y by the solution function $\varphi(t)$, we differentiate it with respect to t, obtaining

$$y'' = 2t + 2yy'.$$

We next replace y' by its value given in (5.56), obtaining

$$\begin{aligned} y'' &= 2t + 2y(t^2 + y^2) \\ &= 2t + 2t^2y + 2y^3. \end{aligned} \tag{5.57}$$

At $(1, -1)$, we find the numerical value of y'' to be

$$2 - 2 - 2 = -2.$$

Repeating this process, we differentiate (5.57) to obtain

$$\begin{aligned} y''' &= 2 + 4ty + 2t^2y' + 6y^2y' \\ &= 2 + 4ty + (2t^2 + 6y^2)(t^2 + y^2) \\ &= 2 + 4ty + 2t^4 + 8t^2y^2 + 6y^4, \end{aligned}$$

and at $(1, -1)$ we have $y''' = 14$.

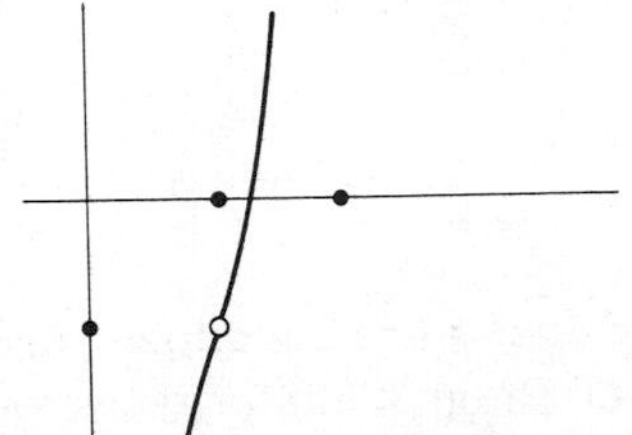

Figure 5–15

Putting together this information about the solution curve $y = \varphi(t)$, we find that $\varphi(1) = -1$, $\varphi'(1) = 2$, $\varphi''(1) = -2$, and $\varphi'''(1) = 14$. The first three of these values translate at once into geometric information about the shape of the curve near the initial point $(1, -1)$, as indicated in Figure 5–15.

It is clear that we could continue this process, obtaining the value of $\varphi^{(k)}(1)$ for arbitrarily large k, although the work involved becomes harder each time due to the increasing complexity of the formulas. Could we make use of such additional information? (Note that we did not give geometric meaning to the information $\varphi'''(1) = 14$, in the example above.)

If we knew all the derivatives of φ at $t = 1$, then we could construct the Taylor series for φ at $t = 1$,

$$\varphi(1) + \varphi'(1)(t-1) + \varphi''(1)\frac{(t-1)^2}{2!} + \varphi'''(1)\frac{(t-1)^3}{3!} + \cdots + \varphi^{(k)}(1)\frac{(t-1)^k}{k!} + \cdots,$$

with the expectation that this series might converge to $\varphi(t)$ for all t in a neighborhood of $t = 1$. (See Appendix 1 for a brief review of series.) If this is true, then the partial sums of the series will be polynomials yielding approximations to $\varphi(t)$ which become more accurate as the degree increases.

With this justification, we are led to consider the polynomial of degree 3 obtained by using the known values of $\varphi^{(k)}(1)$ in our example, and we conclude that

$$\begin{aligned} y = P(t) &= -1 + 2(t-1) + (-2)\frac{(t-1)^2}{2!} + (14)\frac{(t-1)^3}{3!} \\ &= -1 + 2(t-1) - (t-1)^2 + (7/3)(t-1)^3 \end{aligned}$$

is *probably* a good approximation to the solution curve $y = \varphi(t)$ near the given initial point $(1, -1)$.

The conjectural nature of these comments can be removed by a proper application of mathematical theory. For a class of equations including our illustrative example, it can be shown that the solutions are Taylor series; information can be obtained about the rate of convergence and hence about the degree of accuracy of approximation given by the polynomials that are the partial sums of the series. We shall not discuss this aspect further, but shall take it for granted that it is useful to find the first N terms of the Taylor expansion of a solution function.

The method of repeated differentiation is not the only method used. The simplest approach starts by assuming that the solution is of the form of a power series with unknown coefficients, and then these coefficients are found by substituting the series into the equation and solving for them. An example may make this process clear to the reader.

Suppose we wish to solve the equation

$$y' + 2ty = 5t^2 \tag{5.58}$$

with initial conditions $t = 0$, $y = y_0$. (This is a first order linear equation, so it can be solved in the form of an integral by the method explained earlier in this section.) Since $t_0 = 0$, we can assume that the desired solution has the form of a Taylor (power) series about $t = 0$ and write

$$y = c_0 + c_1 t + c_2 t^2 + c_3 t^3 + c_4 t^4 + \cdots = \sum_0^\infty c_k t^k. \tag{5.59}$$

We want to substitute this into the equation (5.58), so we compute

$$y' = c_1 + 2c_2 t + 3c_3 t^2 + 4c_4 t^3 + 5c_5 t^4 + \cdots.$$

Then,

$$2ty = 2c_0 t + 2c_1 t^2 + 2c_2 t^3 + 2c_3 t^4 + 2c_4 t^5 + \cdots$$

and

$$y' + 2ty = c_1 + (2c_2 + 2c_0)t + (3c_3 + 2c_1)t^2 + (4c_4 + 2c_2)t^3 + (5c_5 + 2c_3)t^4 + \cdots.$$

From the equation (5.58), we see that the right side must add exactly to $5t^2$. This can happen only if the coefficient $(3c_3 + 2c_1)$ is 5 while the coefficient of every other power of t is 0. (Here, we have used the uniqueness property of power series; if $\sum_0^\infty a_n t^n$ and $\sum_0^\infty b_n t^n$ are expansions of the same function, then $a_n = b_n$ for every n.)

We are thus led to an infinite system of equations which we need to solve to find the unknown coefficients $\{c_k\}$:

$$\begin{aligned} c_1 &= 0, \\ 2c_2 + 2c_0 &= 0, \\ 3c_3 + 2c_1 &= 5, \\ 4c_4 + 2c_2 &= 0, \\ 5c_5 + 2c_3 &= 0, \\ 6c_6 + 2c_4 &= 0, \text{ etc.} \end{aligned} \tag{5.60}$$

In addition, we want $y = \varphi(t)$ to obey the initial conditions $t = 0$, $y = y_0$. Putting $t = 0$ in (5.59), we find $y = c_0$, so we can start with $c_0 = y_0$. If we then take each of the equations in (5.60), one after the other, it is easy to obtain the following solutions:

$$\begin{aligned} c_0 &= y_0, \\ c_1 &= 0, \\ c_2 &= -c_0 = -y_0, \\ c_3 &= (5 - 2c_1)/3 = \tfrac{5}{3}, \end{aligned}$$

$$
\begin{aligned}
c_4 &= -2c_2/4 = 2y_0/4 = \tfrac{1}{2}y_0,\\
c_5 &= -2c_3/5 = (-\tfrac{10}{3})\tfrac{1}{5} = -\tfrac{2}{3},\\
c_6 &= -2c_4/6 = (-\tfrac{1}{3})(\tfrac{1}{2}y_0) = -\tfrac{1}{6}y_0,\\
c_7 &= -2c_5/7 = (\tfrac{4}{3})\tfrac{1}{7} = \tfrac{4}{21}.
\end{aligned}
$$

If we choose to stop here, then we have a polynomial approximation to the solution, namely

$$
\begin{aligned}
y &= \sum_0^7 c_k t^k\\
&= y_0 - y_0t^2 + \tfrac{5}{3}t^3 + \tfrac{1}{2}y_0t^4 - \tfrac{2}{3}t^5 - \tfrac{1}{6}y_0t^6 + \tfrac{4}{21}t^7\\
&= (\tfrac{5}{3}t^3 - \tfrac{2}{3}t^5 + \tfrac{4}{21}t^7) + y_0(1 - t^2 + \tfrac{1}{2}t^4 - \tfrac{1}{6}t^6).
\end{aligned}
$$

In some cases, such as the second order equation in our next example, it is possible to determine the general form of the coefficients. We thus obtain a series formula for the solution itself, rather than what is at best an approximation to the solution.

Let us solve the equation

$$y'' - 2y' + y = e^{-t} \tag{5.61}$$

with initial conditions $t = 0$, $y = 1$, $y' = 2$. As before, we start with the general power series form for y,

$$y = c_0 + c_1t + c_2t^2 + c_3t^3 + c_4t^4 + \cdots,$$

but we make the replacement $c_n = a_n/n!$ to make the work easier. Thus, we may assume that

$$y = a_0 + a_1t/1! + a_2t^2/2! + a_3t^3/3! + a_4t^4/4! + \cdots.$$

Noting that the derivative of $t^k/k!$ is $t^{k-1}/(k-1)!$, we have

$$
\begin{aligned}
y' &= a_1 + a_2t/1! + a_3t^2/2! + a_4t^3/3! + a_5t^4/4! + \cdots,\\
y'' &= a_2 + a_3t/1! + a_4t^2/2! + a_5t^3/3! + a_6t^4/4! + \cdots.
\end{aligned}
$$

We then calculate the left side of (5.61):

$$
\begin{aligned}
y'' - 2y' + y = {}& (a_0 - 2a_1 + a_2) + (a_1 - 2a_2 + a_3)t/1!\\
&+ (a_2 - 2a_3 + a_4)t^2/2!\\
&+ (a_3 - 2a_4 + a_5)t^3/3! + \cdots.
\end{aligned} \tag{5.62}
$$

We next need a corresponding power series expansion for the right side of (5.61), which is

$$e^{-t} = 1 - t/1! + t^2/2! - t^3/3! + t^4/4! - t^5/5! + \cdots. \tag{5.63}$$

Equating (5.62) and (5.63), we obtain an infinite number of linear equations for the unknown coefficients a_n:

$$\begin{aligned} 1 &= a_0 - 2a_1 + a_2, \\ -1 &= a_1 - 2a_2 + a_3, \\ 1 &= a_2 - 2a_3 + a_4, \\ -1 &= a_3 - 2a_4 + a_5, \\ 1 &= a_4 - 2a_5 + a_6, \text{ etc.} \end{aligned} \tag{5.64}$$

We can start the solution of these equations by using the initial conditions for the differential equation; since $y = 1$ and $y' = 2$ when $t = 0$, we put $t = 0$ into the above series for y and y' to find $a_0 = 1$, $a_1 = 2$. We can then use the first equation in (5.64) to solve for a_2:

$$a_2 = 1 - a_0 + 2a_1 = 1 - 1 + 4 = 4.$$

Using $a_2 = 4$ in the next equation, we solve for a_3:

$$\begin{aligned} a_3 &= -1 - a_1 + 2a_2 = -1 - 2 + 8 \\ &= 5. \end{aligned}$$

The final results are $a_0 = 1$, $a_1 = 2$, $a_2 = 4$, $a_3 = 5$, $a_4 = 7$, $a_5 = 8$, $a_6 = 10$, $a_7 = 11$, etc. If we conjecture that the evident pattern persists, then the desired solution is

$$\begin{aligned} y = 1 + 2t + 4t^2/2! + 5t^3/3! + 7t^4/4! + 8t^5/5! + 10t^6/6! \\ + 11t^7/7! + 13t^8/8! + 14t^9/9! + \cdots. \end{aligned} \tag{5.65}$$

We can produce a formula for these coefficients. We first note that

$$a_n = \begin{cases} \frac{3}{2}n + 1 & \text{when } n \text{ is even,} \\ \frac{3}{2}n + \frac{1}{2} & \text{when } n \text{ is odd.} \end{cases}$$

Using the alternating properties of $(-1)^n$, we can find a single formula: $a_n = (\frac{3}{2})n + (\frac{3}{4}) + (\frac{1}{4})(-1)^n$. To check that the sequence $\{a_n\}$ is indeed the solution of the equations (5.64), we write the single general equation derived from (5.64),

$$(-1)^n = a_n - 2a_{n+1} + a_{n+2}. \tag{5.66}$$

Substituting our proposed general formula for a_n into the right-hand side of (5.66), we obtain

$$\begin{aligned} a_n - 2a_{n+1} + a_{n+2} &= (\tfrac{3}{2}n + \tfrac{3}{4} + \tfrac{1}{4}(-1)^n) - 2(\tfrac{3}{2}(n+1) + \tfrac{3}{4} + \tfrac{1}{4}(-1)^{n+1}) \\ &\quad + (\tfrac{3}{2}(n+2) + \tfrac{3}{4} + \tfrac{1}{4}(-1)^{n+2}) \\ &= \tfrac{3}{2}(n - 2(n+1) + (n+2)) + \tfrac{3}{4}(1 - 2 + 1) \\ &\quad + \tfrac{1}{4}((-1)^n - 2(-1)^{n+1} + (-1)^{n+2}) \end{aligned}$$

$$= \tfrac{3}{2}(0) + \tfrac{3}{4}(0) + (-1)^n \tfrac{1}{4}(1 - 2(-1) + (-1)^2)$$
$$= (-1)^n \tfrac{1}{4}(4) = (-1)^n.$$

Thus (5.66) is an identity. This confirms that the function given in (5.65) is the desired solution.

It is possible in this special case to find the sum of the series for y in terms of familiar functions; using the general formula for a_n, we have

$$y = \sum_0^\infty a_n t^n/n! = \sum_0^\infty (\tfrac{3}{2}n + \tfrac{3}{4} + \tfrac{1}{4}(-1)^n t^n)/n!$$
$$= \frac{3}{2}\sum_0^\infty n\frac{t^n}{n!} + \frac{3}{4}\sum_0^\infty \frac{t^n}{n!} + \frac{1}{4}\sum_0^\infty (-1)^n \frac{t^n}{n!}$$
$$= \tfrac{3}{2}(t + t^2/1! + t^3/2! + t^4/3! + \cdots)$$
$$+ \tfrac{3}{4}(1 + t + t^2/2! + t^3/3! + t^4/4! + \cdots)$$
$$+ \tfrac{1}{4}(1 - t + t^2/2! - t^3/3! + t^4/4! - \cdots)$$
$$= \tfrac{3}{2}te^t + \tfrac{3}{4}e^t + \tfrac{1}{4}e^{-t}$$
$$= (\tfrac{3}{2}t + \tfrac{3}{4})e^t + \tfrac{1}{4}e^{-t}.$$

That this is the correct solution of (5.61) satisfying the given initial conditions can be checked directly by substituting into (5.61).

This has been only a minimal introduction to the use of series in solving differential equations, more to illustrate the method than to give systematic guidance in its use. Any further exploration of this subject ought to be part of a more intensive study of differential equations and ought to be related closely to the use of high-speed computers to facilitate the calculations and to carry out more complex methods for approximating solutions by numerical methods.

Exercises

1 Let $y = \varphi(t)$ be the solution of $y' = t - y$ passing through $(0, 0)$. Use the Euler numerical method as in (5.55) to estimate the value of $\varphi(1)$, using steps of size $h = .1$. Also compare your result with the exact value obtained by solving the equation.

2 Let $y = \varphi(t)$ be the solution to $dy/dt = 2y/t$ passing through $(1, 1)$. Use the Euler method with $h = .2$ to estimate $\varphi(2)$, and compare the estimate with the exact value.

3 (a) Let $y = \varphi(t)$ be the solution of $dy/dt = 4ty - t^3$ such that $y = \frac{1}{8}$ when $t = 0$. Use the Euler method to estimate $\varphi(1)$, first with $h = .2$ and then with $h = .1$.

(b) Knowing that the solution discussed in (a) has the form $y = At^2 + B$, find it and compare the exact value of y at $t = 1$

with the estimates obtained in (a). Can you explain why the errors are so large?

4 Use the method of repeated differentiation to extend the solution of (5.56) to obtain an approximating polynomial of degree 5.

5 Use the method of repeated differentiation on the equation $y' = ty + t^2$ to obtain a polynomial approximation of degree 5 to the solution passing through (1, 1).

6 (a) Use repeated differentiation to find a cubic polynomial approximation for the solution of $dy/dt = y^2$ passing through (0, 1).
(b) Solve the equation exactly, and sketch both the exact solution and the polynomial approximation for $-2 \leq t \leq 2$.

7 Apply repeated differentiation to find a polynomial approximation of degree at most 5 for the equation $d^2y/dt^2 = y^2$ with initial conditions $t = 0, y = -1, y' = 2$.

For each of the next three exercises, use the method of power series to solve the given equation. Then, see if you can recognize the solution from its series and thus obtain a simple formula for the solution.

8 Solve $y' = y$ with initial condition $t = 0, y = 1$.

9 Solve $y' = y + t^2$ with $t = 0, y = 1$.

10 Solve $d^2y/dt^2 = y$ with $y = 1, y' = 2$ when $t = 0$.

11 Find four nonzero terms of the series solution for the equation $d^2y/dt^2 = t^2y$ such that $y = 1, y' = 0$ when $t = 0$.

12 Find a series expansion, at least through the term containing t^4, of the solution of the equation $dy/dt = y^2$ which passes through the point (0, 1). Also compare this with the exact solution.

5.7

Linear Equations

One of the most thoroughly developed portions of the theory of differential equations deals with the general class of linear equations. We have seen that a linear first order equation $dy/dt = a_0(t) + a_1(t)y$ can always be solved as an integral (Section 5.5). The general case of higher order linear equations is not so neatly summarized; we shall sketch part of the complete picture and give an almost complete treatment of the important special case of linear equations with constant coefficients.

It will be easier if we adopt a more modern approach to the subject and talk about linear operators on spaces of functions. Let us use $\mathcal{C}$ to denote the collection of all continuous functions φ defined and having derivatives *of all orders* on the interval $-\infty < t < \infty$. Thus, $\mathcal{C}$ contains functions such as $3t^4 - 5t + 7$, $3 \sin 2t - \cos(t^2 + 1)$, $3t^2e^{-t} + (1 + t^2)^{-1}$, but would not contain $\tan t$ or $1/t$. We also include complex-valued functions such as $(3 + 4i)t^2 -$

$2ie^{-1} \sin t$, but we shall not make much use of them until later in the discussion, when we want to relate the functions $e^{\gamma t}$, $e^{\bar{\gamma} t}$ to the functions $e^{at} \cos bt$ and $e^{at} \sin bt$. (A brief review of complex numbers and functions is in Appendix 2.) More generally, one might wish to work with $\mathcal{C}_I$, the infinitely differentiable functions defined on a fixed interval I, $a < t < b$.

Such classes $\mathcal{C}$ are examples of the important category of mathematical structures called **linear spaces**. This implies that $\mathcal{C}$ has the property that if φ and ψ are in $\mathcal{C}$, then so is the function $C_1\varphi + C_2\psi$ for any choice of the constants C_1 and C_2.

We start from the observation that the process of differentiation carries the class $\mathcal{C}$ into itself. If f is any member of $\mathcal{C}$, then the function f' is also in $\mathcal{C}$. If we denote the differentiation operation d/dt by D, then we may write that $Df \in \mathcal{C}$ for every $f \in \mathcal{C}$. More generally, we write D^2 for the operation of taking the second derivative, and we observe that if $f \in \mathcal{C}$, then $D^n f \in \mathcal{C}$ for any n. We can rephrase this by saying that D^n is a function on the set $\mathcal{C}$ and takes its values in $\mathcal{C}$. However, because the members of $\mathcal{C}$ are themselves functions, we change the terminology and call D^n an **operator**. More generally, we shall use the term "operator" to denote any transformation that can be applied to members of $\mathcal{C}$, sending a member of $\mathcal{C}$ into some other member of $\mathcal{C}$ in a unique way. Examples of such operators T are the following transformations T_1, T_2, T_3, T_4:

(5.67)

T_1 sends a function f into the function $3f'' - 2f' + 5f$,

T_2 sends a function f into the function g, where $g(t) = f''(t) - (3t + 1)f'(t) + (4t^2 - 1)f(t)$,

T_3 sends a function f into the function $f'' + ff' - (f)^2$,

T_4 sends a function f into the function g, where $g(t) = f(-t)$.

To illustrate, T_1 sends t^3 into $18t - 6t^2 + 5t^3$, T_2 sends t^3 into $6t - 3t^2 - 10t^3 + 4t^5$, T_3 sends t^3 into $6t + 3t^5 - t^6$, and T_4 sends t^3 into $-t^3$.

We are not concerned with *all* of the operators on $\mathcal{C}$, but only with a small subset of them, the linear differential operators of finite order, and ultimately with a much smaller subset of those. It is convenient at times to use the standard functional notation to denote the result of applying an operator T to a function. Thus, we write $T_1(t^3) = 18t - 6t^2 + 5t^3$, or $T_4(t^3) = -t^3$. The following definition is basic.

Definition 1 An operator T on $\mathcal{C}$ is said to be linear if and only if it has the following properties:

$$T(f + g) = T(f) + T(g) \quad (T \text{ is additive}),$$
$$T(Cf) = CT(f) \quad (T \text{ is homogeneous})$$

for any constant C and any functions f and g in $\mathcal{C}$.

In the examples given above, T_1, T_2, and T_4 are linear, but T_3 is not. To check the latter fact, observe that $T_3(t) = t - t^2$, whereas $T_3(Ct) = C^2t - C^2t^2 = C^2(t - t^2) \neq CT(t)$.

We can combine operators, either by addition and subtraction or by multiplication (see Exercise 4). Beware of the latter, however, for some of the simple rules of algebra fail; for example, it may happen that multiplication is not commutative. Thus, T_1T_4 is not the same as T_4T_1, nor does $T_1T_2 = T_2T_1$. Using the abbreviation D for d/dt, we may write

$$T_1 = 3D^2 - 2D + 5,$$
$$T_2 = D^2 - (3t + 1)D + (4t^2 - 1).$$

(Strictly speaking, we should have written the last terms in these expressions as $5I$ and $(4t^2 - 1)I$, where I is the identity operator that sends a function f into itself; however, for simplicity we have followed the idiom and dropped the I.)

What does all this have to do with differential equations? To answer this question, let us commence with an example. Suppose we want to solve the differential equation

$$y'' = -3ty + t^2y' + 5t^2 - \sin t. \tag{5.68}$$

We first rewrite this as

$$y'' - t^2y' + 3ty = 5t^2 - \sin t.$$

Then, let T be the differential operator

$$T = D^2 - t^2D + 3t.$$

We see that we can rewrite (5.68) as the equation

$$T(y) = 5t^2 - \sin t,$$

or if we let $y = \varphi(t)$ be the desired solution of (5.68), then we want to find a function $\varphi \in \mathcal{C}$ such that

$$T(\varphi) = 5t^2 - \sin t.$$

In other words, we have a specific operator T, and we want to find all the functions φ in $\mathcal{C}$ which T sends into the specific function $5t^2 - \sin t$.

This simple maneuver is a process peculiar to the discipline of mathematics among all the sciences. We have recast the problem in a new form in which it now fits a pattern that occurs in many branches of abstract mathematics. This pattern is the following: We have a linear space $\mathcal{C}$, a particular member f, and an operator T on $\mathcal{C}$ to $\mathcal{C}$; and we want to solve the equation $T(\varphi) = f$ for φ. The mathematician's approach is to study the general pattern and to develop an abstract theory for the solution of the problem without regard to the specific nature of T, f, or $\mathcal{C}$, and then apply the general theory to the special case arising from (5.68).

This general theory is called functional analysis[1] and is one of the landmarks of mathematics of the 20th century. We cannot go into it much further, but we shall look at several simple aspects, examining their consequences for a special class of differential equations. Our first result is an immediate consequence of the definition of linearity.

[1] *Functional analysis is a merging of analysis (calculus and its later developments) and linear algebra (a branch of algebra having special relevance to analysis). The interactions between calculus and linear algebra are further discussed in the Postlude.*

Theorem 3 *Let T be a linear operator on $\mathcal{C}$. Then the set of all φ in $\mathcal{C}$ that are solutions of the equation $T(\varphi) = 0$ form a class $\mathfrak{N}_T$ that is a linear subspace of $\mathcal{C}$. Stated more simply, if $T(\varphi_1) = 0$ and $T(\varphi_2) = 0$, then $T(c_1\varphi_1 + c_2\varphi_2) = 0$ for any constants c_1 and c_2.*

The class $\mathfrak{N}_T$ is called the null space of the operator T. Its nature depends upon the specific type of operator T involved. For example, if $T = D^3$, then the null space of T consists of all the polynomials $c_0 + c_1 t + c_2 t^2$ of degree at most 2. (Why?)

Theorem 4 *Let T be a linear operator, and let $\mathfrak{N}_T$ be its null space. Let φ_0 be any solution of the equation $T(\varphi) = f$. Then, all the solutions of this equation have the form*

$$\varphi = \varphi_0 + g,$$

where g is some member of $\mathfrak{N}_T$.

Proof First, let g be any member of $\mathfrak{N}_T$. Then, by linearity, if $\varphi = \varphi_0 + g$, $T(\varphi) = T(\varphi_0) + T(g)$. However, $T(\varphi_0) = f$ and $T(g) = 0$, so $T(\varphi) = f + 0 = f$ and φ is a solution of the equation. Conversely, let φ be any solution of the equation $T(\varphi) = f$. Since we also have $T(\varphi_0) = f$, we have

$$T(\varphi - \varphi_0) = T(\varphi) - T(\varphi_0) = f - f = 0.$$

Thus, setting $g = \varphi - \varphi_0$, we see that $T(g) = 0$, so that g is in the null space of T; and thus we have $\varphi = \varphi_0 + (\varphi - \varphi_0) = \varphi_0 + g$.

Together Theorem 3 and Theorem 4 reveal the general abstract structure of the solution of linear equations, which is applicable to the solution of systems of algebraic linear equations and to the solution of a differential equation. The solution of $T(\varphi) = f$ involves two steps:

Step 1 *Find at least one solution φ_0.*

Step 2 *Determine the null space of T.*

Then, all solutions—including therefore the specific solution that you really want—can be obtained simply by setting $\varphi = \varphi_0 + g$ for $g \in \mathfrak{N}_T$.

One further simplification is possible. In many cases, including that of ordinary differential equations, the null space $\mathfrak{N}_T$ of an operator T is **finite dimensional**. This has a very simple meaning; if $\mathfrak{N}$ has dimension k, then there is a special set of functions $g_1, g_2, \ldots, g_k$ with the property that each function is g in $\mathfrak{N}$ of the form

$$g = C_1 g_1 + C_2 g_2 + \cdots + C_k g_k \tag{5.69}$$

for some (unique) choice of the numbers C_i. Such a set of g_i's is called a **basis** for the linear space $\mathfrak{N}$.

As an example, we noted above that the null space of the operator D^3 consists of all the polynomials in t of degree at most 2. The general form of such a polynomial is

$$g(t) = C_1 + C_2 t + C_3 t^2,$$

and this is merely another way to say that a basis for the space $\mathfrak{N}_{D^3}$ consists of the three functions $1, t, t^2$; its dimension is then three. In passing, we note that a space always has an infinite number of different bases. For example, the three functions $t + 1$, $t^2 - t$, and $2t^2 - 1$ also form a basis for the space of quadratic polynomials, since every such polynomial can be represented (uniquely) in the form $C_1(t + 1) + C_2(t^2 - t) + C_3(2t^2 - 1)$. (See Exercise 7.)

With the help of the concept of basis, we can replace Step 2 above (which was *Determine the null space of T*) by the alternate

Step 2′ Determine a set of functions g_i which together form a basis for the null space of T.

For, knowing these functions, we then can construct all the functions g in $\mathfrak{N}_T$ by means of (5.69).

We next apply these remarks to the study of differential equations. An operator T is said to be a linear differential operator of order n if

$$T = D^n + a_1(t)D^{n-1} + a_2(t)D^{n-2} + \cdots + a_{n-1}(t)\, D + a_n(t),$$

where the $a_i(t)$ are functions that are continuous. For any function φ,

$$T(\varphi) = \varphi^{(n)}(t) + a_1(t)\varphi^{(n-1)}(t) + \cdots + a_{n-1}(t)\varphi'(t) + a_n(t)\varphi(t), \tag{5.70}$$

and T is readily seen to be a linear operator. The central fact about such operators, proved in most texts on the theory of differential equations, is the following:

Theorem 5 *The null space $\mathfrak{N}_T$ of the operator given by* (5.70) *has dimension n.*

This means that the task of determining the null space of an operator such as that in (5.70) becomes the problem of finding a set of n solutions to the special equation $T(\varphi) = 0$ which together form a basis for $\mathfrak{N}_T$. Fortunately, a theorem from abstract linear algebra tells us that such a set of n functions will be a basis if and only if the functions are **linearly independent**; this means no identity of the form

$$0 = C_1 g_1 + C_2 g_2 + \cdots + C_n g_n \tag{5.71}$$

holds among the functions g_i except when *all* the C_i are 0.

We will not study the general theory of operators (5.70), but will look at a special important case where things are much simpler. This is the case of linear differential operators with constant coefficients, operators of the form (5.70) where the $a_i(t)$ are constants. An example is

$$T = D^2 - 2D - 3.$$

As predicted by Theorem 5, the null space of T has dimension 2. We claim that the pair of functions e^{-t} and e^{3t} form a basis for $\mathfrak{N}_T$. To prove this, we may check directly that the functions e^{-t} and e^{3t} are solutions of the equation

$$T(\varphi) = \varphi'' - 2\varphi' - 3\varphi = 0,$$

and then check that e^{-t} and e^{3t} are linearly independent. To verify the latter, suppose that there are constants C_1 and C_2 such that

$$C_1 e^{-t} + C_2 e^{3t} = 0$$

for all t. If we put $t = 0$, and then $t = 1$, we obtain two equations for the C_i:

$$C_1 + C_2 = 0,$$
$$e^{-1}C_1 + e^3 C_2 = 0,$$

which are easily seen to have no solutions other than $C_1 = C_2 = 0$. Hence, the null space of T is generated by these two functions. It consists of all the functions $g(t) = C_1 e^{-t} + C_2 e^{3t}$ for any choice of the numbers C_i.

Where did the pair of functions e^{-t} and e^{3t} come from, and can solutions for the general case be obtained in the same way? A partial answer is supplied by the following result.

Theorem 6 *Let T be any constant coefficient linear operator such as*

$$T = D^n + a_1 D^{n-1} + a_2 D^{n-2} + \cdots + a_{n-1} D + a_n,$$

and let P be the ordinary polynomial of degree n

$$P(s) = s^n + a_1 s^{n-1} + a_2 s^{n-2} + \cdots + a_{n-1} s = a_n.$$

Then, if $P(\gamma) = 0$, the function $e^{\gamma t}$ belongs to the null space of T. Each root of $P(s)$ yields a function in $\mathfrak{N}_T$.

Proof The proof is immediate. For, $T(e^{\gamma t})$ is easily seen to be $P(\gamma)e^{\gamma t}$ (see Exercise 8), and thus is 0 if $P(\gamma) = 0$.

We also need the following special fact which can be proved as an exercise in elementary algebra. (See Exercise 10.)

Lemma 1 *If the numbers $\gamma_1, \gamma_2, \gamma_3, \ldots, \gamma_m$ are all different, then the functions $e^{\gamma_1 t}, e^{\gamma_2 t}, \ldots, e^{\gamma_m t}$ are linearly independent. Moreover, this holds whether the numbers γ_i are real or complex.*

Let us see how this pair of results can be used to solve differential equations. Suppose we want to find all the solutions of the equation

$$y'' - 2y' - 3y = 3t^2 + t - 4. \tag{5.72}$$

This has the form $T(\varphi) = f$, where $f(t) = 3t^2 + t - 4$ and T is the linear operator $D^2 - 2D - 3$. The associated polynomial, according to Theorem 6, is $P(s) = s^2 - 2s - 3$. Since $P(s) = (s + 1)(s - 3)$, the roots of this polynomial are -1 and 3, and the two functions e^{-t} and e^{3t} must belong to the null space of T. Since T has order 2, the null space has dimension 2, and therefore any pair of functions that belong to $\mathfrak{N}_T$ and are linearly independent form a basis for $\mathfrak{N}_T$. Since the roots -1 and 3 are distinct, the functions e^{-t} and e^{3t} are—by Lemma 1—linearly independent. Hence, e^{-t} and e^{3t} form a basis for $\mathfrak{N}_T$, and we have completed Step 2; we have found the entire null space for T.

Step 1 requires us to find some solution for the equation (5.72). Suppose we search for it among the polynomials and try $y = \varphi(t) = A_0 + A_1 t + A_2 t^2$. (This is really a special case of the method of power series, discussed in the last section.) We find that $y' = A_1 + 2A_2 t$, $y'' = 2A_2$. Substituting these values into (5.72), we have

$$2A_2 - 2(A_1 + 2A_2 t) - 3(A_0 + A_1 t + A_2 t^2) = 3t^2 + t - 4,$$

or

$$-3A_2 t^2 + (-4A_2 - 3A_1)t + (2A_2 - 2A_1 - 3A_0) = 3t^2 + t - 4.$$

As usual, this yields a set of equations for the unknown coefficients A_0, A_1, A_2, obtained by comparing coefficients of 1, t, and t^2:

$$\begin{aligned} -3A_2 &= 3, \\ -4A_2 - 3A_1 &= 1, \\ 2A_2 - 2A_1 - 3A_0 &= -4, \end{aligned}$$

from which we obtain $A_2 = -1$, $A_1 = 1$, and $A_0 = 0$.

Thus, we have discovered that $y = t - t^2$ is a solution of (5.72). By the fundamental structure theory for solving linear equations, we conclude that the general solution of the equation (5.72) is given by

$$y = t - t^2 + C_1e^{-t} + C_2e^{3t}. \tag{5.73}$$

Suppose the original problem had been to find a solution of (5.72) which satisfied the initial conditions $t = 0$, $y = 1$, $y' = 0$. From (5.73), we have

$$y' = 1 - 2t - C_1e^{-t} + 3C_2e^{3t}. \tag{5.74}$$

We set $t = 0$, $y = 1$, $y' = 0$ in (5.73) and (5.74), obtaining the equations

$$1 = 0 + C_1 + C_2,$$
$$0 = 1 - C_1 + 3C_2.$$

Solving, we find $C_1 = 1$, $C_2 = 0$, and the required solution is $y = t - t^2 + e^{-t}$.

This illustration is typical of differential equations with constant coefficients in which the associated polynomial $P(s)$ of Theorem 6 has no repeated roots. Each root of P gives rise to a function in the null space; they are all independent and thus form a basis. A particular solution can be found, by a power series method if nothing else, and thus the general solution can be obtained. Any initial conditions enable you to determine the numerical values of the constants C_i and thereby single out the unique solution that fits the conditions.

What happens if the associated polynomial $P(s)$ has two or more equal roots? Let us look at the operator

$$T = D^2 - 2D + 1.$$

The associated polynomial is $P(s) = s^2 - 2s + 1 = (s - 1)^2$, whose roots are 1, 1. By Theorem 6, we know that e^t is in the null space of T, but the pair of roots yield only one function, which cannot be a basis for this space, which must have dimension 2. Where can we find another function φ satisfying $T(\varphi) = 0$?

We can guess the correct answer by looking at an even simpler operator, D^2. Here, the associated polynomial is s^2, and the roots are 0 and 0. Theorem 6 gives us only the function $e^{0t} = 1$ as a member of the null space. However, we know that the complete null space of D^2 is generated by the functions 1 and t. Likewise, the associated polynomial for the operator D^3 has roots 0, 0, 0 and its null space is generated by 1, t, and t^2. We are thus led to conjecture that the null space of the operator $T = D^2 - 2D + 1$ contains both e^t and te^t. This is easily checked:

$$
\begin{aligned}
T(te^t) &= \frac{d^2}{dt^2}(te^t) - 2\frac{d}{dt}(te^t) + te^t \\
&= \frac{a}{dt}(e^t + te^t) - 2(e^t + te^t) + te^t \\
&= e^t + e^t + te^t - 2e^t - 2te^t + te^t \\
&= 0.
\end{aligned}
$$

Since e^t and te^t are linearly independent (Exercise 9), the null space of T consists of all the functions $C_1e^t + C_2te^t = (C_1 + C_2t)e^t$. This example extends to the general case.

Theorem 7 *If the associated polynomial $P(s)$ for an operator T has a factor of the form $(s - \gamma)^m$, then the null space of T contains the functions $e^{\gamma t}, te^{\gamma t}, t^2e^{\gamma t}, \ldots, t^{m-1}e^{\gamma t}$.*

Although this has a simple proof, we choose to omit it because it would lead us into another digression. The result can also be proved by induction. In what follows, we shall be concerned mainly with second order operators, and here the proof is trivial. The theorem enables us to complete the general discussion of the theory of linear differential equations with constant coefficients. The polynomial $P(s)$ associated with an operator T of nth order has degree n, and by the fundamental theorem of algebra, $P(s)$ can be expressed as the product of n factors of the form $(s - \gamma)$. Each distinct γ yields a function $e^{\gamma t}$, and repeated factors yield enough additional functions of the form $t^ke^{\gamma t}$ to provide a total of n linearly independent functions. These functions therefore comprise a basis for the null space of T.

As usual, practical questions arise as soon as one starts to use a theoretical result. We shall illustrate this with a brief discussion of some second order equations.

(i) Let us solve the equation

$$y'' - 3y' - 2y = 0.$$

The operator is $D^2 - 3D - 2$, and $P(s) = s^2 - 3s - 2$. The roots of P are $(3 \pm \sqrt{9 + 8})/2$, or 3.56 and $-.56$, to two decimal places. Hence, the solutions consist of all the functions

$$y = C_1e^{3.56t} + C_2e^{-.56t}.$$

(ii) Let us solve the equation

$$y'' + 2y' + 5y = 0. \tag{5.75}$$

The operator is $D^2 + 2D + 5$, and $P(s) = s^2 + 2s + 5$. The roots of P are $(-2 \pm \sqrt{4 - 20})/2 = -1 \pm 2i$. Hence, the solutions consist of all the functions

$$y = C_1 e^{(-1+2i)t} \, C_2 e^{(-1-2i)t}. \tag{5.76}$$

Although it is entirely correct and often easier to leave the solutions in this form, it is also useful to know how to choose another basis not containing complex functions such as $e^{(-1+2i)t}$ and $e^{(-1-2i)t}$. There are two identities satisfied by the exponential function which make this possible.

Lemma 2 (a) *For any real or complex numbers u and v,*

$$e^{u+v} = e^u e^v.$$

(b) *For any real number θ,*

$$e^{i\theta} = \cos\theta + i \sin\theta.$$

These are discussed and proved in Appendix 2.

Returning to (5.76), we apply the lemma to obtain

$$\begin{aligned} e^{(-1+2i)t} &= e^{-t+2it} = e^{-t}e^{2it} \\ &= e^{-t}(\cos 2t + i \sin 2t), \end{aligned}$$

and in the same way,

$$\begin{aligned} e^{(-1-2i)t} &= e^{-t}(\cos(-2t) + i \sin(-2t)) \\ &= e^{-t}(\cos 2t - i \sin 2t). \end{aligned}$$

Hence, the null space of T, which is also the set of solutions of (5.75), can also be described as the set of all functions

$$C_1 e^{-t}(\cos 2t + i \sin 2t) + C_2 e^{-t}(\cos 2t - i \sin 2t).$$

If we take $C_1 = C_2 = \frac{1}{2}$, then this expression becomes $e^{-t}\cos 2t$; if we take $C_1 = -i/2$ and $C_2 = i/2$, it becomes $e^{-t}\sin 2t$. Hence, the null space of T also has a basis consisting of the two functions $e^{-t}\cos 2t$ and $e^{-t}\sin 2t$, and comprises all the functions of the form $g(t) = C_1 e^{-t}\cos 2t + C_2 e^{-t}\sin 2t$. One direct advantage of the use of this basis is that it is often easier to graph real functions such as $e^{-t}\cos 2t$ and to understand their properties than to do this for complex functions such as $e^{(-1+2i)t}$.

This type of reduction is possible whenever a pair of roots of the associated polynomial $P(s)$ are conjugate complex numbers of the form $a + bi$, $a - bi$. In complex form, the corresponding functions in the basis are $e^{(a+bi)t}$ and $e^{(a-bi)t}$, and a similar calculation based on Lemma 2 shows that these functions may be replaced by the real functions $e^{at} \cos bt$ and $e^{at} \sin bt$.

We can give a complete treatment of second degree operators $T = D^2 + pD + q$ with constant *real* coefficients. The nature of the roots of the associated polynomial $P(s) = s^2 + ps + q$ depends upon the discriminant $p^2 - 4q$:

(i) *If* $p^2 - 4q > 0$, *the roots of P are real and distinct, and the null space of T has for a basis two exponential functions of the form* $e^{\gamma_1 t}, e^{\gamma_2 t}$.
(ii) *If* $p^2 - 4q = 0$, *P has a double root* γ, *and the null space of T has* $e^{\gamma t}$ *and* $te^{\gamma t}$ *as a basis.*
(iii) *If* $p^2 - 4q < 0$, *then the roots of P are two conjugate complex numbers* $a \pm bi$, *and a basis for* $\mathfrak{N}_T$ *is the pair of functions* $e^{at} \cos bt, e^{at} \sin bt$.

In the next section, we explore the physical meaning of these cases in the context of a specific model for a physical system, the harmonic oscillator.

Finding the null space of T is equivalent to solving the equation $T(\varphi) = 0$. We have said little about the actual problem of solving $T(\varphi) = f$, except that if we can find one solution, we can then find them all. As a last resort, one can always try the method of power series; however, there are some special cases for which other approaches are much easier. We first need a general observation.

Theorem 8 *If* φ_k *is a solution of the equation* $T(\varphi) = f_k$, $k = 1, \ldots, m$, *and T is linear, then the function* $\varphi_1 + \varphi_2 + \cdots + \varphi_m$ *is a solution of the equation* $T(\varphi) = f_1 + f_2 + \cdots + f_m$.

This follows at once from the meaning of linearity; the result is called by scientists the principle of superposition.

The following theorems and corollaries are easily proved.

Theorem 9 *If* $T = D^2 + pD + q$, *then the equation* $T(\varphi) = t^m$ *has a polynomial solution whose degree is at most m if* $q \neq 0$, *and degree at most* $m + 1$ *if* $q = 0$ *but* $p \neq 0$.

Corollary *The same statement holds for the equation*

$$T(\varphi) = a_0 + a_1 t + a_2 t^2 + \cdots + a_m t^m.$$

Theorem 10 *If* $T = D^2 + pD + q$, $P(s) = s^2 + ps + q$, and $P(\beta) \neq 0$, *then the equation* $T(\varphi) = e^{\beta t}$ *always has a solution of the form* $Ae^{\beta t}$. *If* $P(\beta) = 0$, *then there is always a solution either of the form* $Ate^{\beta t}$ *or* $At^2 e^{\beta t}$.

Corollary *If none of the numbers* $\beta_1, \beta_2, \ldots, \beta_m$ *are roots of* $P(s)$, *then the equation* $T(\varphi) = a_1e^{\beta_1 t} + a_2e^{\beta_2 t} + \cdots + a_me^{\beta_m t}$ *always has a solution of the form*

$$\varphi = A_1e^{\beta_1 t} + A_2e^{\beta_2 t} + \cdots + A_me^{\beta_m t}.$$

An illustration of a very important typical instance of this would be the equation

$$y'' + 9y = \sin t + 3\cos 2t. \tag{5.77}$$

Here, $P(s) = s^2 + 9$, whose roots are $\pm 3i$. Recall that we have

$$\sin t = \frac{e^{it} - e^{-it}}{2i} = \frac{1}{2i}e^{it} - \frac{1}{2i}e^{-it},$$

$$\cos 2t = \frac{e^{2it} + e^{-2it}}{2} = \tfrac{1}{2}e^{2it} + \tfrac{1}{2}e^{-2it}.$$

By the corollary to Theorem 10 we know that there exists a solution of the equation of the form

$$y = A_1e^{it} + A_2e^{-it} + A_3e^{2it} + A_4e^{-2it}, \tag{5.78}$$

which by Lemma 2 can be replaced by the equivalent form

$$y = C_1\sin t + C_2\cos t + C_3\sin 2t + C_4\cos 2t. \tag{5.79}$$

If we substitute either form in (5.77), we can solve for the coefficients. For example, if we use the exponential form (5.78), we have

$$y' = A_1ie^{it} - A_2ie^{-it} + 2A_3ie^{2it} - 2A_4ie^{-2it}$$

and

$$y'' = -A_1e^{it} - A_2e^{-it} - 4A_3e^{2it} - 4A_4e^{-2it}.$$

Hence,

$$y'' + 9y = 8A_1e^{it} + 8A_2e^{-it} + 5A_3e^{2it} + 5A_4e^{-2it}. \tag{5.80}$$

The right side of (5.77), expressed in exponential form, is

$$\sin t + 3\cos 2t = \frac{1}{2i}e^{it} - \frac{1}{2i}e^{-it} + 3(\tfrac{1}{2}e^{2it} + \tfrac{1}{2}e^{-2it}). \tag{5.81}$$

Comparing coefficients of the corresponding basis functions in both (5.80) and (5.81) we read off the following equations:

$$8A_1 = \frac{1}{2i},\ 8A_2 = -\frac{1}{2i},\ 5A_3 = \tfrac{3}{2},\ 5A_4 = \tfrac{3}{2},$$

and the desired solution of (5.77) is therefore

$$y = \frac{1}{16i} e^{it} - \frac{1}{16i} e^{-it} + \frac{3}{10} e^{2it} + \frac{3}{10} e^{-2it}.$$

In trigonometric form, this is

$$y = \tfrac{1}{8} \sin t + \tfrac{3}{5} \cos 2t. \tag{5.82}$$

Finally, therefore, we apply our general results to obtain the general solution of (5.77) by adding the null space of $D^2 + 9$:

$$y = \tfrac{1}{8} \sin t + \tfrac{3}{5} \cos 2t + C_1 \sin 3t + C_2 \cos 3t.$$

We conclude this section by showing how the preceding results can be used to solve simple systems of differential equations. Suppose we are given the equations

$$\begin{cases} \dfrac{dx}{dt} = 3x - y, \\ \dfrac{dy}{dt} = x + y \end{cases} \tag{5.83}$$

with initial conditions $t = 0$, $x = 1$, $y = 2$. We want to find two functions φ, ψ so that $x = \varphi(t)$ and $y = \psi(t)$ satisfy (5.83) and $\varphi(0) = 1$, $\psi(0) = 2$. The procedure we shall use is to eliminate either x or y from the equations, solve the remaining linear differential equation in the other variable, and return to the original equations to recover the first variable, thereby obtaining a general solution for (5.83). The initial conditions will then single out the unique solution desired.

We start by rewriting (5.83) in operator form:

$$(D - 3)x = -y, \tag{5.84}$$

$$(D - 1)y = x. \tag{5.85}$$

Suppose that we apply the operator $D - 3$ to both sides of (5.85), and then use (5.84) to simplify the right side of the result:

$$\begin{aligned} (D - 3)(D - 1)y &= (D - 3)x \\ &= -y. \end{aligned} \tag{5.86}$$

Because we are dealing with operators with constant coefficients, the product of operators is easily computed; $(D - 3)(D - 1) = D^2 - 4D + 3$. Hence, (5.86) becomes $(D^2 - 4D + 3)y = -y$, which is the same as

$$(D^2 - 4D + 4)y = 0.$$

The associated polynomial factors as $P(s) = (s - 2)^2$. Thus, the solution of (5.86) is

$$y = C_1 e^{2t} + C_2 t e^{2t}.$$

To find x, we substitute this formula for y into (5.85):

$$\begin{aligned} x &= (D - 1)y = (D - 1)\{C_1 e^{2t} + C_2 t e^{2t}\} \\ &= C_1\{2e^{2t} - e^{2t}\} + C_2\{e^{2t} + 2te^{2t} - te^{2t}\} \\ &= C_1 e^{2t} + C_2 e^{2t} + C_2 t e^{2t} \\ &= (C_1 + C_2)e^{2t} + C_2 t e^{2t}. \end{aligned}$$

We substitute $t = 0$ and the initial information $x = 1$, $y = 2$ into the above general solutions for x and y to obtain equations for C_1 and C_2:

$$\begin{aligned} 1 &= (C_1 + C_2)(1) + C_2(0), \\ 2 &= C_1(1) + C_2(0). \end{aligned}$$

Solving, we find $C_1 = 2$, $C_2 = -1$, and therefore the solution of the system (5.83) is

$$\begin{aligned} x &= e^{2t} = te^{2t}, \\ y &= 2e^{2t} - te^{2t}. \end{aligned}$$

We note that we could have chosen to eliminate y from the equations (5.84) and (5.85) instead of eliminating x. If we operate on both sides of (5.84) by $D - 1$, we obtain

$$\begin{aligned} (D - 1)(D - 3)x &= (D - 1)(-y) \\ &= -(D - 1)y = -x. \end{aligned}$$

Proceeding as before, we first obtain

$$(D^2 - 4D + 3)x = -x$$

and then

$$(D^2 - 4D + 4)x = 0,$$

whose solutions we can write as

$$x = A_1 e^{2t} + A_2 t e^{2t}.$$

If we substitute into (5.84), we have

$$\begin{aligned} -y &= (D - 3)x \\ &= (D - 3)\{A_1 e^{2t} + A_2 t e^{2t}\} \\ &= A_1\{2e^{2t} - 3e^{2t}\} + A_2\{e^{2t} + 2te^{2t} - 3te^{2t}\} \\ &= (-A_1 + A_2)e^{2t} - A_2 t e^{2t}, \end{aligned}$$

yielding

$$y = (A_1 - A_2)e^{2t} + A_2 t e^{2t}.$$

If again we require that $x = 1$ and $y = 2$ when $t = 0$, we find

$$1 = A_1(1) + A_2(0),$$
$$2 = (A_1 - A_2)(1) + A_2(0),$$

whose solution is $A_1 = 1$, $A_2 = -1$; again, $x = e^{2t} - te^{2t}$, $y = 2e^{2t} - te^{2t}$.

As a final illustration, let us return to Section 5.1 and solve the Two-Tank Problem. The system (5.15) which we used as a model was

$$\frac{dx}{dt} = .02y - .06x, \qquad \frac{dy}{dt} = .06x - .06y. \tag{5.87}$$

The initial conditions require that $x = 0$ and $y = 10$ when $t = 0$, and we want to estimate the value of t for which y is smaller than .1. (This tells us when the water coming out of the tank is sufficiently pure.)

In operator form, (5.87) becomes

$$(D + .06)x = .02y, \qquad (D + .06)y = .06x. \tag{5.88}$$

Hence,

$$\begin{aligned}(D + .06)(D + .06)y &= (D + .06)(.06x)\\ &= (.06)(D + .06)x\\ &= (.06)(.02y),\end{aligned}$$

or

$$(D^2 + .12D + .0036)y = .0012y,$$
$$(D^2 + .12D + .0024)y = 0.$$

(In contrast with artificially simple textbook exercises, the numbers that appear in more realistic situations are seldom integers.) The associated polynomial $s^2 + .12s + .0024$ has roots $-.06 \pm 2\sqrt{.0003}$, or $-.0254$ and $-.0946$. Hence, the general solution for y is

$$y = C_1e^{-.025t} + C_2e^{-.095t}.$$

Substituting this in (5.88), we have

$$\begin{aligned}.06x = (D + .06)y &= C_1\{(-.025)e^{-.025t} + (.06)e^{-.025t}\}\\ &\quad + C_2\{(-.095)e^{-.095t} + (.06)e^{-.095t}\}\\ &= (.035)C_1e^{-.025t} + (-.035)C_2e^{-.095t},\end{aligned}$$

and we obtain

$$x = .58C_1e^{-.025t} - .58C_2e^{-.095t}.$$

When $t = 0$, we must have $x = 0$ and $y = 10$. Hence,

$$0 = 58C_1 - 58C_2,$$
$$10 = C_1 + C_2,$$

giving $C_1 = C_2 = 5$. The solution for y is therefore

$$y = 5e^{-.025t} + 5e^{-.095t}$$
$$= 5e^{-.025t}\{1 + e^{-.07t}\}.$$

To answer the original question posed in the Two-Tank Problem, we want to know when we have $y \leq .1$. To find the exact solution of the equation $5e^{-.025t}\{1 + e^{-.07t}\} = .1$ is difficult; however, it is sufficient if we can find a good approximation to the solution. Note that $e^{-.07t}$ is about $(e^{-.025t})^3$; when $e^{-.025t}$ is small, $e^{-.07t}$ will be much smaller. Thus, to find which value of t will give $y = .1$, it is probably sufficient to find the value of t for which $5e^{-.025t} = .1$, since the factor $1 + e^{-.07t}$ will be very nearly 1. We therefore solve

$$5e^{-.025t} = .1$$

by taking natural logarithms, yielding

$$.025t = 3.91,$$

and thus $t = 156$. Since the unit time was a minute, we can give the desired answer in this form: The water flowing out of the lower tank will be approximately pure in two hours and forty minutes from the time the valves were opened.

Exercises

1 Show that an operator T on $\mathcal{C}$ is linear if and only if $T(c_1 f_1 + c_2 f_2) = c_1 T(f_1) + c_2 T(f_2)$ for all constants c_1, c_2 and all functions $f_1, f_2 \in \mathcal{C}$.

2 (a) For each of the operators described in (5.67), find $T_i(3t + 1)$.
(b) Verify that T_1, T_2, and T_4 are linear operators.

3 Define operators T_1, T_2, T_3 by $T_1(f) = ff'$, $T_2(f) = 3f + f'$, $T_3(f) = g$, where $g(t) = f(t^2)$.
(a) For each of these operators, calculate $T_i(t^2 - t)$.
(b) Which if any of these operators are linear?

4 Multiplication of operators is defined as follows:

$$(T_1T_2)(f) = T_1(T_2(f)).$$

(a) Verify that $D(D + t)(e^{2t}) = (5 + 2t)e^{2t}$.
(b) By calculating $D(D + t)(f)$, show that

$$D(D + t) = D^2 + tD + 1.$$

*(c) Show that $(D + t)(D - t) = D^2 - (t^2 + 1)$, whereas

$$(D - t)(D + t) = D^2 - (t^2 - 1).$$

5 (a) Prove that the null space of D^3 consists of the polynomials of degree at most 2.

(b) What is the null space of D^n?

6 By solving a separable differential equation, find the null space of the operator $D + 2t$.

7 (a) Show that the functions $t + 1$, $t^2 - t$, $2t^2 - 1$ form a basis for the space of polynomials of degree at most 2.

(b) Show that the functions $2t - 1$, $t^2 + t$, $2t^2 + 1$ do *not* form a basis for the space of quadratic polynomials.

8 Verify that $T(e^{\gamma t}) = P(\gamma)e^{\gamma t}$, where T is the operator in Theorem 6.

9 Show that $e^{\gamma t}$ and $te^{\gamma t}$ are linearly independent.

***10** Show that the functions $e^{\alpha t}$, $e^{\beta t}$, $e^{\gamma t}$ are linearly independent if α, β, γ are distinct. (*Hint:* If an identity is differentiated, it remains an identity.)

11 Verify that the null space of $T = D^3 + 3D^2 + 3D + 1$ has a basis consisting of e^{-t}, te^{-t}, t^2e^{-t}.

12 Find the null space of each of the operators:

(a) $D^2 + 3D - 10$, (b) $D^2 + 9$, (c) $D^2 - 4D + 4$.

13 Find the null space of each of the operators:

(a) $D^2 - 2D - 1$, (b) $D^2 + 2D + 3$, (c) $D^2 + 3D + 18$.

14 Knowing that the equation $y' - y = 3e^t$ has one solution of the form $y = Cte^t$, find a solution of the equation such that $y = 1$ when $t = 0$.

15 Knowing that the equation $y'' + y = 2e^{-2t}$ has one solution of the form $y = Ae^{-2t}$, find a solution of the equation such that $y = 0$, $y' = 0$ when $t = 0$.

16 Knowing that the equation $y'' + y = 3e^{2it}$ has a solution of the form $y = Ae^{2it}$, find a solution such that $y = 1$, $y' = 0$ when $t = 0$.

17 Knowing that the equation $y'' + 3y' + 2y = 1 - 4t - 2t^2$ has a polynomial solution, find all the solutions of the differential equation and find the particular solution satisfying the initial conditions $y = 1$, $y' = 0$ when $t = 0$.

18 Obtain a solution of (5.77) by substitution of the trigonometric form (5.79).

19 By the method of this section, find the general solution to

$$y' + 2y = \sin 2t.$$

20 Find the general solution for $y'' + 2y' - 3y = \cos t$.

21 (a) Find the general solution for $y'' + y' = e^{-2t}$.

(b) Find the solution obeying the initial conditions $t = 0$, $y = 2$, $y' = -2$.

22 Find the general solution of $y'' - 4y = 3\cos 2t$.

23 (a) Find the general solution of the linear system

$$\frac{dx}{dt} = x + 2y,$$

$$\frac{dy}{dt} = 8x + y.$$

(b) Find the particular solution obeying the initial conditions $t = 0$, $x = 1$, $y = 2$.

24 (a) Change the initial conditions on the Two-Tank Problem to $t = 0$, $x = 10$, $y = 0$ and find the solution.

(b) When does tank II have the greatest amount of toxic substance, and how much is it?

25 If the roots of the associated polynomial $P(s) = 0$ are $a + bi$ and $a - bi$, $b \neq 0$, show that the general solution φ of $(D^2 + pD + q)\varphi = 0$ is of the form $y = Ce^{at}\sin(bt + \alpha)$, where C and α, the phase angle, are constants to be determined by the initial conditions.

26 Find the general solution for $y'' + 4y = 3\cos 2t$.

27 Find the general solution for $(D^2 - 3D + 2)\varphi = e^{2t}$.

28 Prove Theorem 9, for $m = 3$, by choosing φ to be a general polynomial of degree at most 4.

29 Prove Theorem 10 by substituting $\varphi = Ae^{\beta t}$, or $\varphi = Ate^{\beta t}$, or $\varphi = At^2e^{\beta t}$, depending upon which case is being considered.

5.8

The Harmonic Oscillator

In this final section, we have chosen to discuss in some detail one particular linear differential equation which appears in many different models. We are interested in the interplay between mathematics and reality, and in the ways in which an applied mathematician may use this interplay to explore the possible behavior of a physical system.

The system we shall use, shown in Figure 5–16, consists of a supporting bar B to which is attached a spring S and a mass M. We wish to study four experiments mathematically.

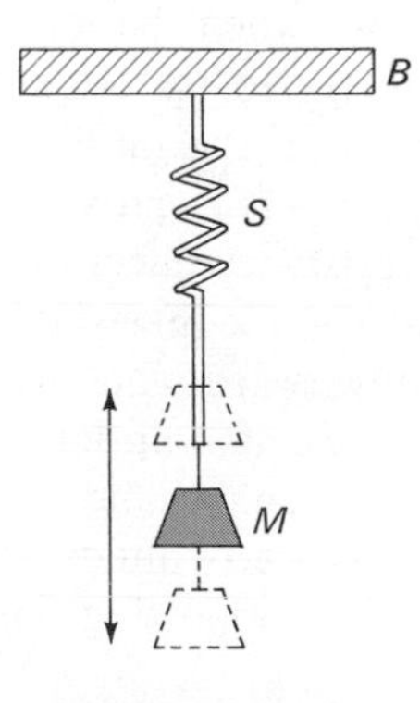

Figure 5–16

(i) *Set the mass in vertical motion by giving it a push.*

(ii) *Set the mass in motion by moving the supporting bar up and down in a periodic manner (Figure 5–17).*

(iii) *Place the mass in a jar of liquid, and then start the mass moving up and down (Figure 5–18).*

(iv) *Combine (ii) and (iii) (Figure 5–19).*

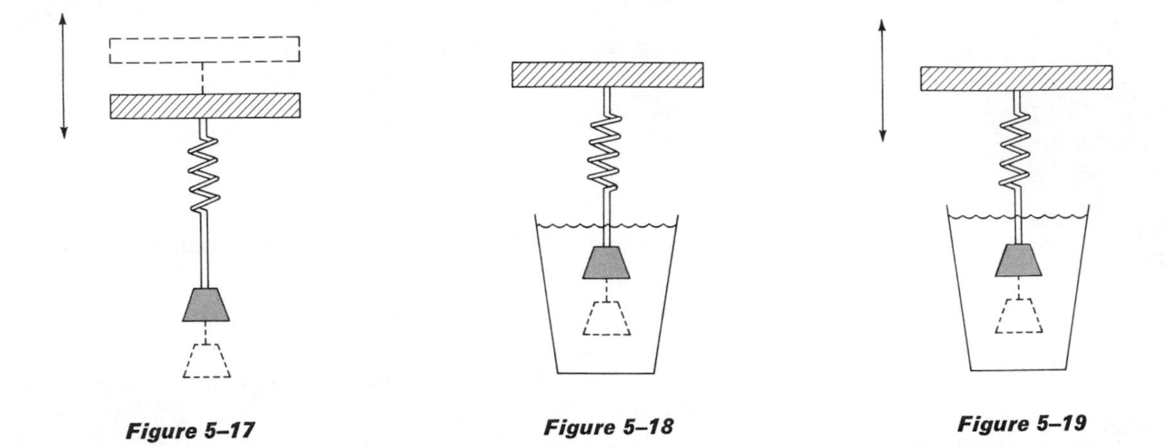

Figure 5–17

Figure 5–18

Figure 5–19

In each case, we want to see if we can "understand" the resulting behavior of the system by showing that a model can be created which predicts the result observed. (*Attention:* To the reader! We cannot control *your* behavior, but we would like you to stop reading at this point and spend some time setting down a description of what you think the motion of the mass will be for each of the four cases above; in (iii) and (iv), be sure to discuss the differences in the motion you would expect to see depending on whether the jar is filled with alcohol or water, or with a thick heavy oil.)

We shall not go into the reasoning which leads to an appropriate model for this system. It can be found in any adequate text on mechanics; the result is the equation

$$M\frac{d^2x}{dt^2} = -A\frac{dx}{dt} - Bx + f(t). \tag{5.89}$$

Here, x specifies the location of the moving mass at time t and is the vertical distance from the mass (regarded as a point mass) to the equilibrium point or rest position of the mass. The number B is positive and measures the strength of the spring; when the mass is displaced from equilibrium, the spring acts to try to restore equilibrium, so that the force exerted by the spring on the mass must be negative when x is positive and positive when x is negative. The number A is also positive and measures the resistance effect of the liquid on the moving mass. Again, the associated force must oppose the direction of motion, so that it must act downward when dx/dt is positive and upward when dx/dt is negative. The assumption that the resistance law is linear is mathematically convenient and seems to fit fairly well for small speeds and certain types of fluids. The function $f(t)$ describes the manner in which the supporting bar B is moved.

We can now describe each of the separate cases (5.89) we want to study. For case (i), with no external forcing motion of the bar and no frictional forces on the moving mass, we take $A = 0$ and $f(t) = 0$. For case (iii) we still have $f(t) = 0$, but now we allow $A > 0$; increasing the value of A corresponds to increasing the viscosity of the liquid, e.g., changing from alcohol to water to light oil to heavy oil. For case (ii) we again set $A = 0$, but this time we take for $f(t)$ a periodic function such as $\sin \beta t$. Finally, in case (iv), we leave $f(t) = \sin \beta t$, but take $A > 0$.

We are not interested in deriving general formulas, so each time we take special numerical values for some of the parameters to simplify the calculations.

Case (i). Solve the equation $d^2x/dt^2 = -Bx$. If we set $B = \omega^2$, then this equation becomes

$$(D^2 + \omega^2)x = 0, \tag{5.90}$$

and by the methods of the preceding section, the general solution is

$$x = C_1 \sin \omega t + C_2 \cos \omega t.$$

If we use Exercise 25 in Section 5.7, we note that an alternate form is

$$x = C \sin (\omega t + \alpha),$$

where C and α are now arbitrary constants. In either form, the resulting solution shows that the motion of the mass is periodic with a constant amplitude determined by the initial conditions. The length of the period is $2\pi/\omega$, since that is the time required for the expression $\omega t + \alpha$ to increase by exactly 2π. A typical graph of $x = \varphi(t)$ is shown in Figure 5–20. We note that the period becomes shorter as ω (and therefore B) increases; this means that the stiffer the spring, the shorter the period of the up and down motion of the moving mass and the higher the frequency $\omega/2\pi$).

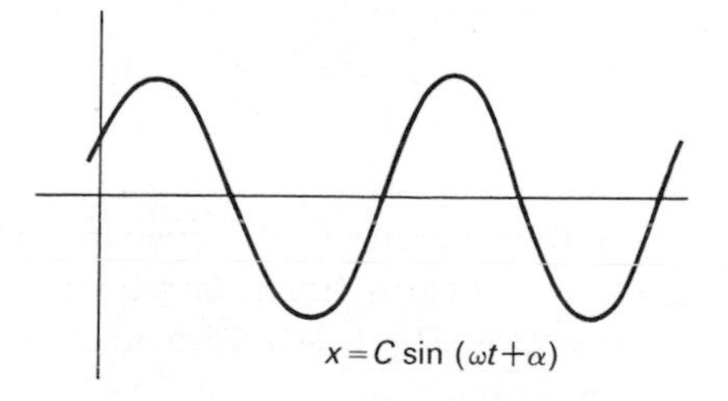

Figure 5–20

Case (ii). We take the equation to be

$$\frac{d^2x}{dt^2} = -Bx + \sin \beta t,$$

where as before $B > 0$. If we set $B = \omega^2$, then this equation becomes

$$(D^2 + \omega^2)x = \sin \beta t. \tag{5.91}$$

The roots of the associated polynomial are ωi and $-\omega i$. Recalling that $\sin \beta t$ can be expressed in terms of $e^{i\beta t}$ and $e^{-i\beta t}$, we have two cases to consider, that in which $\beta \neq \omega$, and that in which $\beta = \omega$. In the first case, we know that (5.91) has a solution $x = A \sin \beta t + B \cos \beta t$. Substituting this in (5.91), we want

$$\begin{aligned}\sin \beta t &= (D^2 + \omega^2)(A \sin \beta t + B \cos \beta t)\\ &= -A\beta^2 \sin \beta t + A\omega^2 \sin \beta t - B\beta^2 \cos \beta t + B\omega^2 \cos \beta t\\ &= A(\omega^2 - \beta^2) \sin \beta t + B(\omega^2 - \beta^2) \cos \beta t.\end{aligned}$$

Thus, we choose $A = 1/(\omega^2 - \beta^2)$, $B = 0$. The general solution of (5.91) is

$$\begin{aligned}x &= C_1 \sin \omega t + C_2 \cos \omega t + \frac{1}{\omega^2 - \beta^2} \sin \beta t\\ &= C \sin (\omega t + \alpha) + \frac{1}{\omega^2 - \beta^2} \sin \beta t.\end{aligned}$$

This is a combination of two sine curves with different periods. Its shape depends on the numerical relationship between ω and β. If we assume that the moving weight is at rest (in equilibrium) at $t = 0$, so that when $t = 0$ we have $x = dx/dt = 0$, then we can solve for the constants C_1 and C_2, obtaining the special solution

$$x = \frac{1}{\omega^2 - \beta^2}\left(\sin \beta t - \frac{\beta}{\omega} \sin \omega t\right). \tag{5.92}$$

A portion of the graph of this motion, for particular values of ω and β, is given in Figure 5–21.

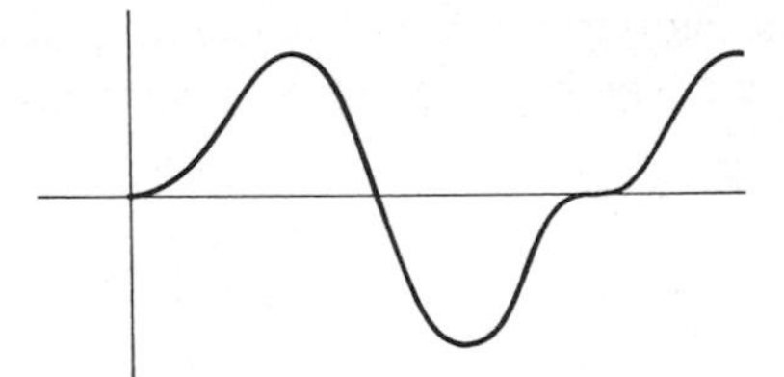

Figure 5–21

When $\beta = \omega$, the equation (5.91) does not have a solution of the form $A \sin \omega t$, since such a function would in fact be in the null space of the operator $D^2 + \omega^2$. Instead, by Theorem 10 in Section 5.7, we can expect a solution of the form $x = C_1 te^{i\omega t} + C_2 te^{-i\omega t}$, or in trigonometric form,

$$x = At \sin \omega t + Bt \cos \omega t.$$

Differentiating this, we have

$$\frac{dx}{dt} = A \sin \omega t + A\omega t \cos \omega t + B \cos \omega t - B\omega t \sin \omega t,$$

$$\begin{aligned}\frac{d^2x}{dt^2} &= 2A\omega \cos \omega t - 2B\omega \sin \omega t - A\omega^2 t \sin \omega t - B\omega^2 t \cos \omega t, \\ &= 2A\omega \cos \omega t - 2B\omega \sin \omega t - \omega^2 x,\end{aligned}$$

so that

$$(D^2 + \omega^2)x = 2A\omega \cos \omega t - 2B\omega \sin \omega t.$$

In order for this to equal $\sin \omega t$, as required by (5.91) with $\beta = \omega$, we must have $2A = 0$ and $-2B\omega = 1$. We have therefore determined that a solution of the equation (5.91) is

$$x = -\frac{1}{2\omega} t \cos \omega t.$$

Adding the functions in the null space of $D^2 + \omega^2$, we arrive at the general solution of (5.91) in the case $\beta = \omega$:

$$x = C_1 \sin \omega t + C_2 \cos \omega t - \frac{1}{2\omega} t \cos \omega t.$$

If we again take the initial conditions to be $t = 0$, $x = 0$, $dx/dt = 0$, then the corresponding solution is

$$x = \frac{1}{2\omega^2} \sin \omega t - \frac{1}{2\omega} t \cos \omega t. \tag{5.93}$$

A portion of the graph is shown in Figure 5–22.

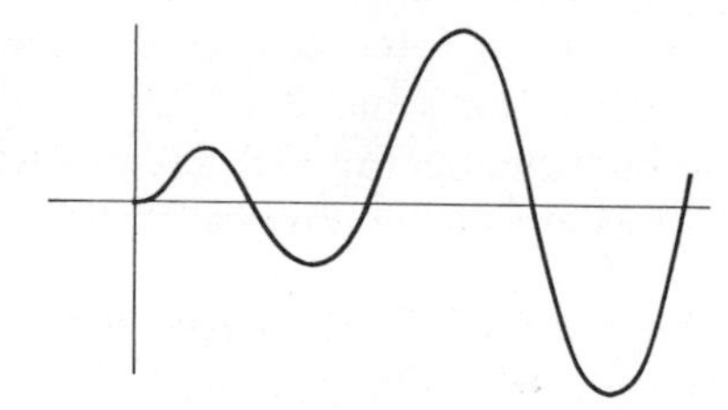

Figure 5–22

Let us now return to the physical system and ask what these mathematical solutions say about the behavior of the system. We describe again our current problem. The mass suspended from the spring is initially at rest. We begin to move the supporting bar to which the spring is attached up and down in a periodic way described by the function $f(t) = \sin \beta t$. This transfers energy to the mass-spring system, which also starts to move. The mathematical solution suggests that there is a profound difference in behavior between the case when $\beta = \omega$ and the case when $\beta \neq \omega$. We have seen that the number ω, which is derived from the physical properties of the spring and which is large if the spring is stiff, determines the "natural" frequency and period of the mass-spring system. If β and ω are different, so that the equation of motion of the mass is (5.92), then we see that the mass oscillates up and down in a complex way. The maximum and minimum positions can be found from (5.92). If we observe that each of the sine terms can take on only values between -1 and 1, then we know that for all t,

$$|x| \leq \frac{1 + \beta/\omega}{|\omega^2 - \beta^2|} = \frac{1}{\omega|\omega - \beta|}. \tag{5.94}$$

On the other hand, when $\beta = \omega$ and the corresponding solution is given by (5.92), we see that the cosine term is multiplied by t, and accordingly the function describing the motion of the moving mass is unbounded. Such a function obviously cannot describe the true motion of the oscillating mass, since the values of x would quickly exceed the distances allowed by the geometry of the physical system itself.

Do these mathematical conclusions correspond to observable physical phenomena? If we were to carry out the actual experiment and gradually allow the frequency β of the external driving form to approach the natural frequency ω of the mass-spring system, would there be a change in behavior? The answer is affirmative; as suggested by (5.93), as β approaches ω and $1/|\omega - \beta|$ becomes large, the extent of oscillation of the moving mass becomes wider and wider. When the frequencies β and ω coincide, the phenomenon called **resonance** occurs, sometimes with destructive effects. In practice, of course, no physical system is entirely without resistance, and no external supplied force can have exactly the form $f(t) = \sin \beta t$ with β exactly the same real number as ω. However, the relatively crude mathematical model has predicted the existence of a very important aspect of the physical system.

Case (iii) We now include the effect of a simple resistance law by studying the equation

$$\frac{d^2x}{dt^2} = -2k\frac{dx}{dt} - x.$$

The number k is positive and measures the amount of resistance to the motion of the moving weight: When $k = 0$, the resistance is absent and we have the undamped oscillator of case (i), with natural period 2π. Rewrite the general equation as

$$(D^2 + 2kD + 1)x = 0, \tag{5.95}$$

and examine the associated polynomial $P(s) = s^2 + 2ks + 1$. Its roots are

$$\frac{-2k \pm \sqrt{4k^2 - 4}}{2} = -k \pm \sqrt{k^2 - 1}.$$

There are three special cases to be studied: (a) $0 < k < 1$, (b) $k = 1$, (c) $k > 1$.

In case (a), let $\gamma = \sqrt{1 - k^2}$. Then, the roots of the associated polynomial are $-k \pm i\gamma$ and the corresponding solutions of the equation (5.95) will be

$$x = C_1 e^{-kt} \sin \gamma t + C_2 e^{-kt} \cos \gamma t,$$

which can also be written as

$$x = Ce^{-kt} \sin (\gamma t + \alpha).$$

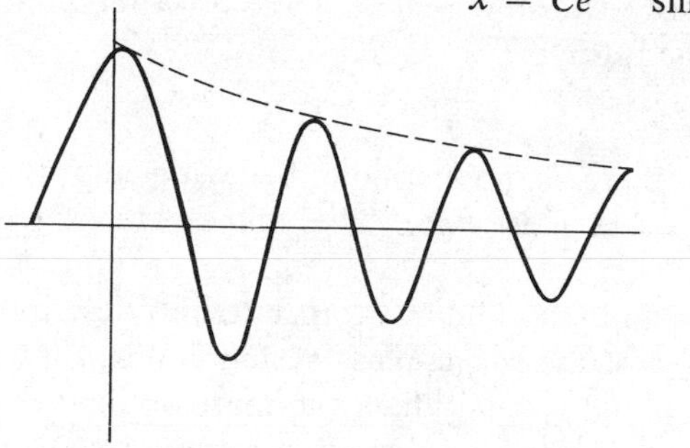

Figure 5–23

This has the shape of a sine curve whose amplitude is decreasing rapidly as t increases. (See Figure 5–23.) When k is quite small, the effect of the damping factor e^{-kt} may be initially small, but since $\lim_{t \to 0} e^{-kt} = 0$, the range of oscillation of the moving mass will eventually decrease toward zero. When k is nearly 1, the damping effect will be much stronger, and the period will be very long. (See Figure 5–24.)

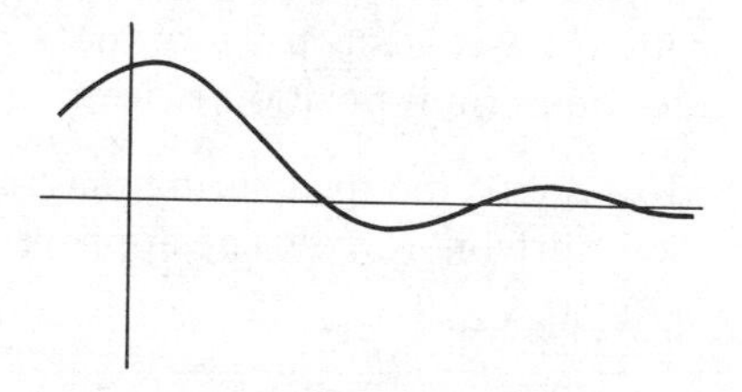

Figure 5–24

In case (b), $k = 1$, and the roots of $P(s)$ are $-1, -1$. As explained in Section 5.7, the corresponding solutions of (5.95) will be

$$\begin{aligned} x &= C_1 e^{-t} + C_2 t e^{-t} \\ &= (C_1 + C_2 t)e^{-t}. \end{aligned}$$

The shape of such a curve is profoundly different from the solution found in case (a), shown in Figure 5–23. For $t \geq 0$, the graph will cross the t-axis at most once, and there is no periodic behavior. In Figure 5–25 we show several examples of possible shapes.

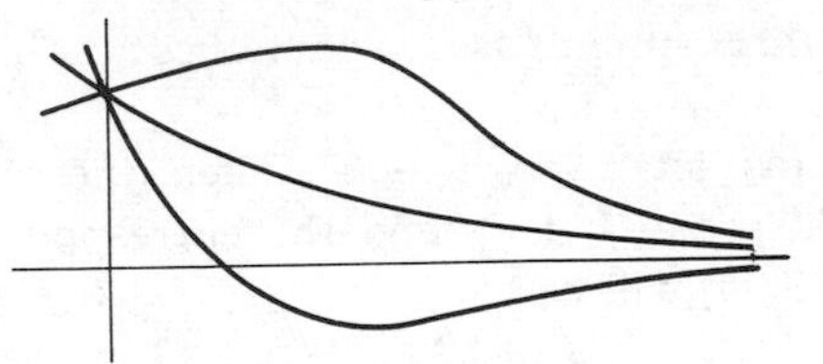

Figure 5–25

In case (c), $k > 1$, and the roots of $P(s)$ are real and distinct, $-k + \sqrt{k^2 - 1}$ and $-k - \sqrt{k^2 - 1}$. Since $\sqrt{k^2 - 1}$ is smaller than k, both of these roots are negative. For example, if $k = 3$, the roots are $-.17$ and -5.83. The corresponding solutions of (5.95) have the form

$$x = C_1 e^{-at} + C_2 e^{-bt}.$$

Again, there is no periodic behavior and the graph will cross the t-axis at most once. The shapes are very similar to those in Figure 5–25.

How do these solutions translate into descriptions of the physical behavior of the spring-mass system? When the resistance supplied by the liquid is small, then the mass oscillates with a gradually decreasing amplitude and with a period somewhat longer than the natural period. As the resistance is increased, the damping effect is accentuated, and the period increases. However, at a certain point ($k = 1$ in our example) the behavior alters, and from this point on the behavior of the system is much simpler. If the mass is lifted and released, it will merely move downward toward the equilibrium position slowly and approach it asymptotically. If it is given a large enough vertical velocity, it might pass through the equilibrium position, but it will then move up toward it in the same asymptotic manner. No matter how it is started into motion, the mass will not pass through the equilibrium position twice.

Case (iv). How does a damped spring-mass system behave under an external periodic driving force? The appropriate equation is now

$$\frac{d^2x}{dt^2} = -2k\frac{dx}{dt} - x + \sin \beta t,$$

which we rewrite as

$$(D^2 + 2kD + 1)x = \sin \beta t. \tag{5.96}$$

Since $k > 0$, we know that none of the functions in the null space of the operator are of the form $e^{i\beta t}$, so that we can be sure that (5.96) has a solution of the form $x = A \sin \beta t + B \cos \beta t$. Substituting this for x in (5.96) and solving, we find $A = (1 - \beta^2)/\Delta$ and $B = -2k\beta/\Delta$, where $\Delta = (1 - \beta^2)^2 + (2k\beta)^2$. As we have seen above, the null space of the operator $D^2 + 2kD + 1$ has for a basis functions of the form $e^{-kt} \sin(\gamma t + \alpha)$, or $(C_1 + C_2 t)e^{-t}$, or e^{-at}, e^{-bt}, depending on the relative size of k. Each of these functions tends to zero as t becomes large. Thus, every solution of the equation (5.96) has the form

$$x = \left\{\frac{1 - \beta^2}{\Delta} \sin \beta t - \frac{2k\beta}{\Delta} \cos \beta t\right\} + \{\text{something tending to zero}\}.$$

The second term is called transient because it is effective only for a short while, after which the motion of the moving mass is given essentially by the first term alone.

Let us therefore examine the behavior of the function

$$x_0 = \varphi(t) = \frac{1 - \beta^2}{\Delta} \sin \beta t = \frac{2k\beta}{\Delta} \cos \beta t.$$

If we assume that this can be written as $C \sin(\beta t - \theta) = C \cos \theta \sin \beta t - C \sin \theta \cos \beta t$, we find that

$$C = \frac{1}{\sqrt{\Delta}} = \frac{1}{\sqrt{(\beta^2 - 1)^2 + (2k\beta)^2}}. \tag{5.97}$$

Thus, the nontransient motion of the oscillating mass has the form

$$x_0 = \frac{1}{\sqrt{\Delta}} \sin(\beta t - \theta)$$

and is therefore a sine curve of amplitude $1/\sqrt{\Delta}$, displaced by a certain phase shift θ.

What does this mathematical analysis of the equation (5.96) predict about the behavior of the physical system? First, it shows that if we move the supporting bar up and down as specified by the function $f(t) = \sin \beta t$, then the mass will start to move in a complicated manner, eventually settling down to a periodic harmonic motion with the same frequency as that of the motion of the supporting bar, but not necessarily in phase. The amplitude of the motion of the mass—meaning the value of the maximum displacement of the mass from equilibrium—is given by the number $1/\sqrt{\Delta}$, as in (5.97). Now, observe the effect of altering the frequency of the external driving force. The greatest amplitude of the motion will occur when Δ is least. It is easy to see that the minimum of Δ occurs when β is $\beta_0 = \sqrt{1 - 2k^2}$, assuming that $2k^2 < 1$, and occurs for $\beta = 0$ when $2k^2 \geq 1$. If we first consider the behavior of such an oscillating system with a very small resistance k, then we again see the phenomenon of resonance as in case (ii). As the frequency of the driving force is altered, the maximum response of the moving mass will change and will reach a peak when β is $\beta_0 = \sqrt{1 - 2k^2}$. Since k is small, this does not differ much from the natural frequency $\omega = 1$ of the undamped system nor in fact from the actual frequency $\sqrt{1 - k^2}$ of the freely moving damped system of case (iii). This is an instance of resonance, but because of the presence of the damping terms, the maximum amplitude of the moving mass is finite. For example, if we take $\beta = \omega = 1$ in (5.97), then we see that

$$|x| \leq \frac{1}{\sqrt{\Delta}} = \frac{1}{2k}.$$

What happens now if the resistance is increased? When k becomes large enough so that $2k^2 > 1$, then the minimum value of Δ occurs for $\beta = 0$ and is 1; moreover, Δ increases steadily as β increases. This tells us that the range of oscillation of the moving mass is at most 1, and that it decreases as the frequency of the driving force increases, approaching zero as a limit. This is precisely the theory behind any type of a shock absorber. A rapid movement of the supporting bar through a considerable distance will result in only a small movement of the mass M—which indeed is smaller for high frequency external motions than it is for low.

Without going into the physics involved, we point out the fact that another important physical system also leads to the same basic differential equation as does the spring-mass system. A simple LCR circuit consists of a resistor R, a capacitor C, and an inductor L, driven by an alternating voltage source $V(t)$ (see Figure 5–26) leads to the following equation for the current I:

$$L\frac{d^2I}{dt^2} + R\frac{dI}{dt} + \frac{1}{C}I = V(t).$$

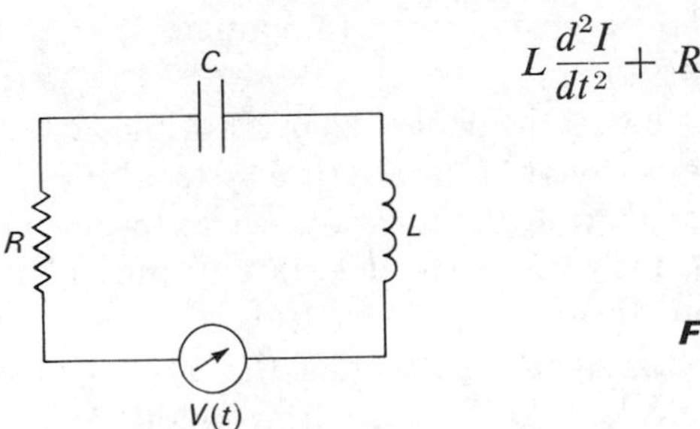

Figure 5–26

Note that the inductance L is the analogue of mass, the resistance R is akin to the damping resistance produced by the oil dashpot, and the reciprocal of the capacitance plays the role of the spring constant in the mechanical problem.

Since the equation is fundamentally the same, all the conclusions again apply. If $V(t)$ is identically zero and R is absent or very small, the system behaves like a simple oscillator; set into activity, it will produce a lightly damped oscillating current I. If the resistance is large, then it will not support an alternating current. Much more interesting results happen if we choose $V(t)$ to be a simple alternating voltage like $\sin \beta t$. In the mechanical system, resonance could be pictured as resulting in violent unbounded oscillations; the classic story is of the corps of soldiers whose exactly timed marching, in resonance with the natural frequency of a bridge, led to its collapse. In the electrical system, resonance is what makes it possible to tune a receiver to match the frequency of a transmitter. Thus it has made possible our entire communications industry.

In the exercises that follow, we have included a variety of word problems which lead to differential equations and systems of equations. All can be solved by the methods given in this chapter, but many are of considerable difficulty. In some cases, the central problem will be to devise an appropriate model. We have also included some of the classical problems. It may be better to use these exercises as a basis for group research projects than as individual assignments. If nothing else, they illustrate the variety of areas to which one can apply elementary methods of analysis.

Exercises

1 The rate at which a body changes temperature is, according to Newton, approximately proportional to the difference between its temperature and that of the surrounding media. A hot meteor weighing thirty pounds falls into a rapidly flowing stream whose temperature is 40°. Ten minutes later, the temperature of the meteor is down to 200°, and in ten more minutes, it is 60°. What was the initial temperature of the meteor, and what will be the temperature half an hour after it has landed?

2 A special braking system for a racing car supplies a frictional force that is proportional to the square root of the velocity of the car. In a test, the moving car is brought to a stop in 3 seconds after traveling 100 feet. How fast was it going when the brakes were applied?

3 Let P be the population of the United States at time t. A frequently used model for population growth assumes that there is a maximum size M that can be tolerated, and that the rate of population growth dP/dt is proportional to the product of P and $M - P$. Some census data is given below. Use the data for 1900, 1920, and 1940 to estimate the value of M, and compare this with the value of M which can be obtained from the data for 1940, 1950, 1960. What does this suggest about the suitability of the proposed model? What value of M would you obtain from the data for 1930, 1940, 1950?

year of census	1900	1920	1930	1940	1950	1960
population/(10^6)	76	106	123	132	151	180

4 In a simple chemical reaction, two grams of A and one gram of B combine to yield 3 grams of C. The rate at which C is produced is proportional to the product of the amounts of A and B that are left *uncombined* at the moment. If we start with 20 grams of A, 10 grams of B, and no C, it is found that 10 grams of C is produced at the end of 5 minutes. How long will it take to obtain a total yield of 25 grams of C? Would it save much time to have added 10 more grams of A at the start?

5 A tank with a capacity of 100 gallons is half full of fresh water. A pipe is opened to admit treated sewage into the tank at 4 gal/min at the same time that the drain valve is opened to allow 3 gal/min to leave the tank. If the treated sewage contains 10 grams of usable potassium per gallon, what is the concentration of potassium in the tank when it is full?

6 It began to snow some time in the evening. At midnight the plow started out and by 1 a.m. it had plowed a distance of 2 miles down the main road. During the next hour, however, it could only cover another mile. When did it start to snow? Assume it continues to snow at a constant rate and that the plow removes snow at a constant rate.[2]

[2] *This exercise comes from Prof. R. P. Agnew, Cornell University.*

7 The setup for a complicated mixing system is shown in Figure 5–27. We have three tanks interconnected as indicated by pipes whose capacity of flow is 5 gallons/min. Initially, tank I contains 20 gallons of red paint, tank II contains 30 gallons of yellow paint, and tank III contains 40 gallons of blue paint. What will be the mixtures of paints in each tank at the end of 5 minutes?

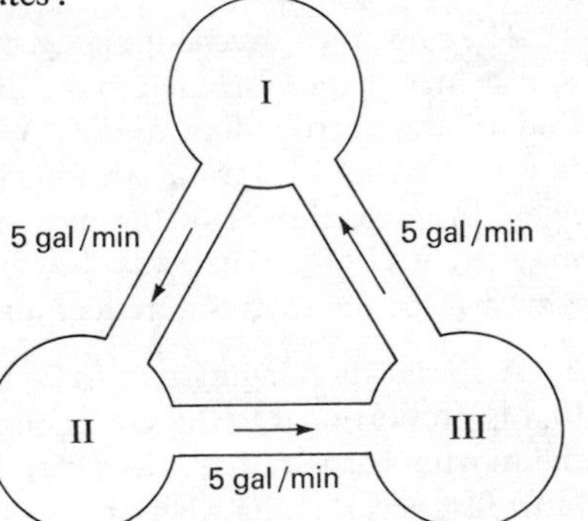

Figure 5–27

8 The speed with which water will issue through a hole in the bottom of a vessel is $\sqrt{2gh}$, where $g = 32$ ft/sec^2 and h is the depth of water above the hole. Compare the length of time it takes a conical tank to empty through a hole in the apex of the cone and through a hole of the same size in the flat top. (See Figure 5–28.)

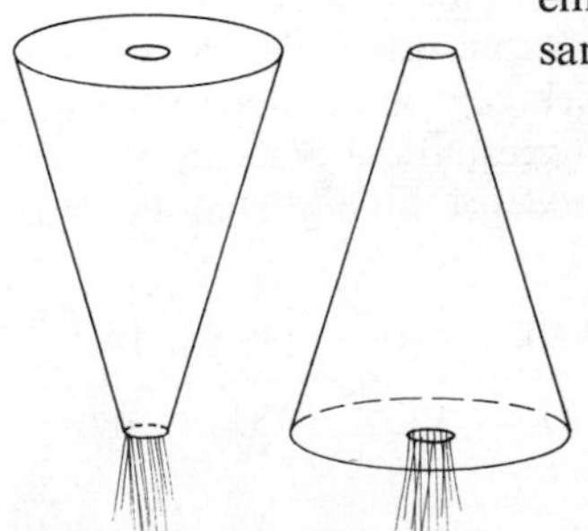

Figure 5–28

9 A certain parasite is hatched only from eggs deposited in a host, and the host always dies. If x and y are the number of hosts and parasites alive at time t, then we assume that the number of eggs deposited is proportional to the product xy. Take account of the birth and (normal) death rates of the host and the death rates of the parasites, and show that a model for this situation is the system

$$\frac{dx}{dt} = ax - kxy,$$

$$\frac{dy}{dt} = kxy - by.$$

Assume that $a > 0$, $b > 0$, and solve the system. Then, plot the point (x, y) in the plane and examine the path of this point as t increases. What does this mean about the biological behavior of the system?

---

series

appendix 1

An infinite series is usually written as $\sum a_n$ or as

$$a_1 + a_2 + a_3 + \cdots.$$

It consists of two sequences of numbers, the sequence of terms $\{a_n\}$ and the sequence of partial sums $\{A_n\}$, which are related by the identities

$$\begin{aligned} A_n &= a_1 + a_2 + a_3 + \cdots + a_n, & n &= 1, 2, 3, \ldots, \\ a_1 &= A_1, & & \\ a_n &= A_n - A_{n-1}, & n &= 2, 3, 4, \ldots . \end{aligned}$$

For example, the first five terms of the series

$$1 + \tfrac{1}{2} + \tfrac{1}{4} + \tfrac{1}{8} + \tfrac{1}{16} + \tfrac{1}{32} + \cdots$$

are 1, $\frac{1}{2}$, $\frac{1}{4}$, $\frac{1}{8}$, and $\frac{1}{16}$, whereas the first five partial sums are 1, $\frac{3}{2}$, $\frac{7}{4}$, $\frac{15}{8}$, and $\frac{31}{16}$.

Definition 1 The series $\sum a_n$ is said to be convergent, or to converge to the sum A, if $\lim_{n\to\infty} A_n$ exists and is A. If the sequence of partial sums $\{A_n\}$ is not a convergent sequence, the series is said to diverge.

The series

$$\tfrac{1}{2} + \tfrac{1}{6} + \tfrac{1}{12} + \tfrac{1}{20} + \cdots,$$

which is

$$\frac{1}{(1)(2)} + \frac{1}{(2)(3)} + \frac{1}{(3)(4)} + \frac{1}{(4)(5)} + \cdots$$

has as its general term $a_n = 1/(n)(n+1)$. Calculation shows that its partial sums are given by $A_1 = \frac{1}{2}$, $A_2 = \frac{2}{3}$, $A_3 = \frac{3}{4}$, $A_4 = \frac{4}{5}$.

One may show (for example by mathematical induction) that $A_n = n/(n+1)$. Since $\lim_{n\to\infty} n/(n+1) = 1$, the given series converges and has sum 1.

Another familiar example is the geometric series

$$a + ar + ar^2 + ar^3 + \cdots = a\sum_0^\infty r^n,$$

whose partial sums are $A_1 = a$, $A_2 = a(1+r) = a(1-r^2)/(1-r)$, and in general $A_n = a(1-r^n)/(1-r)$, assuming that the number r is not 1. (If $r = 1$, then $A_n = an$.) Since $\lim_{n\to\infty} r^n$ exists if and only if $-1 < r \leq 1$, we may conclude that the geometric series converges only when $-1 < r < 1$. The series then has sum $a/(1-r)$.

If two series $\sum a_n$ and $\sum b_n$ are convergent, then the series $\sum(a_n + b_n)$, whose terms are the sums of the terms of the other two series, is also convergent, and its sum is the sum of the other sums; we write

$$\sum(a_n + b_n) = \sum a_n + \sum b_n.$$

For example, the series

$$(\tfrac{1}{2} + \tfrac{1}{3}) + (\tfrac{1}{4} + \tfrac{1}{9}) + (\tfrac{1}{8} + \tfrac{1}{27}) + \cdots = \tfrac{5}{6} + \tfrac{13}{36} + \tfrac{35}{216} + \cdots$$

is convergent and has sum

$$\frac{1}{2}\,\frac{1}{1-\frac{1}{2}} + \frac{1}{3}\,\frac{1}{1-\frac{1}{3}} = \frac{3}{2}.$$

The discovery of tests to show that a given series is either convergent or divergent is central to the theory of series. Of such tests, some can be applied to the terms of the series; others may also involve the partial sums.

Theorem 1 *If $a_n \geq 0$, then $\sum a_n$ is convergent if and only if the partial sums form a bounded sequence.*

Proof Since $a_n \geq 0$, the sequence $\{A_n\}$ is monotonic increasing. If it is bounded, then $\lim_{n\to\infty} A_n$ exists,[1] and the series $\sum a_n$ is convergent. If the sequence $\{A_n\}$ is unbounded, then $\{A_n\}$, and hence the series $\sum a_n$, is divergent.

[1] *This follows from the basic theorem, often cited in an elementory calculus course:* Every bounded monotonic sequence of real numbers converges to a limit.

Theorem 2 *The series $\sum a_n$ is convergent if and only if*

$$\sum_{n+1}^{m} a_k = a_{n+1} + a_{n+2} + \cdots + a_{m-1} + a_m$$

converges to 0 as n and m independently increase.

Proof The sum in question can be expressed in terms of the partial sums as $A_m - A_n$. Thus, the stated condition is equivalent to the statement that the sequence $\{A_k\}$ is a Cauchy sequence. This, in turn, is known to be a necessary and sufficient condition for any sequence of real or complex numbers to be convergent.

There is also a very useful test which can be used only to prove that a series is divergent.

Theorem 3 *If it is* not *true that* $\lim_{n\to\infty} a_n = 0$, *then the series* $\sum a_n$ *is divergent.*

Proof Suppose that $\sum a_n$ is convergent, with sum A. Then, $\lim_{n\to\infty} A_n = A$. It must also be true that $\lim_{n\to\infty} A_{n-1} = A$ and therefore that $\lim_{n\to\infty} (A_n - A_{n-1}) = A - A = 0$. However, $A_n - A_{n-1} = a_n$, so that $\lim_{n\to\infty} a_n = 0$. This is therefore a necessary condition for convergence; without it, $\sum a_n$ diverges.

There is a strong analogy between infinite series and integrals of the form $\int_a^\infty f(x)\,dx$, usually called "improper" integrals.

Theorem 4 *Let $f(x)$ be a positive continuous monotonically decreasing function on the interval $a \le x \le \infty$. Let $a_n = f(n)$ for all $n > a$. Then the series $\sum a_n$ is convergent if and only if the improper integral $\int_a^\infty f(x)\,dx$ is finite.*

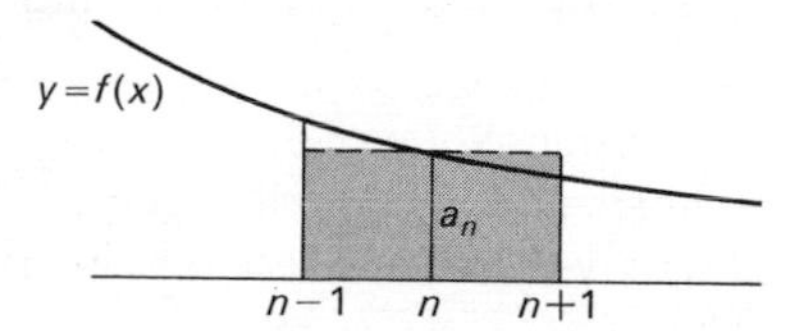

Figure A1–1

Proof From Figure A1–1, we see that

$$a_n \le \int_{n-1}^{n} f(x)\,dx,$$

and so, if k is the first integrel greater than a,

$$A_m - A_{k-1} = \sum_k^m a_n \le \int_a^k f + \int_k^{k+1} f + \cdots + \int_{m-1}^{m} f \le \int_a^\infty f.$$

If the integral on the right is finite, then the partial sums $\{A_m\}$ form a bounded sequence. Since $a_n \ge 0$, we see by Theorem 1 that the series $\sum a_n$ is convergent.

Also, from Figure A1–1, we observe that

$$a_n \ge \int_n^{n+1} f,$$

and so $$\sum_k^m a_n \geq \int_k^{m+1} f.$$

If the integral $\int_a^\infty f$ is infinite (divergent), then the number $\int_k^{m+1} f$ will grow without bound as m increases, and so must the number $A_m = \sum_1^m a_n$. Since the partial sums of the series $\sum a_n$ are therefore unbounded, the series diverges. Hence $\sum a_n$ convergent implies that the improper integral is finite.

If we take $f(x) = 1/x^p$, we find that $\int_1^\infty f(x)\,dx$ is finite when $p > 1$ and infinite if $0 \leq p \leq 1$. Accordingly, we find that the series

$$\sum_1^\infty \frac{1}{n^p} = 1 + \frac{1}{2^p} + \frac{1}{3^p} + \cdots$$

is convergent when $p > 1$, and diverges when $p \leq 1$.

It is sometimes possible to test a series with positive terms by comparing it with another positive series whose behavior is known.

Theorem 5 *Let $0 \leq a_n \leq b_n$. Then, if $\sum b_n$ converges, $\sum a_n$ converges.*

Proof If the corresponding partial sums are A_n and B_n, then we have $A_n \leq B_n$. Since $\sum b_n$ converges and has positive terms, $\{B_n\}$ is a bounded sequence. So must be $\{A_n\}$, and $\sum a_n$ must converge.

A consequence of this theorem is worth noting; if the same relation holds and $\sum a_n$ is divergent, so is $\sum b_n$.

A very similar test that is often easy to apply is the following.

Theorem 6 *Let $0 < a_n$, $0 < b_n$, and $\lim_{n\to\infty} a_n/b_n = L$, with $0 < L < \infty$. Then, the two series $\sum a_n$ and $\sum b_n$ are either both convergent or both divergent.*

When two sequences $\{a_n\}$ and $\{b_n\}$ of positive numbers are related as in Theorem 6, so that $\lim a_n/b_n = L \neq 0$ exists, then one may write $a_n \sim b_n$, read "a_n is asymptotically like b_n." For example, $n/(n^3 + 4) \sim 1/n^2$. Since $\sum 1/n^2$ converges, so does $\sum n/(n^3 + 4)$. Likewise, the series $\sum \dfrac{\sqrt{n+1}}{(2 + \sqrt{n^3 + n + 1})}$ is divergent, since its terms behave like $1/n$.

When the terms of a series $\sum a_n$ are not all positive, the simplest indicator of convergence is the convergence of the series $\sum |a_n|$, whose terms are the absolute values of the terms of the series $\sum a_n$.

Theorem 7 *A series $\sum a_n$ is convergent if the series $\sum |a_n|$ is convergent.*

Proof Suppose $\sum |a_n|$ is convergent. We give two proofs. For the first, we observe that

$$\begin{aligned}\left|\sum_{n+1}^{m} a_k\right| &= |a_{n+1} + a_{n+2} + \cdots + a_m| \\ &\leq |a_{n+1}| + |a_{n+2}| + \cdots + |a_m| \\ &= \sum_{n+1}^{m} |a_k| \to 0\end{aligned}$$

(by Theorem 2), so that (by Theorem 2), $\sum a_n$ converges.

For the second proof, we observe that

$$0 \leq a_n + |a_n| \leq 2|a_n|.$$

Since $\sum 2|a_n|$ is convergent, so is the series $\sum(a_n + |a_n|)$ (by Theorem 5). Noting that

$$\sum a_n = \sum(a_n + |a_n|) - \sum |a_n|,$$

the difference of two convergent series, we conclude that $\sum a_n$ is convergent.

For example, each of the following series is convergent:

$$1 - \tfrac{1}{4} + \tfrac{1}{9} - \tfrac{1}{16} + \tfrac{1}{25} + \cdots + (-1)^{n+1}\frac{1}{n^2} + \cdots,$$

$$\tfrac{1}{2} + \tfrac{2}{4} - \tfrac{3}{8} + \tfrac{4}{16} + \tfrac{5}{32} - \tfrac{6}{64} + \tfrac{7}{128} + \cdots.$$

(Compare the latter series with $\sum n/2^n$.)

It is possible to have the series $\sum a_n$ convergent while the positive series $\sum |a_n|$ is divergent. This is the case for the "harmonic" series

$$1 - \tfrac{1}{2} + \tfrac{1}{3} - \tfrac{1}{4} + \tfrac{1}{5} + \cdots + (-1)^{n+1}\frac{1}{n} + \cdots.$$

The convergence of this series is a consequence of the following useful result, usually called the "alternating series test."

Theorem 8 *Let the sequence $\{c_n\}$ be monotonic decreasing with $\lim_{n\to\infty} c_n = 0$. Then the series $\sum(-1)^{n+1}c_n$ is convergent.*

Proof The series has the form

$$c_1 - c_2 + c_3 - c_4 + c_5 - c_6 + \cdots,$$

where all the numbers c_k are positive, but decrease with limit 0. The partial sums of such a series behave in a special way; they obey the inequalities

$$A_2 \le A_4 \le A_6 \le \ldots \le A_{2n} \le \ldots,\ A_{2n+1} \le \ldots \le A_5 \le A_3 \le A_1,\ A_{2n} \le A_{2n+1}.$$

To see this, note that $A_{2n+1} = A_{2n} + c_{2n+1}$, so that $A_{2n+1} \ge A_{2n}$. Also, note that $A_{2n+2} = A_{2n} + (c_{2n+1} - c_{2n+2})$ and that $c_{2n+1} - c_{2n+2} \ge 0$, so that $A_{2n} \le A_{2n+2}$. In the same way, we see that $A_{2n+1} \le A_{2n-1}$.

Thus, the even partial sums form a bounded increasing sequence, and the odd partial sums form a bounded decreasing sequence. Both must therefore converge. Since $A_{2n+1} - A_{2n} = c_{2n+1}$, which has limit 0, we have $\lim A_{2n+1} = \lim A_{2n}$, and the original series converges.

This test can be used to show that each of the following series converges:

$$1 - \frac{1}{\sqrt{2}} + \frac{1}{\sqrt{3}} - \frac{1}{\sqrt{4}} + \cdots + \frac{(-1)^{n+1}}{\sqrt{n}} + \cdots,$$

$$\frac{2}{(1)(3)} - \frac{(2)(4)}{(1)(3)(5)} + \frac{(2)(4)(6)}{(1)(3)(5)(7)} - \frac{(2)(4)(6)(8)}{(1)(3)(5)(7)(9)} + \cdots.$$

Another useful test, which can be applied to series with both positive and negative terms, is the ratio test.

Theorem 9 *If $a_n \ne 0$ and $\lim |a_{n+1}|/|a_n| = L$, $L < 1$, then the series $\sum a_n$ converges. If $L > 1$, then the series diverges. (If $L = 1$, the series might do either.)*

Proof Clearly, if $L > 1$, then for large n, $|a_{n+1}|/|a_n| > 1$ and the sequence $\{|a_k|\}$ is monotonic increasing. Certainly then, we do not have $\lim a_n = 0$, and the series diverges (by Theorem 3). Suppose then that $L < 1$. Choose r with $L < r < 1$. Then, for large n, we must have $|a_{n+1}|/|a_n| \le r$. This can be rewritten as

$$\frac{|a_{n+1}|}{|a_n|} \le \frac{r^{n+1}}{r^n},$$

or as

$$\frac{|a_{n+1}|}{r^{n+1}} \le \frac{|a_n|}{r^n}.$$

This shows that the sequence $\{|a_n|/r^n\}$ is positive and decreasing. It must therefore be bounded, so that $|a_n|/r^n \le B$ for all n. Accordingly, we have $|a_n| \le Br^n$, and since $\sum Br^n$ converges when $r < 1$, the comparison test (Theorem 5) applies and $\sum |a_n|$ converges. By Theorem 7, so does $\sum a_n$.

The importance of infinite series lies largely in their use as representations for functions, especially by power series.

Definition 2 A series of the form $\sum_0^\infty a_n x^n$, or more generally $\sum_0^\infty a_n(x - c)^n$, is called a power series. The convergence set is the set of all x for which such a series converges. On this set, the series defines a function $f(x)$.

Theorem 10 *The set of convergence of a power series is either a bounded interval, possibly with one or both of its endpoints, or the infinite interval $-\infty < x < \infty$. Specifically, a power series $\sum a_n x^n$ will converge for all x such that $-R < x < R$, where $R \leq \infty$ is called the radius of convergence. It will diverge for x satisfying $|x| > R$. It may converge or it may diverge at the points $x = R$, $x = -R$.*

Proof The series $\sum_0^\infty a_n x^n$ clearly converges when $x = 0$. Suppose that it also converges for a value $x = x_0$. By Theorem 3, the sequence of terms $\{a_n {x_0}^n\}$ must converge to 0 and must therefore be bounded: $|a_n {x_0}^n| \leq B$ for all n. Let x be any number with $|x| < |x_0|$. Then,

$$\begin{aligned} |a_n x^n| &= |a_n {x_0}^n| r^n \\ &\leq B r^n, \end{aligned}$$

where $r = |x|/|x_0| < 1$. By the comparison test (Theorem 5), we conclude that $\sum |a_n x^n|$ converges, and x belongs to the convergence set for $\sum a_n x^n$. Let R be the least upper bound of the set of all numbers $x_0 > 0$ for which the series $\sum a_n x^n$ converges. If R is infinite, so that the set of x_0 has no upper bound, then the previous argument shows that the series converges for all numbers x, and the convergence set is $-\infty < x < \infty$. If R is finite, then the argument shows that the series converges for all x with $-R < x < R$ and that it diverges for all $x > R$ and for all $x < -R$. The argument gives no information about the behavior of the series for $x = R$ or for $x = -R$, where the series may converge or diverge, depending on its detailed nature.

The ratio test is often a convenient method for finding R. For example, consider the series

$$\sum_0^\infty n x^n = x + 2x^2 + 3x^3 + 4x^4 + \cdots.$$

Applying the test, we examine

$$\begin{aligned} \lim_{n\to\infty} \frac{|a_{n+1}x^{n+1}|}{|a_n x^n|} &= \lim_{n\to\infty} \frac{|(n+1)x^{n+1}|}{|nx^n|} \\ &= \lim_{n\to\infty} \frac{n+1}{n}|x| = |x|. \end{aligned}$$

Hence, with $|x| = L$ as above, we see that the series converges for all x with $|x| < 1$ and diverges for all x with $|x| > 1$ $(R = 1)$. Testing the series at $x = 1$, or at $x = -1$, we obtain the series

$1 + 2 + 3 + \cdots$ and $-1 + 2 - 3 + 4 - 5 + \cdots$, both of which diverge. Hence, the convergence set is $-1 < x < 1$.

The following result is an example of the usefulness of power series.

Theorem 11 *Power series can be differentiated termwise. That is, if*

$$f(x) = a_0 + a_1(x - c) + a_2(x - c)^2 + \cdots$$

and if this series converges for all x with $-R < x - c < R$, then the function f is continuous on this set and is of class C^∞ (i.e., has continuous derivatives of all orders), and

$$f'(x) = a_1 + 2a_2(x - c) + 3a_3(x - c)^2 + \cdots,$$
$$f''(x) = 2a_2 + (2)(3)a_3(x - c) + (3)(4)a_4(x - c)^2 + \cdots,$$

etc. Each of these power series also converges for those x with $-R < x - c < R$.

We omit the proof of this theorem, but suggest any text on advanced calculus or the theory of series for a proof. One immediate consequence is that the coefficients in the power series for f can be determined at once by setting $x = c$ in each of the resulting series for the derivatives of f. The result is Taylor's series for f:

$$f(x) = f(c) + \frac{f'(c)}{1!}(x - c) + \frac{f''(c)}{2!}(x - c)^2 + \cdots$$
$$= \sum_{k=0}^{\infty} f^{(k)}(c)(x - c)^k/k!$$

It is therefore also possible to start with a function f and construct its associated Taylor series by this formula. One may then test this series for convergence, and in some cases, show that the function it converges to is indeed the function f. For example, let $f(x) = e^x$. Since $f'(x) = e^x$, we find $f^{(k)}(0) = 1$ for all k, and the resulting series is

$$1 + x + \frac{x^2}{2!} + \frac{x^3}{3!} + \cdots + \frac{x^n}{n!} + \cdots.$$

The ratio test shows that this converges for all x, $-\infty < x < \infty$. If we differentiate this series, the result is the same series. Since e^x is the only function $f(x)$ (see Theorem 1, Chapter 5) that has the property that $f'(x) = f(x)$ for all x and that $f(0) = 1$, we know that the series must converge to the function e^x.

Power series can be regarded as generalized polynomials. If we are careful to use only those values of x for which the series in question converges, we can add and multiply power series as we would polynomials. For example,

$$\begin{aligned} e^x + e^{-x} &= 1 + x + x^2/2! + x^3/3! + x^4/4! + \cdots \\ &\quad + 1 - x + x^2/2! - x^3/3! + x^4/4! - \cdots \\ &= 2 + 2(x^2)/2! + 2(x^4)/4! + 2(x^6)/6! + \cdots, \end{aligned}$$

so that

$$\begin{aligned} \cosh x = \frac{e^x + e^{-x}}{2} &= 1 + x^2/2! + x^4/4! + \cdots \\ &= \sum_0^\infty \frac{x^{2n}}{(2n)!}. \end{aligned}$$

Again, we have (from the familiar geometric series)

$$\frac{1}{1-x} = 1 + x + x^2 + x^3 + x^4 + \cdots.$$

Then,

$$\begin{aligned} \frac{1+x}{1-x} &= (1+x)(1 + x + x^2 + x^3 + x^4 + \cdots) \\ &= 1 + x + x^2 + x^3 + x^4 + \cdots \\ &\quad + x + x^2 + x^3 + x^4 + x^5 + \cdots \\ &= 1 + 2x + 2x^2 + 2x^3 + 2x^4 + \cdots. \end{aligned}$$

Finally,

$$\begin{aligned} \frac{e^{-x}}{1-x} &= \frac{1}{1-x} e^{-x} \\ &= (1 + x + x^2 + x^3 + x^4 + \cdots) \\ &\quad \times (1 - x + x^2/2! - x^3/3! + \cdots) \\ &= 1 - x + x^2/2! - x^3/3! + x^4/4! - \cdots \\ &\quad + x - x^2 + x^3/2! - x^4/3! + \cdots \\ &\quad + x^2 - x^3 + x^4/2! - x^5/3! + \cdots \\ &\quad + x^3 - x^4 + x^5/2! - \cdots \\ &\quad + x^4 - x^5 + \cdots \\ &= 1 + 0x + x^2/2 + x^3/3 + (\tfrac{3}{8})x^4 + \cdots. \end{aligned}$$

Lest the reader get the impression that common sense can totally replace rigor when we are dealing with infinite series, we append two instructive examples. Let us define two functions by means of series:

$$\begin{aligned} f(x) &= x + x(1-x) + x(1-x)^2 + x(1-x)^3 + \cdots, \\ g(x) &= 1 + x^2 + (2x-1)x^2 + (3x-2)x^3 + (4x-3)x^4 + \cdots. \end{aligned}$$

We want to calculate two limits, $\lim_{x \downarrow 0} f(x)$, and $\lim_{x \uparrow 1} g(x)$. In the first series, each term becomes 0 when $x = 0$, so that it would seem plausible that

$$\lim_{x \downarrow 0} f(x) = 0 + 0 + 0 + \cdots = 0.$$

In the same way, each term in the second series becomes 1 when $x = 1$, so that it would seem plausible that

$$\lim_{x \uparrow 1} g(x) = 1 + 1 + 1 + \cdots = \infty.$$

However, by writing out the terms and cancelling, it can be seen that the sum of each series on the interval $0 < x < 1$ is 1. Since $f(x) = 1$ for all such x, $\lim_{x \downarrow 0} f(x) = 1$, and since $g(x) = 1$ for all such x, $\lim_{x \uparrow 1} g(x) = 1$. (To check these assertions, calculate the partial sums of each series, and then compute their limits.)

Finally, we give one more illustration to suggest caution in working with infinite sums. Given the task of adding all the numbers in the infinite display shown below, one might try several different procedures.

(i) One might add each row, and then add the results.
(ii) One might add each column, and then add these results.
(iii) One might start at the upper left corner and move downward along the diagonal, at each stage adding all the entries in the upper lefthand square. (For example, at the third step, our partial sum would be obtained by adding all the entries in the 3 by 3 square

$$\begin{bmatrix} 1 & -2 & 0 \\ -3 & 5 & -2 \\ 0 & -3 & 5 \end{bmatrix},$$

which gives us 1.)

Please observe that in the array below, each procedure gives a different answer!

$$\begin{bmatrix} 1 & -2 & 0 & 0 & 0 & 0 & 0 & \cdot & \cdot \\ -3 & 5 & -2 & 0 & 0 & 0 & 0 & \cdot & \cdot \\ 0 & -3 & 5 & -2 & 0 & 0 & 0 & \cdot & \cdot \\ 0 & 0 & -3 & 5 & -2 & 0 & 0 & \cdot & \cdot \\ 0 & 0 & 0 & -3 & 5 & -2 & 0 & \cdot & \cdot \\ 0 & 0 & 0 & 0 & -3 & 5 & -2 & \cdot & \cdot \\ \cdot & \cdot & \cdot & \cdot & \cdot & \cdot & \cdot & \cdot & \cdot \\ \cdot & \cdot & \cdot & \cdot & \cdot & \cdot & \cdot & \cdot & \cdot \end{bmatrix}$$

Exercises

Discuss the convergence or divergence of each of the following series.

1 $\frac{3}{10} + \frac{4}{20} + \frac{5}{30} + \frac{6}{40} + \cdots$ (*Ans.* diverges).

2 $\frac{3}{4} + \frac{5}{8} + \frac{7}{16} + \frac{9}{32} + \frac{11}{64} + \frac{13}{128} + \cdots$ (*Ans.* converges).

3 $\frac{3}{4} + \frac{5}{9} + \frac{7}{16} + \frac{9}{25} + \frac{11}{36} + \frac{13}{49} + \frac{15}{64} + \cdots$ (*Ans.* diverges).

4 $\frac{1}{6} + \frac{\sqrt{2}}{12} + \frac{\sqrt{3}}{20} + \frac{\sqrt{4}}{30} + \frac{\sqrt{5}}{42} + \frac{\sqrt{6}}{35} + \cdots$ (*Ans.* converges).

5 $\sum_1^\infty \frac{6n}{1+n^3}$ (*Ans.* converges).

6 $\sum_1^\infty \frac{7+5n+n^2}{1+n^2+n^4}$ (*Ans.* converges).

7 $\sum_1^\infty \frac{2+5n}{3+4n+5n^2}$ (*Ans.* diverges).

8 $\sum_1^\infty \frac{(-1)^n}{1+3n}$ (*Ans.* converges).

9 $\frac{1}{2} - \frac{3}{4} + \frac{1}{8} - \frac{3}{16} + \frac{1}{32} - \frac{3}{64} + \cdots$ (*Ans.* converges).

10 $\frac{1}{3} - \frac{2}{5} + \frac{3}{7} - \frac{4}{9} + \frac{5}{11} - \frac{6}{13} + \cdots$ (*Ans.* diverges).

For which values of x do the following series converge?

11 $\frac{1}{2} + \frac{1\cdot 3}{2\cdot 4}x + \frac{1\cdot 3\cdot 5}{2\cdot 4\cdot 6}x^2 + \frac{1\cdot 3\cdot 5\cdot 7}{2\cdot 4\cdot 6\cdot 8}x^3 + \frac{1\cdot 3\cdot 5\cdot 7\cdot 9}{2\cdot 4\cdot 6\cdot 8\cdot 10}x^4 + \cdots$ (for all x, $-1 \le x < 1$).

12 $\sum_1^\infty \frac{x^n}{n^2}$ (for all x, $-1 \le x \le 1$).

13 $\sum_0^\infty (2^n + 1)x^n$ (for all x, $-\frac{1}{2} < x < \frac{1}{2}$).

14 $\sum_1^\infty n!x^n$ (only for $x = 0$).

15 $\sum_1^\infty \frac{(x-1)^n}{n^2+n}$ (for all x, $0 \le x \le 2$).

16 $\sum_1^\infty \frac{(x+3)^n}{(2n)!}$ (this converges for all x).

17 Granted that $\sum_0^\infty x^n = 1/(1-x)$, show that $\sum_1^\infty nx^n = x/(1-x)^2$ and that $\sum_1^\infty n/2^n = 2$.

18 Continuing, study the series $\sum_1^\infty n^2x^n$, convergent for all x, $-1 < x < 1$, and show that $\sum_1^\infty n^2/2^n = 6$.

19 Show that

$$\frac{e^{-x}}{1-x^2} = 1 - x + \tfrac{3}{2}x^2 - \tfrac{7}{6}x^3 + \tfrac{37}{24}x^4 - \tfrac{141}{120}x^5 + \cdots.$$

20 Show that

$$\left(\frac{1\cdot 3}{4}\right)\left(\frac{2\cdot 4}{9}\right)\left(\frac{3\cdot 5}{16}\right)\left(\frac{4\cdot 6}{25}\right)\left(\frac{5\cdot 7}{36}\right)\left(\frac{6\cdot 8}{49}\right)\cdots = \frac{1}{2}.$$

complex numbers and functions

appendix **2**

A student's first introduction to complex numbers has usually been given by a teacher who said something like this: We know that x^2 can never be negative—the square of a negative number is positive and the square of a positive is positive. So, the equation $x^2 = -1$ doesn't have any solutions. Let us invent a number i with the property that $i^2 = -1$. Then, numbers like $a + bi$, where a and b are ordinary numbers, will be called complex numbers. The laws for working with these new numbers are the same ones that we already know from algebra; for example, $(2 - 3i)(4 + i) = 8 - 12i + 2i - 3i^2 = 11 - 10i$.

This approach, while aping the historical process, leaves unanswered many reasonable questions: What is i, really? What justification is there for creating something like this? How do we know it obeys all the rules of arithmetic?

The more modern approach, by now quite common, was introduced by Hamilton (c. 1840). The field of complex numbers can be taken to be the ordinary two-dimensional plane, all points of which are given as pairs (a, b) with a special definition of the operations of addition and multiplication. The sum of two points P and Q is defined in the usual vector way: If $P = (a, b)$ and $Q = (c, d)$, then $P + Q = (a + c, b + d)$. The product of P and Q is the point whose coordinates are $(ac - bd, ad + bc)$. Using these definitions, it is merely a tedious calculation to verify that all the rules of arithmetic that describe a field are satisfied. A crucial step is to verify that if (a, b) is not the origin $(0, 0)$, then there is a (unique) point

(x, y) such that $(a, b)(x, y) = (c, d)$ for any choice of (c, d). Using the definition of multiplication, one must solve the equations

$$ax - by = c,$$
$$bx + ay = d,$$

obtaining $x = (ac + bd)/(a^2 + b^2)$, $y = (ad - bc)/(a^2 + b^2)$, which are defined, since $a^2 + b^2 \neq 0$.

With this approach, a correct answer to the question, *What is i?* is given by the answer "$i = (0, 1)$." In passing, it should be mentioned that other models for the complex numbers can also be given. Even though they look entirely different, all the models are isomorphic—meaning that they can be put in a one-to-one structure-preserving correspondence with the points of the plane. One such additional model consists of all the 2 by 2 matrices of the form

$$\begin{bmatrix} a & -b \\ b & a \end{bmatrix},$$

where a and b are ordinary (real) numbers.

From a naive viewpoint, the "number" i was created to enable us to solve the equation $x^2 + 1 = 0$. It is a happy accident that the complex numbers which resulted then allowed us to solve any quadratic equation of the form $ax^2 + bx + c = 0$, where the coefficients a, b, c are real numbers. One might have expected that a new collection of numbers would have to be invented to solve an equation like $x^2 = i$; however, this is solved by

$$x = \pm\{(1/\sqrt{2}) + (1/\sqrt{2})i\}.$$

It was a pleasant surprise to learn indeed that one did not have to go outside the complex numbers to solve *any* quadratic equation with arbitrary complex coefficients, such as $x^2 + (3 - 7i)x + (4 + 5i) = 0$. But it was astounding to discover that this extension of the number system made it possible to find solutions of every polynomial equation of any degree. This result, called by tradition the "fundamental theorem of algebra"—which it certainly is not—does not yet have a proof that is sufficiently easy to present in an elementary class.

Another simple algebraic aspect of the field of complex numbers is the fact that it has a self-isomorphism, the mapping that sends the complex number $a + bi$ into its **conjugate**, the complex number $a - bi$. Thus, the conjugate of $-2 + 4i$ is $-2 - 4i$ and the conjugate of $3 - 7i$ is $3 + 7i$; of course, the conjugate of the real number 5 is 5, since $5 = 5 + 0\mathrm{i}$. If z is any complex number, its conjugate is usually denoted by $\bar{z}$.

The reason for the importance of this concept is contained in the word "isomorphism," which means that the mapping preserves the algebraic structure of the field of complex numbers. If we write $\bar{z} = \mathbf{h}(z)$, then it can be checked easily that $\mathbf{h}$ satisfies the identities

$$\mathbf{h}(\alpha - \beta) = \mathbf{h}(\alpha) - \mathbf{h}(\beta),$$
$$\mathbf{h}(\alpha\beta) = \mathbf{h}(\alpha)\mathbf{h}(\beta),$$
$$\mathbf{h}\left(\frac{1}{\gamma}\right) = \frac{1}{\mathbf{h}(\gamma)}$$

for any complex numbers α, β, γ with $\gamma \neq 0$.

There is one immediate corollary of this isomorphism which is important to the solution of polynomial equations and therefore to the elements of calculus. If $P(x) = a_0x^n + a_1x^{n-1} + \cdots + a_n$ is any polynomial with all its coefficients *real* and γ is any root of the equation $P(x) = 0$, then the complex number $\bar{\gamma}$ is also a root. This is proved by noting that in this case $\mathbf{h}(P(\gamma)) = P(\bar{\gamma})$, so that if $P(\gamma) = 0$, then $\mathbf{h}(P(\gamma)) = \mathbf{h}(0) = 0$. Therefore $\bar{\gamma}$ is also a root.

It is customary to concentrate on real-valued functions in developing the elements of calculus. However, it takes very little work to extend everything needed to make it possible to deal with functions that yield complex values instead. For example, the function defined by

$$f(x) = (2 - 3i)x^2 + 5ix + (7 - i) \tag{1}$$

is a well-defined function, defined for every (real) value of x, and the values it yields are complex numbers. We note that we could rewrite this formula as

$$f(x) = 2x^2 + 7 + i\{-3x^2 + 5x - 1\}. \tag{2}$$

In the same way, any complex-valued function of one variable can be written in the form

$$f(x) = g_1(x) + ig_2(x),$$

where g_1 and g_2 are ordinary real-valued functions of x.

Differentiation and integration can be carried out as usual. For example, we have either

$$f'(x) = 2(2 - 3i)x + 5i$$

using the equation (1), or

$$f'(x) = g_1'(x) + ig_2'(x) = 4x + 0 + i\{-6x + 5\}$$

using formula (2).

Integration can be carried out the same way, either leaving the complex-valued function as it is, or separating it into real and imaginary parts. Thus, we have

$$\int_{-1}^{2} \{(2-3i)x^2 + 5ix + (7-i)\}\, dx$$
$$= \left\{(2-3i)\frac{x^3}{3} + 5i\frac{x^2}{2} + (7-i)x\right\}\Bigg|_{-1}^{2}$$
$$= (2-3i)\left(\frac{8}{3} - \frac{-1}{3}\right) + 5i\left(\tfrac{4}{2} - \tfrac{1}{2}\right) + (7-i)(2-(-1))$$
$$= 6 - 9i + \tfrac{15}{2}i + 21 - 3i$$
$$= 27 - \tfrac{9}{2}i.$$

It is also useful to develop a theory to handle infinite series whose terms are complex numbers. Very little of the theory of series of real numbers need be modified. If α_n is the complex number $a_n + ib_n$, then the series

$$\sum_1^\infty \alpha_n = \sum_1^\infty (a_n + ib_n) = (a_1 + ib_1) + (a_2 + ib_2) + \cdots$$

is convergent if and only if the two series $\sum_1^\infty a_n$ and $\sum_1^\infty b_n$ are convergent, and then one has

$$\sum_1^\infty \alpha_n = \sum_1^\infty a_n + i\sum_1^\infty b_n.$$

The definition of the general exponential function is a very important special application of the above. If z is any complex number, then

$$\exp(z) = 1 + z + \frac{z^2}{2!} + \frac{z^3}{3!} + \frac{z^4}{4!} + \cdots. \tag{3}$$

A special identity, originally discovered by De Moivre (c. 1700), can be obtained from (3) by setting $z = i\theta$. Observe that $(i\theta)^n$ is always either θ^n, $i\theta^n$, $-\theta^n$, or $-i\theta^n$, depending upon the remainder after dividing n by 4; then

$$\exp(i\theta) = 1 + i\theta + \frac{(i\theta)^2}{2!} + \frac{(i\theta)^3}{3!} + \frac{(i\theta)^4}{4!} + \frac{(i\theta)^5}{5!} + \cdots$$
$$= 1 + i\theta - \frac{\theta^2}{2!} - i\frac{\theta^3}{3!} + \frac{\theta^4}{4!} + i\frac{\theta^5}{5!} - \frac{\theta^6}{6!} - \cdots$$
$$= \left\{1 - \frac{\theta^2}{2!} + \frac{\theta^4}{4!} - \frac{\theta^6}{6!} + \cdots\right\} + i\left\{\theta - \frac{\theta^3}{3!} + \frac{\theta^5}{5!} - \frac{\theta^7}{7!} + \cdots\right\}$$
$$= \cos\theta + i\sin\theta,$$

where we have replaced these well-known power series by their sums. If we now replace θ by $-\theta$, we obtain

$$\begin{aligned}\exp(-i\theta) &= \cos(-\theta) + i\sin(-\theta)\\ &= \cos\theta - i\sin\theta.\end{aligned}$$

One at once obtains the familiar identities

$$\cos\theta = \frac{\exp(i\theta) + \exp(-i\theta)}{2},$$

$$\sin\theta = \frac{\exp(i\theta) - \exp(-i\theta)}{2i}.$$

Another interesting identity can be obtained from the series (3) for the exponential function. Suppose we calculate the product $\exp(u)\exp(v)$, using the series for each. Thus,

$$\exp(u) = 1 + u + \frac{u^2}{2!} + \frac{u^3}{3!} + \frac{u^4}{4!} + \frac{u^5}{5!} + \cdots,$$

$$\exp(v) = 1 + v + \frac{v^2}{2!} + \frac{v^3}{3!} + \frac{v^4}{4!} + \frac{v^5}{5!} + \cdots,$$

$$\begin{aligned}\exp(u)\exp(v) &= 1 + (u+v) + \left(\frac{u^2}{2!} + uv + \frac{v^2}{2!}\right) + \left(\frac{u^3}{3!} + \frac{u^2v}{2!} + \frac{uv^2}{2!} + \frac{v^3}{3!}\right)\\ &\quad + \left(\frac{u^4}{4!} + \frac{u^3v}{3!1!} + \frac{u^2v^2}{2!2!} + \frac{uv^3}{1!3!} + \frac{v^4}{4!}\right) + \cdots\\ &= 1 + (u+v) + \frac{1}{2!}(u^2 + 2uv + v^2)\\ &\quad + \frac{1}{3!}(u^3 + 3u^2v + 3uv^2 + v^3)\\ &\quad + \frac{1}{4!}(u^4 + 4u^3v + 6u^2v^2 + 4uv^3 + v^4) + \cdots\\ &= 1 + (u+v) + \frac{(u+v)^2}{2!} + \frac{(u+v)^3}{3!} + \frac{(u+v)^4}{4!} + \cdots\\ &= \exp(u+v).\end{aligned}$$

It is this formula, $\exp(u)\exp(v) = \exp(u+v)$, which corresponds to the law of exponents and which justifies our adoption of the notation e^z for $\exp(z)$. For, the identity just proved thus becomes

$$e^u e^v = e^{u+v},$$

holding now for any real or complex numbers u and v.

An interesting special case casts light on the mathematical origins of the trigonometric addition formulas. We have $e^{i\alpha}e^{i\beta} = e^{i(\alpha+\beta)}$. Rewritten in the De Moivre form, it becomes

$$(\cos\alpha + i\sin\alpha)(\cos\beta + i\sin\beta) = \cos(\alpha+\beta) + i\sin(\alpha+\beta).$$

Calculating the product on the left, we have

$$\cos\alpha\cos\beta - \sin\alpha\sin\beta + i(\sin\alpha\cos\beta + \cos\alpha\sin\beta) = \cos(\alpha+\beta) + i\sin(\alpha+\beta),$$

from which we deduce at once that

$$\cos(\alpha+\beta) = \cos\alpha\cos\beta - \sin\alpha\sin\beta,$$
$$\sin(\alpha+\beta) = \sin\alpha\cos\beta + \cos\alpha\sin\beta.$$

Exercises

1 Show that one value for $\sqrt{3+4i}$ is $2+i$.

2 Show that one value of $\sqrt{9-12i}$ is $\sqrt{12}-\sqrt{3}\,i$.

3 Find a value for $\sqrt{1+3i}$ $\left(Ans.\ \sqrt{\frac{\sqrt{10}+1}{2}} + i\sqrt{\frac{\sqrt{10}-1}{2}}\right)$.

4 Solve the equation $x^2+(3-i)x-(5i+10)$ (*Ans.* -5 and $2+i$).

5 Solve $x^2+3x-10i$ (*Ans.* $1+2i$, $-4-2i$).

6 Let $f(x) = (3x+7i)(x^2-5ix+1)$. Find $f'(x)$ and $f''(x)$ (*Ans.* $f'(x) = 9x^2-16ix+38$, $f''(x) = 18x-16i$).

7 Evaluate $\int_0^1 (2x-i)(x+i)\,dx$ (*Ans.* $\frac{5}{3}+\frac{1}{2}i$).

8 Evaluate $\int_0^1 (t-i)(t+i)\,dt$ (*Ans.* $\frac{4}{3}$).

9 Evaluate $\int_0^{\pi/2} \{(\theta-i)\sin\theta + 2i\cos\theta\}\,d\theta$ (*Ans.* $1+i$).

10 Verify that $|e^{i\theta}| = 1$ for all real θ.

11 Show that $|e^{iy+x}| = e^x$.

12 Let $\omega = -\frac{1}{2}+(\sqrt{3}/2)i$. Verify that $\omega^3 = 1$ and that $1+\omega+\omega^2 = 0$.

13 With ω as above, show that

$$\frac{e^x+e^{\omega x}+e^{\omega^2 x}}{3} = 1+\frac{x^3}{3!}+\frac{x^6}{6!}+\frac{x^9}{9!}+\frac{x^{12}}{12!}+\cdots.$$

14 With ω as above, show that

$$\frac{1}{3}\left\{\frac{1}{1-x}+\frac{1}{1-\omega x}+\frac{1}{1-\omega^2 x}\right\} = 1+x^3+x^6+x^9+\cdots.$$

15 Show that $\sin(x+iy) = \sin x\cosh y + i\cos x\sinh y$.

calculus revisited

postlude

It is often worthwhile, at the end of a long climb, to look back over the route traveled. From a vantage point, it is sometimes easier to recognize landmarks and to grasp the pattern as a whole. We shall attempt to do this to some extent, hoping that it will help in understanding calculus; at the same time, we shall show part of the landscape that lies ahead in the direction usually labeled "advanced calculus" or "introductory analysis."

The subject matter of calculus is the study of functions. Specifically, it is the study of various linear spaces of functions or mappings, and of some of the techniques for exploring their properties. Let $\mathbb{R}$ denote the real numbers, and use $\mathbb{R}^n$ for the n-fold Cartesian product of $\mathbb{R}$ by itself; thus, $\mathbb{R}^3$ is the set of all ordered triples $x = (x_1, x_2, x_3)$ and $\mathbb{R}^n$ is the set of all $x = (x_1, x_2, \ldots, x_n)$. Geometrically, $\mathbb{R}^2$ is the plane, $\mathbb{R}^3$ is space, and $\mathbb{R}^n$ is n-space. The functions that form the central concern of the calculus are defined on subsets of $\mathbb{R}^n$ and take values in $\mathbb{R}^m$. The formula for such a function is $y = F(x)$, where $x \in \mathbb{R}^n$, $y \in \mathbb{R}^m$, and

$$
\begin{aligned}
y_1 &= f_1(x) = f_1(x_1, x_2, \ldots, x_n),\\
y_2 &= f_2(x) = f_2(x_1, x_2, \ldots, x_n),\\
\vdots\ & \quad\ \vdots \qquad\qquad \vdots\\
y_m &= f_m(x) = f_m(x_1, x_2, \ldots, x_n),
\end{aligned}
\tag{1}
$$

and where each of the functions f_j is a real-valued function defined on the same subset $D \subset \mathbb{R}^n$. Since such a family of equations defining a general function F can be viewed as displaying a transformation which sends each point x in the set D into some point y in $\mathbb{R}^m$, we also call F a transformation mapping D into $\mathbb{R}^m$. To indicate this, one sometimes writes $D \xrightarrow{F} \mathbb{R}^m$, where $D \subset \mathbb{R}^n$.

For special choices of n and m, the mapping F becomes a very familiar object. Thus, if $n = m = 1$, then F is nothing more than a real-valued function of one variable defined on some set $D \subset \mathbb{R}$; more than half the time devoted to elementary calculus is spent in learning to work with such functions. When $m = 1$ and n is unrestricted, then F is merely a real-valued function of n variables; the study of such functions has been one of the central concerns of the present text. Again, if $n = 1$ and $m = 3$, so that F is a mapping from $\mathbb{R}$ into 3-space, then F becomes the mathematical model for a point moving along a curve in space. (See Sections 1.7 and 3.1.) Finally, a sample of a mapping from $\mathbb{R}^2$ into $\mathbb{R}^2$ is given by the transformation between polar coordinates and Cartesian coordinates $T(r, \theta) = (x, y)$, where

$$T: \begin{cases} x = r\cos\theta, \\ y = r\sin\theta. \end{cases}$$

Economy is a constant characteristic of mathematics. A concept or definition ought to be selected to apply to a large number of contexts. The concept of continuity, for example, ought to be described in such a way that it makes sense for functions from $\mathbb{R}^n$ to $\mathbb{R}^m$ for any pair n and m. Thus, looking at formula (1), we would want continuity of F to mean that each of the ordinary real-valued functions f_j is continuous. Such comprehensiveness can be obtained by formulating a general notion of continuity which applies to all mappings. Intuitively, a continuous function is one that is approximately constant over a small region; if F is continuous and $y = F(x)$, then we feel that small changes in x should yield only small changes in y. One way to make this idea precise is to introduce the concept of neighborhood. A function F is continuous at a point x^* if and only if given any neighborhood $\mathcal{W}$ about the point $y^* = F(x^*)$, there exists a neighborhood $\mathcal{U}$ about x^* such that $F(x)$ lies in $\mathcal{W}$ for every choice of x in $\mathcal{U}$. Expressed in more geometric language, the mapping F must carry the entire set $\mathcal{U}$ about the point x^* into a subset of $\mathcal{W}$; in mathematical notation, $F(\mathcal{U}) \subset \mathcal{W}$. A reexamination of the above definition makes it possible to describe continuity entirely in terms of the basic notion of an open set, i.e., a set consisting only of points interior to it.

[1] ***A topological space** is any set of "points" in which neighborhoods of points, and therefore open sets, have been appropriately defined.*

Definition 1 A function F from one topological space[1] X into a topological space Y is said to be continuous if and only if the inverse image $F^{-1}(\mathcal{W})$ of every open set $\mathcal{W}$ in Y is in turn an open set in X.

Here, $F^{-1}(\mathcal{W})$ is the set of all points x in X such that $F(x)$ lies in $\mathcal{W} \subset Y$, as pictured in Figure P-1. (As a test of your understanding of this universal definition of continuity, see if you can show that it

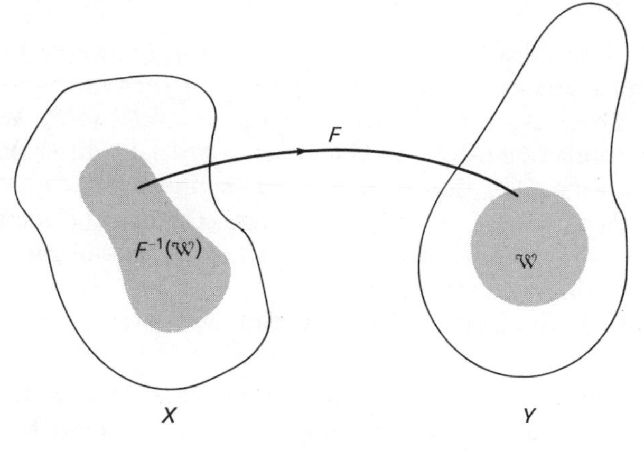

Figure P–1

is equivalent to the definition you learned earlier for functions from $\mathbb{R}$ into $\mathbb{R}$.)

As a second illustration of the search for universal concepts, let us turn to differentiation. We would like to obtain a definition which would permit us to talk about the derivative of any suitable mapping F between $\mathbb{R}^n$ and $\mathbb{R}^m$, yet which would reduce to the usual derivative when we take $n = 1, m = 1$. A key idea here is approximation. The slope of a plane curve at a point can be found by using the line which best fits the curve in a specific sense at this point. In the case of functions of one variable, one seeks a function G of the form $G(x) = a + bx$ whose graph best fits the graph of $F(x)$ at $x = x^*$. This leads to the following general definition. Give two continuous functions F and G both defined on a set $D \subset \mathbb{R}^n$ and taking values in $\mathbb{R}^m$, and a point $x^* \in D$, we say that F and G have contact at x^* of order k if

$$\lim_{x \to x^*} \frac{|F(x) - G(x)|}{|x - x^*|^k} = 0.$$

It is easy to see that contact of order 0 merely means that $F(x^*) = G(x^*)$. If F and G are real-valued functions of one variable and $G(x) = a + bx$, then F and G have contact of order 1 at x^* if and only if $F(x^*) = G(x^*)$ and $F'(x^*)$ exists and is b. Thus, in this case, the line $y = G(x) = a + bx$ is the tangent to the curve $y = F(x)$ corresponding to $x = x^*$. Using this special case as a guide, we are led to a general definition of **differentiation** of mappings as follows.

The simplest class of functions on $\mathbb{R}^n$ to $\mathbb{R}^m$ are those of degree one, also called **affine** transformations. The typical formula is

$$\begin{aligned}
y_1 &= a_1 + b_{11}x_1 + b_{12}x_2 + \cdots + b_{1n}x_n,\\
y_2 &= a_2 + b_{21}x_1 + b_{22}x_2 + \cdots + b_{2n}x_n,\\
y_3 &= a_3 + b_{31}x_1 + b_{32}x_2 + \cdots + b_{3n}x_n,\\
&\vdots \\
y_m &= a_m + b_{m1}x_1 + b_{m2}x_2 + \cdots + b_{mn}x_n.
\end{aligned}$$

(In matrix form, this might be written $y = a + bx$, where x and a are column matrices and b is the matrix (b_{ij}) with m rows and n columns.) Given a general function F from $D \subset \mathbb{R}^n$ to $\mathbb{R}^m$, we look for an affine function G which has contact of order 1 with $F(x)$ at x^*. If there is such a $G(x)$, then we say that F is differentiable at x^* and that the derivative of F at x^* is b, the matrix that is the coefficient of x in the formula for G. The graph of $G(x)$ can be thought of as a hyperplane tangent to the graph of F.

Restating this in the form of a definition, we have

Definition 2 If F is a continuous function defined on a region $D \subset \mathbb{R}^n$ with values in $\mathbb{R}^m$ and x^* is an interior point of D, then F is differentiable at x^* and has derivative b there if and only if the affine function $G(x) = a + bx$ satisfies

$$\lim_{x \to x^*} \frac{|F(x) - G(x)|}{|x - x^*|} = 0.$$

Since this also requires that F and G have contact of order 0 at x^* and that therefore $F(x^*) = G(x^*)$, it is easy to see that an equivalent formulation of this limit relation is

$$\lim_{h \to 0} \frac{|F(x^* + h) - F(x^*) - bh|}{|h|} = 0.$$

The immediate practical result of this definition is the following. If the function F is given in the form $F(x) = y$, where

$$y_j = f_j(x_1, x_2, \ldots, x_n), \quad j = 1, 2, \ldots, m,$$

as in formula (1), then the derivative of F at a point x^* is the matrix

$$dF = \begin{bmatrix} \frac{\partial y_1}{\partial x_1} & \frac{\partial y_1}{\partial x_2} & \cdots & \frac{\partial y_1}{\partial x_n} \\ \frac{\partial y_2}{\partial x_1} & \frac{\partial y_2}{\partial x_2} & & \frac{\partial y_2}{\partial x_n} \\ \vdots & \vdots & & \vdots \\ \frac{\partial y_m}{\partial x_1} & \frac{\partial y_m}{\partial x_2} & \cdots & \frac{\partial y_m}{\partial x_n} \end{bmatrix},$$

where all entries are evaluated at the point x^*. The notation is not yet standardized. Some mathematicians use dF for this matrix, and some use $F'(x^*)$, by analogy with elementary calculus; the matrix is also called by various names, such as the **differential** of F or the **total derivative** of F.

To relate this to the concepts that have been studied in this book, let us examine two special cases. First, consider a single function of three variables, $y = f(x_1, x_2, x_3)$, which we can regard as a mapping from $\mathbb{R}^3$ into $\mathbb{R}$. Then, the differential of the mapping f is a matrix with one row and three columns:

$$df = \left[\frac{\partial y}{\partial x_1}, \frac{\partial y}{\partial x_2}, \frac{\partial y}{\partial x_3}\right].$$

If we rewrite the mapping as $u = f(x, y, z)$, which merely relabels the variables, then we have

$$df = [f_x, f_y, f_z],$$

and we recognize that the differential of f (or the total derivative of f) is exactly the same as the gradient of f.

As a second special case, consider the mapping F from $\mathbb{R}$ into $\mathbb{R}^3$ defined by the equations

$$\begin{aligned} y_1 &= f_1(t), \\ y_2 &= f_2(t), \\ y_3 &= f_3(t). \end{aligned}$$

Then the differential of F (or total derivative of F) is a matrix with three rows and one column:

$$dF = \begin{bmatrix} \dfrac{dy_1}{dt} \\ \dfrac{dy_2}{dt} \\ \dfrac{dy_3}{dt} \end{bmatrix} = \begin{bmatrix} f_1'(t) \\ f_2'(t) \\ f_3'(t) \end{bmatrix}.$$

If we view the mapping F as yielding a curve in space, then we see that dF yields the tangent vector to the curve.

The study of the total derivative of a transformation and its relationship to properties of the transformation is one of the central topics in advanced analysis, and it is not possible to give an outline here. However, one example can be given that may help to show how this more general approach simplifies the treatment of some topics discussed in this text.

Consider the following equations:

$$\begin{gathered} w = F(u, v), \\ u = f(x, y), \quad v = g(x, y). \end{gathered}$$

The combination of these equations yields a new equation which expresses w in terms of x and y, namely

$$\begin{aligned} w &= F(f(x, y), g(x, y)) \\ &= G(x, y). \end{aligned}$$

In fact, if we define a transformation T from $\mathbb{R}^2$ into $\mathbb{R}^2$ by $T(x, y) = (u, v)$, where

$$T: \begin{cases} u = f(x, y), \\ v = g(x, y), \end{cases}$$

then we can write $G(x, y) = F(T(x, y))$, or even $G = F \circ T$, where $\circ$ is the operation of composition of functions.

In elementary calculus, the "function of a function rule" or "chain rule" gave us a formula for calculating the derivative of a composition of two or more functions. It is natural to ask if there is such a rule that works as simply in the general case. Can we find $dG = d(F \circ T)$ in terms of dF and dT? Let us first use what we know to find dG.

From the basic definition of total derivative, we have

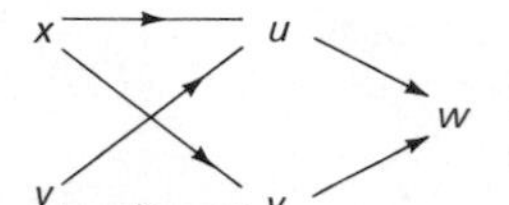

Figure P–2

$$dG = [G_x, G_y] = \left[\frac{\partial w}{\partial x}, \frac{\partial w}{\partial y}\right].$$

Now, the relationship between the variables x, y, u, v, w can be displayed as in Figure P–2. By the techniques given in Section 3.4, we can find the partial derivatives of w with respect to x and y in terms of the partial derivatives of w with respect to u and v. For example,

$$\frac{\partial w}{\partial x} = \frac{\partial w}{\partial u}\frac{\partial u}{\partial x} + \frac{\partial w}{\partial v}\frac{\partial v}{\partial x}.$$

The final result is

$$dG = \left[\frac{\partial w}{\partial u}\frac{\partial u}{\partial x} + \frac{\partial w}{\partial v}\frac{\partial v}{\partial x}, \frac{\partial w}{\partial u}\frac{\partial u}{\partial y} + \frac{\partial w}{\partial v}\frac{\partial v}{\partial y}\right].$$

Suppose that we now write the expressions for dF and dT. Since $w = F(u, v)$, we have

$$dF = \left[\frac{\partial w}{\partial u}, \frac{\partial w}{\partial v}\right].$$

Since $T(x, y) = (u, v)$, we have

$$dT = \begin{bmatrix} \dfrac{\partial u}{\partial x} & \dfrac{\partial u}{\partial y} \\ \dfrac{\partial v}{\partial x} & \dfrac{\partial v}{\partial y} \end{bmatrix}.$$

Comparing these formulas with the formula above for dG, we observe a very simple relationship, given by the matrix equation

$$dG = dF\,dT = \left[\frac{\partial w}{\partial u}, \frac{\partial w}{\partial v}\right]\begin{bmatrix} \dfrac{\partial u}{\partial x} & \dfrac{\partial u}{\partial y} \\ \dfrac{\partial v}{\partial x} & \dfrac{\partial v}{\partial y} \end{bmatrix}.$$

In this particular example, we have verified that if $G = F \circ T$, then $dG = dF\,dT$. This, indeed, is the general chain rule for differentiation! It embraces all the special cases which we discussed in Section 3.4, and many others. All fall under the simple formula $d(F \circ T) = dF\,dT$, where the usual rule for multiplying matrices is used.

One more illustration of this formula may be helpful. Suppose we consider the same transformation T mapping (x, y) into (u, v) and consider a new function H on $\mathbb{R}$ to $\mathbb{R}^2$ given by $H(t) = (x, y) = (\varphi(t), \psi(t))$. The composite $T \circ H = K$ is the mapping sending t into (u, v) defined by

$$K(t) = T(H(t)) = T(\varphi(t), \psi(t)) = (u, v),$$

or

$$u = f(x, y) = f(\varphi(t), \psi(t)),$$
$$v = g(x, y) = g(\varphi(t), \psi(t)).$$

From this pair of equations, we can calculate dK, using the rules in Section 3.4:

$$dK = \begin{bmatrix} \dfrac{du}{dt} \\ \dfrac{dv}{dt} \end{bmatrix} = \begin{bmatrix} \dfrac{\partial u}{\partial x}\dfrac{dx}{dt} + \dfrac{\partial u}{\partial y}\dfrac{dy}{dt} \\ \dfrac{\partial v}{\partial x}\dfrac{dx}{dt} + \dfrac{\partial v}{\partial y}\dfrac{dy}{dt} \end{bmatrix}.$$

Comparing this result with the expressions for dT and dH, we have

$$dT = \begin{bmatrix} \dfrac{\partial u}{\partial x} & \dfrac{\partial u}{\partial y} \\ \dfrac{\partial v}{\partial x} & \dfrac{\partial v}{\partial y} \end{bmatrix},$$

$$dH = \begin{bmatrix} \dfrac{dx}{dt} \\ \dfrac{dy}{dt} \end{bmatrix};$$

and as before, we find that $dK = dT\,dH$, where we have multiplied the matrices in the usual way.

The use of total derivatives of general functions to obtain a simple and universal form of the chain rule for differentiation is one way in which the more modern approach to mathematics has brought unity and coherence to a chaotic mélange of many seemingly unrelated special theories. There are many other examples, all derived from the fact that an affine transformation which approximates a general mapping F must also share many other local properties of F. Some have to do with the way in which mappings alter areas and volumes of sets or the shape of curves and surfaces. One important illustration concerns the technique for making changes of variable in multiple integrals. Recall again the formula for polar coordinates:

$$x = r\cos\theta,$$
$$y = r\sin\theta.$$

We know that "$dx\,dy$" is to be replaced by "$r\,dr\,d\theta$" in changing a double integral from Cartesian coordinates to polar coordinates; the need for the factor "r" is usually explained in terms of geometric diagrams. However, there is in fact another purely analytic reason. Let us calculate the total derivative of the mapping T sending (r, θ) into (x, y). We have

$$dT = \begin{bmatrix} \dfrac{\partial x}{\partial r} & \dfrac{\partial x}{\partial \theta} \\ \dfrac{\partial y}{\partial r} & \dfrac{\partial y}{\partial \theta} \end{bmatrix} = \begin{bmatrix} \cos\theta & -r\sin\theta \\ \sin\theta & r\cos\theta \end{bmatrix}.$$

Then, note that the determinant of this 2 by 2 matrix is

$$\det(dT) = J = r\cos^2\theta + r\sin^2\theta = r,$$

the mysterious factor required for the change to polar coordinates. The above extends to the general case. A change of coordinates given by a mapping T is accompanied by a replacement in the integration formula, with a multiplicative factor J which is the determinant of the matrix dT; the factor J is also called the Jacobian of T. As an example, the substitution

$$T\colon \begin{cases} x = u^2 - v^2, \\ y = uv \end{cases}$$

requires that one replace $dx\,dy$ by $2(u^2 + v^2)\,du\,dv$, because

$$dT = \begin{bmatrix} 2u & -2v \\ v & u \end{bmatrix}$$

and $J = \det(dT) = 2u^2 + 2v^2$.

There is yet another way to arrive at the correct formula in such cases. We use a tiny portion of a vast generalization of the concept of integration, called the theory of **differential forms**. Using this last illustration, suppose we write $x = u^2 - v^2$, and then $dx = 2u\,du - 2v\,dv$, by analogy with what is done in elementary calculus. Similarly, from $y = uv$, we obtain $dy = v\,du + u\,dv$. Then, proceeding as in ordinary algebra, we have

$$\begin{aligned} dx\,dy &= (2u\,du - 2v\,dv)(v\,du + u\,dv) \\ &= 2uv\,du\,du - 2v^2\,dv\,du + 2u^2\,du\,dv - 2uv\,dv\,dv. \end{aligned}$$

Suppose now that we impose two algebraic rules: (i) $du\,du = 0$ and $dv\,dv = 0$, (ii) $dv\,du = -du\,dv$. Then we have

$$\begin{aligned} dx\,dy &= 2uv(0) - 2v^2(-du\,dv) + 2u^2\,du\,dv - 2uv(0) \\ &= (2u^2 + 2v^2)\,du\,dv, \end{aligned}$$

and we once again obtain the correct Jacobian factor. This mysterious process works in the general case. If $x = f(u, v)$, $y = g(u, v)$, then

$$dx = \frac{\partial x}{\partial u}du + \frac{\partial x}{\partial v}dv \quad \text{and} \quad dy = \frac{\partial y}{\partial u}du + \frac{\partial y}{\partial v}dv,$$

and the algebraic process yields

$$\begin{aligned} dx\,dy &= \left(\frac{\partial x}{\partial u}du + \frac{\partial x}{\partial v}dv\right)\left(\frac{\partial y}{\partial u}du + \frac{\partial y}{\partial v}dv\right) \\ &= 0 + \frac{\partial x}{\partial u}\frac{\partial y}{\partial v}du\,dv + \frac{\partial x}{\partial v}\frac{\partial y}{\partial u}dv\,du + 0 \\ &= \frac{\partial x}{\partial u}\frac{\partial y}{\partial v}du\,dv - \frac{\partial x}{\partial v}\frac{\partial y}{\partial u}du\,dv \\ &= \left(\frac{\partial x}{\partial u}\frac{\partial y}{\partial v} - \frac{\partial x}{\partial v}\frac{\partial y}{\partial u}\right)du\,dv. \end{aligned}$$

This is the same result $dx\,dy = J\,du\,dv$ that is obtained by computing the Jacobian of the transformation T, since

$$dT = \begin{bmatrix} \dfrac{\partial x}{\partial u} & \dfrac{\partial x}{\partial v} \\ \dfrac{\partial y}{\partial u} & \dfrac{\partial y}{\partial v} \end{bmatrix}.$$

We must leave further discussion of this and many other applications of the theory of differential forms to texts in advanced calculus and beyond.

In this brief postlude we have looked back at some of the ideas developed in this book with the objective of organizing them into a more unified pattern. Throughout the book we have seen that the process of generalizing basic ideas from elementary calculus to functions from $\mathbb{R}^n$ to $\mathbb{R}^m$ introduces some new and often strange complexities. A relatively straightforward idea about functions of one variable becomes more complicated in the more general setting; a simple formula becomes a complex set of interrelated formulas. However, in the more general setting much of the additional complexity in handling ideas like continuity, differentiability, and integrability, or processes like differentiation with the chain rule and integration by substitution can be simplified by means of a more careful and systematic study of the basic algebra and the basic topology in the higher dimensional spaces.

The algebra in $\mathbb{R}^n$, which generalizes the basic algebraic tools in the real number system, is called *linear algebra*. The basic *topology* of $\mathbb{R}^n$ generalizes the kind of neighborhood manipulations in $\mathbb{R}$ which are central to a discussion of continuity and limits. A careful study of linear algebra and the basic topology of $\mathbb{R}^n$ can provide tools with which almost all of the basic notions and formulas from elementary calculus can be extended to functions from $\mathbb{R}^n$ to $\mathbb{R}^m$ with startling ease and remarkably little change in appearance.

Some ideas from linear algebra and topology have been introduced very rapidly in this postlude, but we hope that their tremendous unifying ability has shown through. A more systematic study of linear algebra and elementary topology should now become a part of your further excursions into advanced calculus.

solutions to selected exercises

chapter 1

page 4

1.1

1 (a) inside, (b) outside, (c) inside,
(d) inside, since we are considering the closed ball.

3 Tangent at the point (5/3, 4/3, 2/3).

5 (a) 1,
(b) 0. (They intersect at (10/3, −10/3, 5/3).)

7 Open the box up flat and draw a straight line from A to B. Distance $AB = \sqrt{(6+2)^2 + (5+2)^2} = \sqrt{113}$.

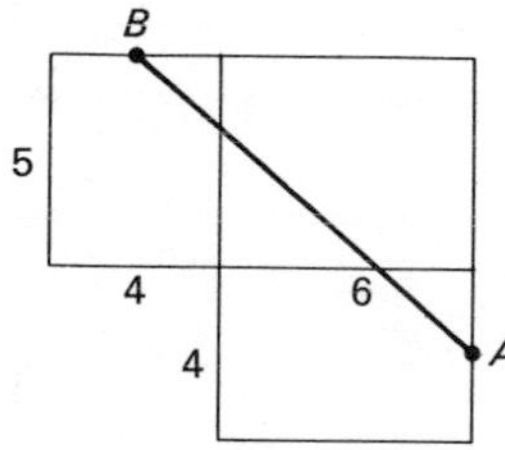

9 Yes.

11 (a) $(x+2)^2 + (y-1)^2 + (z-1)^2 = 4$,
(b) $(x-1)^2 + (y+2)^2 + (z+3)^2 \leq 9$,
(c) $(x-1)^2 + y^2 + (z+1)^2 < \frac{1}{4}$.

13 Let $r = |P_0|$. Then $r < 1$. Take any point Q within the sphere centered at P_0 with radius $(1-r)/2$. Then Q lies in S, since $|Q| = |Q - P + P| \leq |Q - P| + |P| \leq \frac{1}{2}(1-r) + r = \frac{1}{2}(1+r) < 1$.

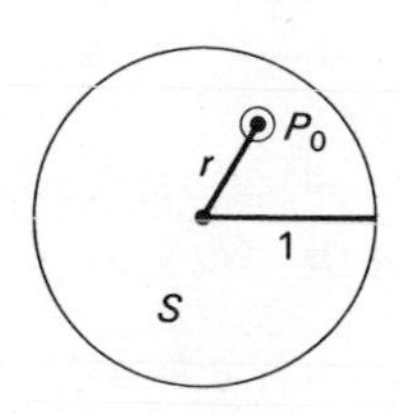

page 12

1.2

1 (a) $(-1, 1, 1)$, (b) $(3, -2, 0)$, (c) $\sqrt{38}$.

3 $A = (3, -2, 0)$,
$B = (2, -1, -1)$.

5 (a) $(-4/5, 0, 3/5)$,
(c) $(2/3, -\sqrt{3}/3, \sqrt{2}/3)$,
(e) $(2/11, -9/11, -6/11)$.

7 (a) $2 \text{ dist } (A, C) = 2\left|A - \left(\dfrac{2A + B}{3}\right)\right| = \dfrac{2}{3}|B - A|.$
Likewise, $\text{dist } (B, C) = (2/3)|B - A|$.
(b) $\text{dist } (A, C) = |A - C| = |A - (\lambda A + (1 - \lambda)B)|$
$= (1 - \lambda)|A - B| = (1 - \lambda) \text{ dist } (A, B);$
$\text{dist } (B, C) = |B - C| = |B - (\lambda A + (1 - \lambda)B)|$
$= \lambda|B - A| = \lambda \text{ dist } (A, B).$

9 (a) $A' = (-1, 1, 6)$, $B' = (0, -3, 8)$;
(c) $\text{dist } (A, B) = |A - B| = |(A + v) - (B + v)| = |A' - B'|$
$= \text{dist } (A', B')$.

11 $\dfrac{p_2 - p_1}{|p_2 - p_1|} = \dfrac{1}{7}(2, 5, -2, 4).$

13 (i) $|u| \geq 0$, since $\sqrt{x^2 + y^2 + z^2}$ is always nonnegative. If $|u| = 0$ then clearly $x = y = z = 0$, i.e., $u = \mathbf{0}$.

(ii) $|\alpha u| = \sqrt{(\alpha x)^2 + (\alpha y)^2 + (\alpha z)^2} = \sqrt{\alpha^2(x^2 + y^2 + z^2)}$
$= |\alpha|\,|u|.$

(iii) Refer to Figure 1–11 and notice that the length of the diagonal is less than the sum of the lengths of the two adjacent sides.

15 It suffices to show that $(M_1 + M_3)/2 = (M_2 + M_4)/2$ (notation as in the proof of Theorem 3). But this is clearly true, since each expression is equal to $(A + B + C + D)/4$.

17 If AC is a diagonal of a parallelogram with vertices A, B, C, and D, then $C - B = D - A$. Using this in the identity $C = A + (B - A) + (C - B)$, we see that $C = A + (B - A) + (D - A) = B + D - A$; that is, $A + C = B + D$. The other two cases, corresponding to parallelograms with diagonals AB and AD respectively, are proved similarly.

---

page 17

1.3

1 (a) $\theta = 45°$, (b) $\theta = 120°$,
(c) $\theta = 90°$, (d) $\theta = 30°$.

3 Let θ be the angle between P and $P + Q$, and let φ be the angle between Q and $P + Q$. Then, recalling that $|P| = |Q|$,

$$\cos\theta = \frac{P \cdot (P + Q)}{|P|\,|P + Q|} = \frac{|P|^2 + P \cdot Q}{|P|\,|P + Q|}$$
$$= \frac{|Q|^2 + P \cdot Q}{|Q|\,|P + Q|} = \frac{Q \cdot (Q + P)}{|Q|\,|Q + P|} = \cos\varphi.$$

Hence $\theta = \varphi$.

5 (a) on the surface, (b) outside, (c) inside.

7 The center is $(2, -1, 0, 2)$, and the radius is 2.

9 Let the unit vectors in the direction of the coordinate axes be $e_1 = (1, 0, 0)$, $e_2 = (0, 1, 0)$, and $e_3 = (0, 0, 1)$. Let $v = (v_1, v_2, v_3)$. Then

$$\cos\theta_x = \frac{v \cdot e_1}{|v|\,|e_1|} = \frac{v \cdot e_1}{|v|} = \frac{v_1}{|v|}.$$

Similarly, $\cos\theta_y = v_2/|v|$ and $\cos\theta_z = v_3/|v|$. Hence

$$(\cos\theta_x, \cos\theta_y, \cos\theta_z) = \frac{1}{|v|}(v_1, v_2, v_3) = \frac{v}{|v|}.$$

Since $\left|\frac{v}{|v|}\right| = 1$, this implies that $(\cos\theta_x)^2 + (\cos\theta_y)^2 + (\cos\theta_z)^2 = 1$.

11 If $A = (0, 0, z)$ is the vertex on the Z-axis and D is the fourth vertex, then there are four correct answers for the pair A, D. They are $A = (0, 0, 3)$, $D = (1, 2, -3)$; $A = (0, 0, -3)$, $D = (1, 2, 3)$; $A = (0, 0, 27/2)$, $D = (-1, -6, 23/2)$; $A = (0, 0, -7)$, $D = (1, 6, -5)$.

---

page 21

1.4

1 hyperbolic cylinder.

3 The graph is the union of the two planes $x - y - z = 0$ and $x - y + z = 0$ through the origin, since the equation is $z^2 - (x - y)^2 = 0$, which factors as $[z - (x - y)][z + (x - y)] = 0$.

5 right circular cylinder perpendicular to the YZ-plane.

7 hyperboloid of two sheets.

9 (c) The points satisfy both equations and hence lie on the intersection of a plane and a bowl-shaped paraboloid.

11

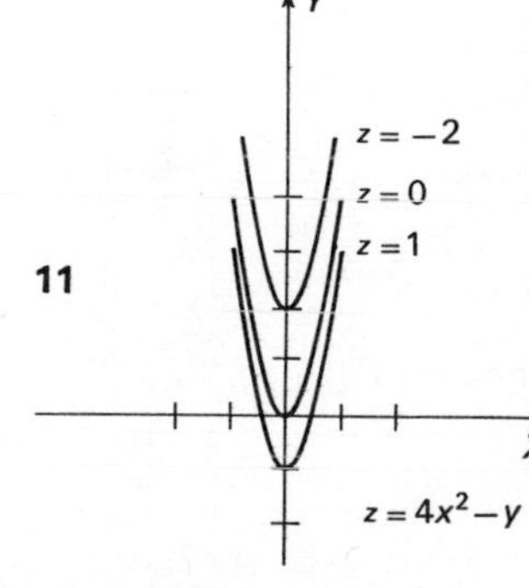

13 The surface is swept out by a horizontal line that rotates as it moves upward along the z-axis and is therefore a ruled surface.

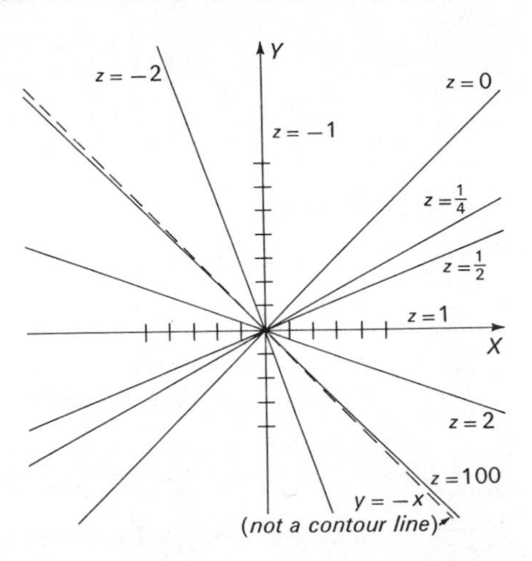

15

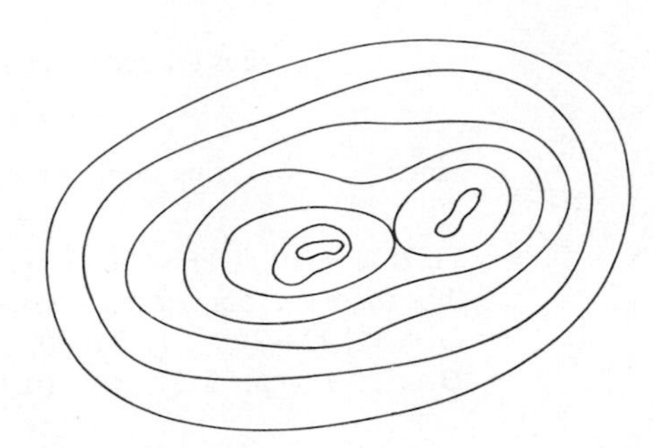

(Note that one peak is higher than the other.)

17 (a) two parallel planes.

(b) a rippled surface which is 0 on the line $x = y$; the equation of the surface in the XZ-plane is $z = \sin x$; the equation of the surface in the YZ-plane is $z = -\sin y$.

(c) concentric spherical surfaces with center at the origin and radii $\sqrt{n\pi}$, where n is a nonnegative integer.

19 hyperbolic paraboloid.

---

page 25

1.5

1 $(6, 0, 0)$, $(0, 4, 0)$; it does not intersect the z-axis; $u = (2, 3, 0)$.

3 $(4, 0, 0)$, $(0, 7, 0)$, $(0, 0, 7)$; $u = (7, 4, 4)$.

5 $(-2, 0, 0)$, $(0, -1, 0)$, $(0, 0, 1)$; $u = (1, 2, -2)$.

7 The determinant

$$\begin{vmatrix} V_1 & V_2 & V_3 \\ V_1 & V_2 & V_3 \\ W_1 & W_2 & W_3 \end{vmatrix}$$

is 0. Evaluate the determinant by minors along the first row:

$$\begin{aligned} 0 &= V_1\begin{vmatrix} V_2 & V_3 \\ W_2 & W_3 \end{vmatrix} - V_2\begin{vmatrix} V_1 & V_3 \\ W_1 & W_3 \end{vmatrix} + V_3\begin{vmatrix} V_1 & V_2 \\ W_1 & W_2 \end{vmatrix} \\ &= V \cdot U. \end{aligned}$$

Similar computations yield $0 = W \cdot U$.

9 A vector U perpendicular to the plane determined by A, B, and C must be perpendicular to the vectors $B - A$ and $C - A$. By Exercise 7,

$$U = (-9, 12, 6).$$

11 (a) The three points A, B, and C determine a pencil of planes. For example, $(3c - 2b)x + by + cz - (b + c) = 0$ could express the pencil. If $b = 0$ and $c = 1$, this is $3x + z - 1 = 0$; if $c = 0$ and $b = 1$, this is $-2x + y - 1 = 0$.
(b) $A - B = (3, 6, -9)$, $A - C = (1, 2, -3)$. The two vectors have the same direction, and hence the three points are collinear.

15 (a) $D = 3$, (b) $D = 7/3$, (c) $D = 3$.
Using the solution of Exercise 13, $u = (1, -2, 2)$ and $u \cdot P_0 + d = \alpha|u|^2$ yields $\alpha = 1$. Since $P_0 - P_1 = \alpha u$, we have $P_1 = P_0 - \alpha u = (3, -1, 1)$.

17 $3x - 4y + 2z - w - 5 = 0$.

---

page 29

1.6

1 The Cartesian coordinates are (using (1.16)) $A = (2, 2\sqrt{3}, -3)$ and $B = (-\sqrt{2}, -\sqrt{2}, 2)$.

3 Using (1.17), the Cartesian coordinates are $A = (1, \sqrt{3}, 2\sqrt{3})$ and $B = (\sqrt{2}/2, -\sqrt{6}/2, -\sqrt{2})$.

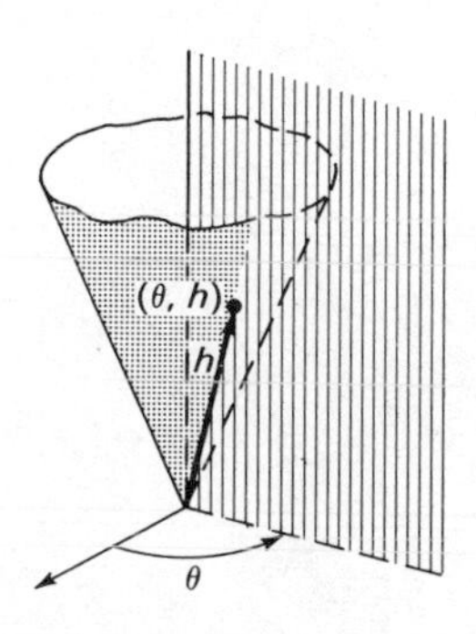

5 (a) A point on the upper half of a cone can be uniquely determined by giving an angle θ and a nonnegative distance h. θ indicates which half plane through the axis of the cone to use, and h indicates how far to move from the vertex along the intersection of the half plane and the cone.
(b) For the full cone, use negative h for points on the other nappe below the vertex.

7 (a) cylinder (b) half plane, (c) upper half of a cone,
(d) cylinder based on spiral,
(e) circular cylinder based on $(x - \frac{1}{2})^2 + y^2 = \frac{1}{4}$.

9 Points on the line through A and B cannot be located, since $\theta = \varphi = 0$ for all such points.

11 To locate a point, given the distances d_1 and d_2 and the angle θ, we proceed thus. The spheres centered at P_1 and P_2 with radii d_1 and d_2 meet in a circle that encloses the line P_1P_2. The choice of θ determines uniquely which point on this circle is P. Note that θ is not uniquely defined if P lies on the line P_1P_2. Also note that no point P results if d_1 and d_2 are both too small.

---

page 35

1.7

1 The lines intersect at the point corresponding to $s = 1$ and $t = 0$, namely $(2, 3, -1)$.

3 $(3, 0, -6)$.

5 One set of parametric equations for the line is $x = t + 2$, $y = 3t - 1$, $z = -t + 3$. Another set is $x = 1 + t$, $y = -4 + 3t$, $z = 4 - t$.

7 A set of parametric equations for the line is $x = t$, $y = t - 1$, $z = -t + 5$.

9

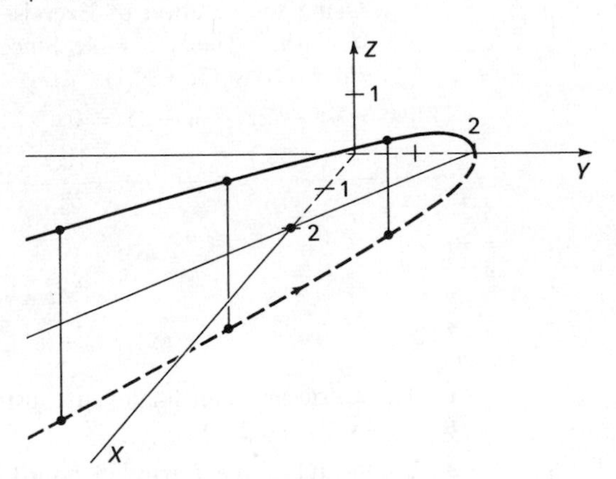

11 One such equation is $x = (1/2) \sin t$, $y = \cos t$.

13

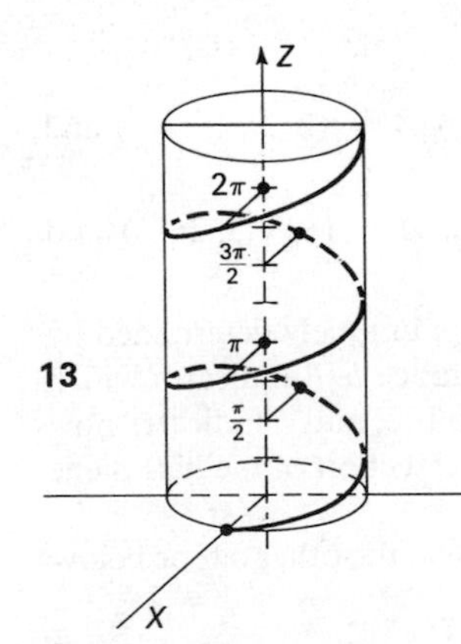

15 The correct equation is (c).

---

page 40

1.8

1 $P = (1, 0, 0)$.

3 If p_0 is outside the closed ball, then $|p_0| > r$. A suitable neighborhood of p_0 would be the open ball with center at p_0 and radius $(|p_0| - r)/2$. This clearly cannot touch the original closed ball.

5 Open, connected, simply connected, bounded. The boundary is the union of the two concentric circles with centers at the origin at radii 1 and $\sqrt{2}$.

7 Closed, disconnected, and unbounded. Each component is simply connected. The boundary is a hyperboloid of revolution of two sheets, namely the set itself.

9 The sequence $\{p_n\} = \{(1/n, 1/n^2, 1/n^3)\}$ is an example.

---

chapter 2

page 48

2.1

1 (a)

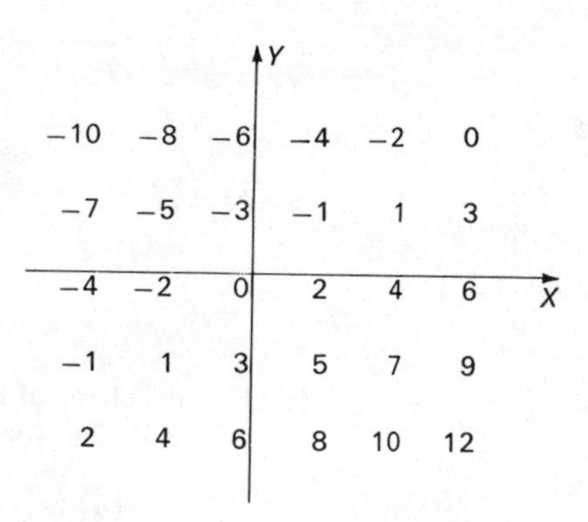

(b) Looking at the chart, one can conclude that as $|p| \to \infty$, p in the second quadrant, the temperature $T \to -\infty$; as $|p| \to \infty$, p in the fourth quadrant, $T \to +\infty$; and as $|p| \to \infty$, p spiraling about the origin, T takes on all possible values, $-\infty < T < \infty$.

3 For the function in Exercise 1, the equithermal curves are the lines

$$y = \frac{2}{3}x - \frac{C}{3},$$

where C is the temperature. For the function in Exercise 2, the equithermal curves are parabolas with vertex at $(-C/2, 0)$ which open toward the right, where C is the temperature.

5 (a) half plane above (and including) $y = -x$.

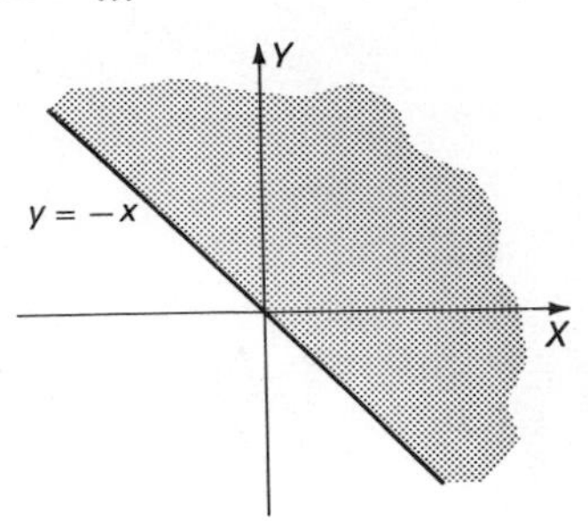

(b) Entire XY-plane except for the lines $x = y$ and $x = -y$.
(c) All of 3-space except for the cone $x^2 + y^2 - z^2 = 0$.
(d) It is necessary that $(x - z)(2x - y) \geq 0$, which means that $x \geq z$ and $2x \geq y$ or $x \leq z$ and $2x \leq y$. The domain of f is a pair of opposing regions formed by the intersecting planes $x = z$ and $2x = y$.

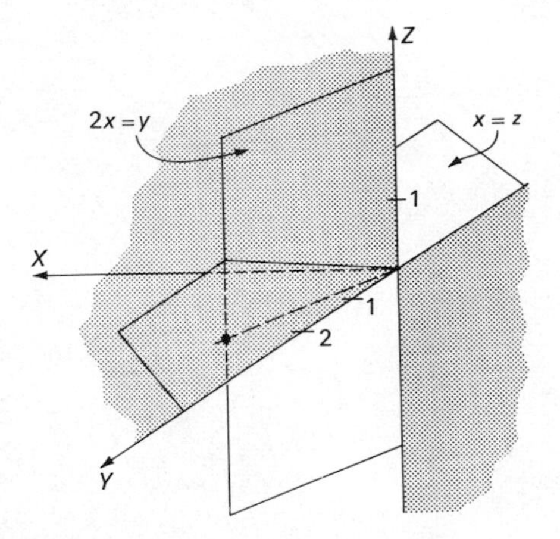

(e) Parallel strips of the plane lying between $y = x + (2n - 1)\pi$ and $y = x + 2n\pi$ for all integers n.

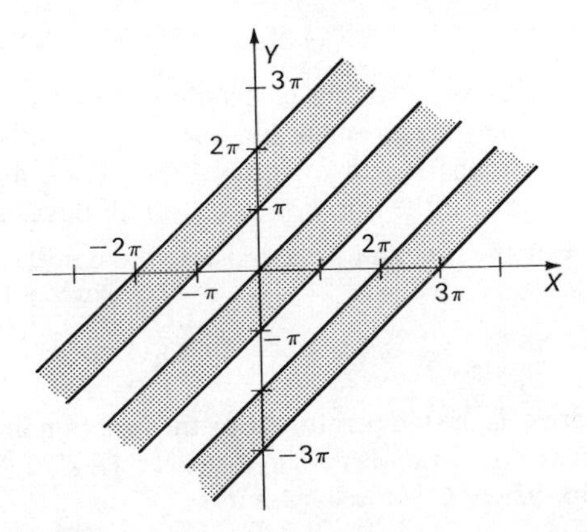

9 (a) Yes.

(b) Yes, since $F(x, y, z, w) = |p|^2 + w$, so that for fixed w, $F \to \infty$ as $|p| \to \infty$.

(c) Yes. Fix a temperature C and suppose that $C = |p|^2 + t$ for all $t \leq C$. Then $C - t = |p|^2 = x^2 + y^2 + z^2$. Approach the origin along an axis, for example along $y = 0$, $z = 0$. Then $x^2 = C - t$ and $2x \dfrac{dx}{dt} = -1$. Thus $\dfrac{dx}{dt}$ is not a constant.

---

page 51

2.2

1 Note that all the vectors end at points on the line $y = x$.

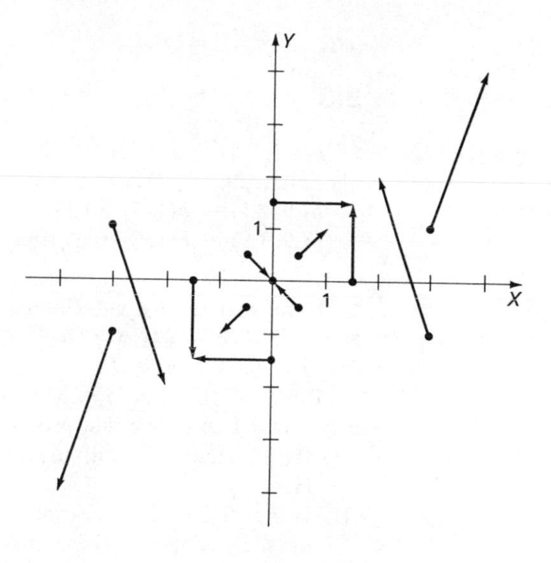

3 (a)

(b) The graph of the function consists of all pairs $(t, f(t))$, which turn out to be triples $(t, 2 \cos t, 3 \sin t)$. The graph of f is therefore a curve in 3-space, an elliptical helix.

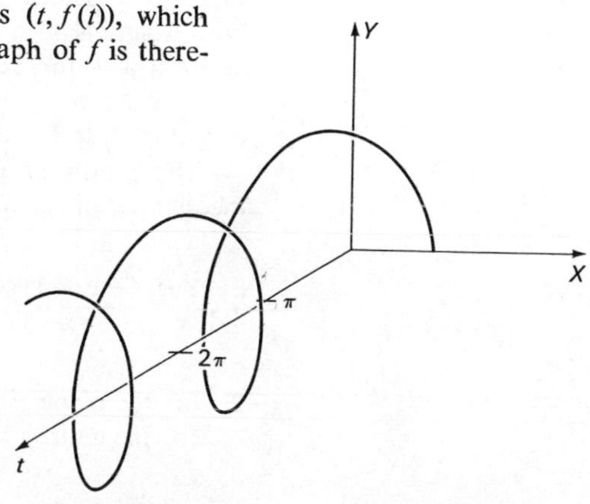

7

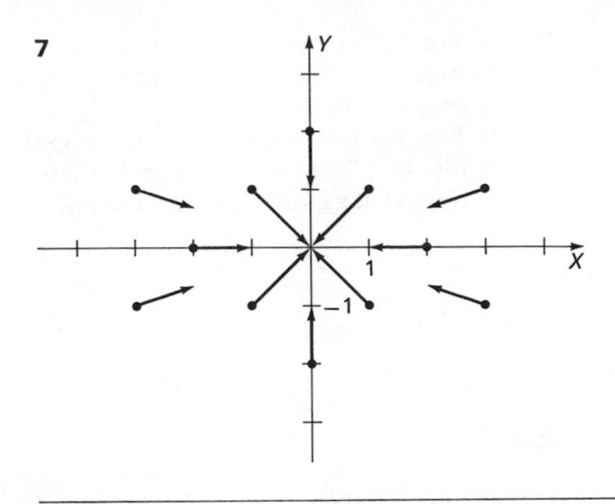

page 59

2.3

1 $F(1.1, 2.1) - F(1, 2) = .21,$
$F(1.1, 1.9) - F(1, 2) = -.21,$
$F(.9, 2.1) - F(1, 2) = .19,$
$F(.9, 1.9) - F(1, 2) = -.19.$

3 Proof of (i): Any $x \in I$ and $y \in J$ obey $a \leq x \leq b$ and $c \leq y \leq d$. Hence $a + c \leq x + y \leq b + d$, so $x + y \in [a + c, b + d]$. Thus $I + J = [a + c, b + d]$.

Proof of (ii): Let $a \leq x \leq b$ and $c \leq y \leq d$. Since I is positive, $a > 0$ and $x > 0$. We first prove four easy results.

I. If $c < 0$ then $b \geq x$ implies that $bc \leq xc$, and $x > 0$ implies $xc \leq xy$. Hence $bc \leq xy$.

II. If $c > 0$ then $c \leq y$ implies that $ac \leq ay$, and $y > 0$ implies that $ay \leq xy$. Together these imply that $ac \leq xy$.

III. If $d > 0$ then $xd \leq bd$ and since $x > 0$, $xy \leq xd$: hence $xy \leq bd$.

IV. If $d < 0$ then $xd \leq ad$ and since $x > 0$, $xy \leq xd$; hence $xy \leq ad$.

Now when J is positive, $d \geq c > 0$, so II and III hold. Thus $ac \leq xy \leq bd$ and $I \times J = [ac, bd]$. When J is posineg, $c < 0$ and $d > 0$, so I and III hold. Thus $bc \leq xy \leq bd$ and $I \times J = [bc, bd]$. When J is negative, $c \leq d < 0$, so I and IV hold. Thus $bc \leq xy \leq ad$ and $I \times J = [bc, ad]$.

The proofs of (iii) and (iv) are similar in that they are based on inequalities of real numbers.

5 Yes; $[-1, -1] \times J = \{\text{all } -y \text{ where } y \in J\}$, so $I + ([-1, -1] \times J) = \{\text{all } x + (-y), \text{ where } x \in I \text{ and } y \in J\} = I - J$.

7 (a) the plane except for the lines $y = x$ and $y = -x$;
(b) the entire plane (since $2x^2 + 3y^2 + 4 > 0$ for all (x, y));

(c) all points in the plane below and on the parabola $y = x^2$, i.e., whenever $y \leq x^2$;
(d) all of 3-space;
(e) all points in 3-space not on the parabolic cylinder $z^2 = x$.

9 If $p = (x, y, z, w)$ is close to $p_0 = (x_0, y_0, z_0, w_0)$, then $A(p)$ is close to the number $A_0 = A(p_0)$, $B(p)$ is close to $B_0 = B(p_0)$, and $C(p)$ is close to $C_0 = C(p_0)$, since A, B, and C are continuous functions. Because f is continuous, $f(A(p), B(p), C(p))$ is close to $f(A_0, B_0, C_0)$. Therefore, $h(p) = h(x, y, z, w)$ is close to $h(p_0) = f(A_0, B_0, C_0)$ when p is close to p_0.

11 (a) Recall that the rational numbers and the irrational numbers are each dense in the real numbers; that is, any interval, no matter how small, contains infinitely many of each. Applied to the function f, this tells us that its values oscillate between 1 and 0 in any small neighborhood of any point, so that f is continuous nowhere.
(b) The function g has the value 1 or 0 on the line $y = x - r$, depending on whether r is rational or irrational. Again, g oscillates between 0 and 1 in any small neighborhood of any point, no matter how small, and hence cannot be continuous at any point.

---

page 66

2.4

1 (a) 1; (b) $\sqrt{2}, -\sqrt{2}$;
(c) ± 1, depending on whether the approach to (0, 0) is from the right or the left.

3 (a) On the Y-axis $x = 0$, so the limit is $\sin 0 = 0$.
(b) If $y = x$, $\sin(x/y) = \sin 1$, so the limit is $\sin 1$.
(c) If $y = \sqrt{x}$ then $x/y = x/\sqrt{x} = \sqrt{x}$, so the limit is

$$\lim_{x \to 0} \sin \sqrt{x} = 0.$$

(d) If $x = \sqrt{y}$ then $y = x^2$, so $x/y = 1/x$ and $\lim_{x \to 0} \sin(1/x)$ does not exist.

5 1.

7 0.

9 (a) 1/3, (b) 2/3, (c) 2/9.

11 Think of $f(x, y) = x^y$ as defined on the open first quadrant $x > 0$, $y > 0$. Limits exist for points on the axes except (0, 0), so that $f(x, 0) = 1$ and $f(0, y) = 0$ are defined. Since $0 \neq 1$, this shows that we do not get a unique limit value as $(x, y) \to (0, 0)$ and that f is not continuous there.

---

page 72

2.5

1 (a) Look at the function g defined by $g(p) = f(p) - C$. This new function is continuous on D, and $g(p_0) > 0$. By Theorem 8 there is a neighborhood $\mathfrak{N}$ about p_0 such that $g(p) > 0$ for all $p \in \mathfrak{N}$ that are in D. Hence $f(p) > C$ for such p. To show that we can take $\mathfrak{N}$ so that $\mathfrak{N} \subset D$, observe that there must be a ball with center p_0 lying in D (since D is open) and also one lying in $\mathfrak{N}$. The smaller of these has the desired property. (Draw a picture!)

(b) Let $p_0 \in \mathcal{O}$. We have to show that some neighborhood of p_0 is contained in $\mathcal{O}$. Since $p_0 \in \mathcal{O}$, $f(p_0) > C$, so by (a) there is a neighborhood $\mathfrak{N} \subseteq D$ about p_0 such that $f(p_0) > C$ for all $p \in \mathfrak{N}$. But this says precisely that $\mathfrak{N} \subseteq \mathcal{O}$, so $\mathfrak{N}$ will serve as the required neighborhood.

(c) Let $g = -f$, $C' = -C$.

3 Fix p_0 and let $\epsilon > 0$ be given. Let $a = f(p_0) - \epsilon$ and $b = f(p_0) + \epsilon$. Then by hypothesis there exists a neighborhood $\mathfrak{N}$ of p_0 such that $f(p_0) - \epsilon < f(p) < f(p_0) + \epsilon$ for all $p \in \mathfrak{N}$, so by Definition 1, f is continuous at p_0.

5 Observe that on the X- and Y-axes, $f(x, y) = 2$ and $f(1, -1) = -1$. Thus by the intermediate value theorem there is a point p_0 where $f(p_0) = 0$ on *every* line segment in the fourth quadrant joining $(1, -1)$ to a point on either axis.

7 (a) $|f(x_1) - f(x_2)| = |x_1 - x_2|\,|x_1 + x_2| \leq 8|x_1 - x_2|$, since $2 \leq |x_1 + x_2| \leq 8$.

(b)
$$\begin{aligned} |f(x, y) - f(x_0, y_0)| &= |(x - x_0)(x + x_0) + (y_0 - y)(y_0 + y)| \\ &\leq 2|x - x_0| + 2|y_0 - y| \\ &\leq 4|(x, y) - (x_0, y_0)|, \end{aligned}$$

since

$$|x - x_0| \leq |(x, y) - (x_0, y_0)|$$

and

$$|y_0 - y| \leq |(x, y) - (x_0, y_0)|.$$

13 Let $B = \text{lub}\,\{f(p) \mid p \in D\}$. ($B$ exists, since any bounded set of real numbers has a least upper bound.) Then $f(p) \leq B$ for all $p \in D$. Suppose $f(p) \neq B$ for all $p \in D$. Then $g(p) = 1/(B - f(p))$ is a continuous real-valued function on D. By hypothesis g is bounded on D, so there exists a real number $M > 0$ such that $g(p) \leq M$ for all $p \in D$. But then $0 < 1/M \leq B - f(p)$, or $f(p) \leq B - (1/M)$ for all $p \in D$, where $B - (1/M)$ is strictly less than B. This implies that B is not the least upper bound of f, which is a contradiction, so we must have $f(p) = B$ for some $p \in D$.

---

chapter 3

page 80

3.1

1 (a) A particle, at first stationary, moves forward, stops awhile, returns to its initial position and stops.

(b) Two moving particles approaching each other collide and cease to exist.

(c) A stolen car trapped on a street between two police cars each initially at rest attempts to escape and is hemmed in and captured.

3 The velocity $F'(t) = (3, 3t^2 - 2t, 2t - 4)$, the speed $|F'(t)| = \sqrt{9 + (3t^2 - 2t)^2 + (2t - 4)^2} = \sqrt{25 - 16t + 8t^2 - 12t^3 + 9t^4}$, and the acceleration $F''(t) = (0, 6t - 2, 2)$. At $(-3, -2, 5)$, $t = -1$, so the velocity is $v = (3, 5, -6)$, the speed is $|v| = \sqrt{70}$, and the acceleration is $a = (0, -8, 2)$. At $(6, 4, -4)$, $t = 2$, so $v = (3, 8, 0)$, the speed is $|v| = \sqrt{73}$, and $a = (0, 10, 2)$.

7 (a) No.

(b) Yes, when $t = -1$, which corresponds to the point $(-1, 1, -1)$.

9 The point $(0, 0, 0)$ corresponds to $t = 0$ for C_1 and C_2, and to $t = 1$ for C_3. $T_1 = (1, -1, 2)$, $T_2 = (-1, 1, 1)$, $T_3 = (1, 1, 0)$. Then $T_1 \cdot T_2 = T_2 \cdot T_3 = T_3 \cdot T_1 = 0$.

11 No.

---

page 85

3.2

1 $$\frac{\partial f}{\partial x} = 3x^2y - 4xy^2 + 5y - 2, \qquad \frac{\partial f}{\partial y} = x^3 - 4x^2y + 5x,$$
$$\frac{\partial^2 f}{\partial x^2} = 6xy - 4y^2, \qquad \frac{\partial^2 f}{\partial x\,\partial y} = 3x^2 - 8xy + 5,$$
$$\frac{\partial^2 f}{\partial y\,\partial x} = 3x^2 - 8xy + 5, \qquad \frac{\partial^2 f}{\partial y^2} = -4x^2.$$

3 $$\frac{\partial f}{\partial x} = 2xz - 3yz + 2y^2, \frac{\partial f}{\partial y} = -3xz + 4xy, \frac{\partial f}{\partial z} = x^2 - 3xy;$$
$$\frac{\partial^2 f}{\partial x\,\partial y} = -3z + 4y, \frac{\partial^2 f}{\partial y\,\partial z} = -3x, \frac{\partial^2 f}{\partial x\,\partial z} = 2x - 3y;$$
$$\frac{\partial^2 f}{\partial x^2} = 2z, \frac{\partial^2 f}{\partial y^2} = 4x, \frac{\partial^2 f}{\partial z^2} = 0.$$

5 $$\frac{\partial f}{\partial t} = -xe^{-t} + xye^{xt}, \frac{\partial^2 f}{\partial x\,\partial t} = -e^{-t} + ye^{xt} + x^2ye^{xt},$$
$$\frac{\partial^2 f}{\partial t^2} = xe^{-t} + x^2ye^{xt}.$$

7 $\dfrac{\partial f}{\partial x} = 3x^2 - 3y^2$, $\dfrac{\partial^2 f}{\partial x^2} = 6x$, $\dfrac{\partial f}{\partial y} = -6xy$, and $\dfrac{\partial^2 f}{\partial y^2} = -6x$. Hence
$$\frac{\partial^2 f}{\partial x^2} + \frac{\partial^2 f}{\partial y^2} = 6x + (-6x) = 0.$$

9 $\dfrac{\partial f}{\partial z} = \dfrac{\partial B(y, z)}{\partial z} + \dfrac{\partial C(x, z)}{\partial z}$, so $\dfrac{\partial^2 f}{\partial y\,\partial z} = \dfrac{\partial^2 B(y, z)}{\partial y\,\partial z}$, since $C(x, z)$ is a function of x and z only. Therefore $\dfrac{\partial^3 f}{\partial x\,\partial y\,\partial z} = 0$, since $\dfrac{\partial^2 B(y, z)}{\partial y\,\partial z}$ is a function of y and z only.

---

page 93

3.3

1 $\nabla f = (2y^2 - 6x, 4xy + 5)$, $\nabla f(p_0) = (2, 13)$. Using Theorem 1,
(a) $D_u f(p_0) = u \cdot \nabla f(p_0) = 2/\sqrt{2} - 13/\sqrt{2} = -11/\sqrt{2}$,
(b) $D_u f(p_0) = 6/5 + 52/5 = 58/5$,
(c) $D_u f(p_0) = 10/13 - 13(12/13) = -146/13$.

3 $\nabla f = (2xy - 3z^2, x^2 + 2z, 2y - 6xz)$, $\nabla f(1, 2, -1) = (1, -1, 10)$. Using unit vectors and applying Theorem 1,
(a) $D_u(1, 2, -1) = u \cdot \nabla f(1, 2, -1) = 1/3 + 2/3 + 20/3 = 23/3$,
(b) $D_u(1, 2, -1) = -1/9 - 4/9 + 80/9 = 25/3$,
(c) $D_u(1, 2, -1) = 4/9 + 4/9 - 70/9 = -62/9$.

5 $\nabla f = (2y - yzw, 2x - xzw, w - xyw, z - xyz)$;
$\nabla f(1, 1, -1, 2) = (4, 4, 0, 0)$, $\nabla f(1, 2, -1, 1) = (6, 3, -1, 1)$,
$\nabla f(2, 0, -1, 2) = (0, 8, 2, -1)$.

7

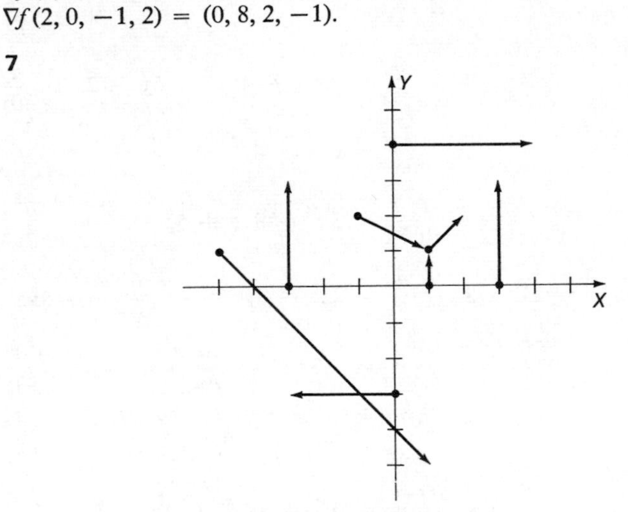

9 (a) Let f be defined and have first order partial derivatives on a spherical neighborhood of the point $p_0 = (x_0, y_0, z_0)$. Let p be another point in this neighborhood, and write $p = p_0 + \Delta p$, where $\Delta p = (\Delta x, \Delta y, \Delta z)$. Then there are three points q_1, q_2, and q_3 with $|p_0 - q_i| \le |\Delta p|$, $i = 1, 2, 3$, such that

$$f(p_0 + \Delta p) - f(p_0) = \Delta x \frac{\partial f}{\partial x}(q_1) + \Delta y \frac{\partial f}{\partial y}(q_2) + \Delta z \frac{\partial f}{\partial z}(q_3).$$

(b) We write

$$\begin{aligned} f(p_0 + \Delta p) - f(p_0) &= f(x_0 + \Delta x, y_0 + \Delta y, z_0 + \Delta z) - f(x_0, y_0, z_0) \\ &= [f(x_0 + \Delta x, y_0 + \Delta y, z_0 + \Delta z) \\ &\quad - f(x_0 + \Delta x, y_0 + \Delta y, z_0)] \\ &\quad + [f(x_0 + \Delta x, y_0 + \Delta y, z_0) \\ &\quad - f(x_0 + \Delta x, y_0, z_0)] \\ &\quad + [f(x_0 + \Delta x, y_0, z_0) - f(x_0, y_0, z_0)]. \end{aligned}$$

Applying the Mean Value Theorem for one variable to each of these differences gives the existence of points q_1, q_2, and q_3, as shown in Figure 3–17, such that

$$f(p_0 + \Delta p) - f(p_0) = \Delta x \frac{\partial f}{\partial x}(q_1) + \Delta y \frac{\partial f}{\partial y}(q_2) + \Delta z \frac{\partial f}{\partial z}(q_3).$$

The inequalities $|p_0 - q_i| \leq |\Delta p|$, $i = 1, 2, 3$, are clear from the figure.

---

page 102

3.4

1 (a) $\dfrac{\partial w}{\partial x} = 2xy \cos(x^2y)$, $\dfrac{\partial^2 w}{\partial x^2} = 2y \cos(x^2y) - (2xy)^2 \sin(x^2y)$;

(b) $\dfrac{\partial w}{\partial x} = 6x(x^2 - y^2)^2$, $\dfrac{\partial^2 w}{\partial x^2} = 6(x^2 - y^2)^2 + 24x^2(x^2 - y^2)$;

(c) $\dfrac{\partial w}{\partial x} = 1/y$, $\dfrac{\partial^2 w}{\partial x^2} = 0$.

3 $\dfrac{dw}{dt} = f_1(x, y)(2t - 1) + f_2(x, y)(3 + 2t)$,

$$\begin{aligned} \frac{dw}{dt} &= y(2t - 1) + x(3 + 2t) = 2t^3 + 5t^2 - 3t + 2t^3 + t^2 - 3t \\ &= 4t^3 + 6t^2 - 6t. \end{aligned}$$

Directly, we see that $w(t) = (t^2 - t)(3t + t^2) = t^4 + 2t^3 - 3t^2$.

5
$$\begin{aligned} \frac{\partial w}{\partial x} &= \frac{\partial w}{\partial u}\frac{\partial u}{\partial x} + \frac{\partial w}{\partial v}\frac{\partial v}{\partial x} + \frac{\partial w}{\partial t}\frac{\partial t}{\partial x} = (1)\frac{\partial w}{\partial u} + (0)\frac{\partial w}{\partial v} + (2x)\frac{\partial w}{\partial t} \\ &= F_1 + 2xF_3; \\ \frac{\partial w}{\partial y} &= \frac{\partial w}{\partial u}\frac{\partial u}{\partial y} + \frac{\partial w}{\partial v}\frac{\partial v}{\partial y} + \frac{\partial w}{\partial t}\frac{\partial t}{\partial y} = (-1)\frac{\partial w}{\partial u} + z\frac{\partial w}{\partial v} + (0)\frac{\partial w}{\partial t} \\ &= -F_1 + zF_2; \\ \frac{\partial w}{\partial z} &= (0)\frac{\partial w}{\partial u} + y\frac{\partial w}{\partial v} - 2z\frac{\partial w}{\partial t} = yF_2 - 2zF_3. \end{aligned}$$

7 Here $X(t) = at$, $Y(t) = bt$, $Z(t) = ct$ and $\dfrac{d^2X}{dt^2} = \dfrac{d^2Y}{dt^2} = \dfrac{d^2Z}{dt^2} = 0$, so the general formula for $\dfrac{\partial^2 w}{\partial t^2}$ becomes

$$\frac{\partial^2 w}{\partial t^2} = a^2\frac{\partial^2 w}{\partial x^2} + b^2\frac{\partial^2 w}{\partial y^2} + c^2\frac{\partial^2 w}{\partial z^2} + 2ab\frac{\partial^2 w}{\partial x\,\partial y} + 2bc\frac{\partial^2 w}{\partial y\,\partial z} + 2ac\frac{\partial^2 w}{\partial z\,\partial x}\cdot$$

9 Formula (a) corresponds to diagram (i), formula (c) corresponds to diagram (ii), and formula (b) corresponds to neither diagram.

11 (a) Since u and v are functions of t alone, we can think of x, y, and z as functions of t alone. Consequently, $w = F(x, y, z)$ can be regarded as a function of t alone.

(b)

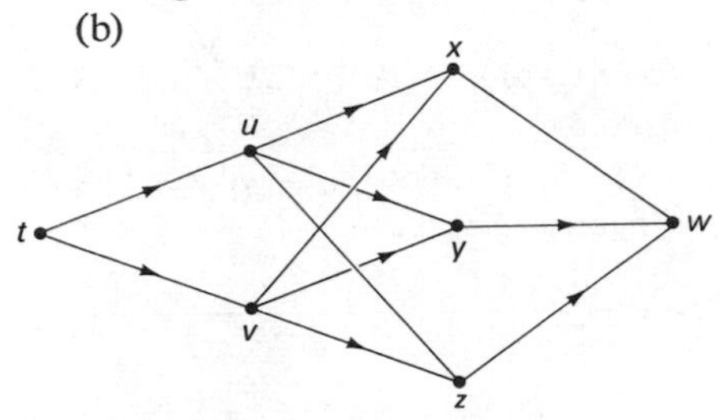

(c) There are 6 paths. The diagram below shows three of these, two having the path from t to v in common, and two having the path from y to w in common.

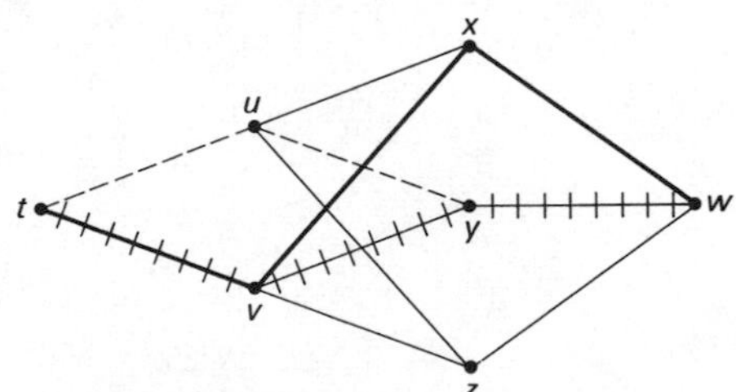

(d)
$$\begin{aligned}\frac{dw}{dt} &= \frac{\partial w}{\partial x}\frac{dx}{dt} + \frac{\partial w}{\partial y}\frac{dy}{dt} + \frac{\partial w}{\partial z}\frac{dz}{dt}\\ &= \frac{\partial w}{\partial x}\left(\frac{\partial x}{\partial u}\frac{du}{dt} + \frac{\partial x}{\partial v}\frac{dv}{dt}\right) + \cdots\\ &= \frac{\partial w}{\partial x}\frac{\partial x}{\partial u}\frac{du}{dt} + \frac{\partial w}{\partial x}\frac{\partial x}{\partial v}\frac{dv}{dt} + \frac{\partial w}{\partial y}\frac{\partial y}{\partial u}\frac{du}{dt} + \frac{\partial w}{\partial y}\frac{\partial y}{\partial v}\frac{dv}{dt}\\ &\quad + \frac{\partial w}{\partial z}\frac{\partial z}{\partial u}\frac{du}{dt} + \frac{\partial w}{\partial z}\frac{\partial z}{\partial v}\frac{dv}{dt}\cdot\end{aligned}$$

page 113

3.5

1 (a) Solving the equations $f_x = 6x - 4y - 2 = 0$ and $f_y = -4x + 10y + 3 = 0$, we see that the only critical point is $(2/11, -5/22)$.
(b) Five critical points: $(0, 0)$, $(\pm 1, \pm 2)$.
(c) Two critical points: $(2, 2)$ and $(-2, -2)$.
(d) f has no critical points in D.
(e) All points except $(0, 0)$ on the lines $y = \pm x$.

3 (a) $f_x = 3y - 2x = 0$ and $f_y = 3x - 2y = 0$, and $f_{xx} = f_{yy} = -2$.
(b) In the plane $x = y$, the function becomes $f(x, x) = x^2$.

5 $(0, 0)$ is a local maximum; each of the four points $(\pm 1, \pm 2)$ is a saddle point.

7 $f_x = 2x - 6y + 1$, $f_y = -2y - 6x + 2$,
$f_{xx} = 2$, $f_{yy} = -2$, $f_{xy} = -6$.
$(1/4, 1/4)$ is a saddle point; the maximum and minimum for f on S must occur on the boundary of S. The minimum value for f over S is $f(1, 1) = -3$, and the maximum value is $f(1, 0) = 2$.

9 The only critical point of f is $(1, 2)$, which is not in S. The maximum value of f over S is $f(1, 0) = 3$, and the minimum value is $f(1/2, 1) = -3/4$.

11 $\nabla F = (yz, xz, xy)$, $\nabla F(3, -1, -2) = (2, -6, -3)$.

13 $F(x, y, z) = (x - y)(1 - x - y)$, so that the surface $F(x, y, z) = 0$ is the union of two perpendicular planes, each of which is perpendicular to the XY-plane.

15 $t = 1$ or $t = -1/2$, and the corresponding points are $(0, 2, -3)$ and $(3/2, -1, 0)$. The line is orthogonal to the surface at $(0, 2, -3)$.

17 The graph of f is $z - f(x, y) = 0$. With $F(x, y, z) = z - f(x, y)$, the general normal is $(-f_x, -f_y, 1)$ so that the normal above a critical point for f is $(0, 0, 1)$, which is a vertical vector. (Note that the tangent plane is therefore horizontal at any critical point.)

---

page 122

3.6

3 At $(1, 1, 1)$, $\dfrac{\partial z}{\partial x} = -\dfrac{2}{7}$ and $\dfrac{\partial z}{\partial y} = -\dfrac{3}{7}$.

5 (a) $\dfrac{\partial z}{\partial x} = \dfrac{-(2xy + yz)}{6yz^2 + xy} = \dfrac{-2x - z}{6z^2 + x}$, $\dfrac{\partial z}{\partial y} = \dfrac{-(x^2 + 2z^3 + xz)}{6yz^2 + xy}$.

(b) $\dfrac{\partial x}{\partial z} = -\dfrac{6yz^2 + xy}{2xy + yz} = -\dfrac{6z^2 + x}{2x + z}$, $\dfrac{\partial x}{\partial y} = -\dfrac{x^2 + 2z^3 + xz}{2xy + yz}$.

(c) $\dfrac{\partial z}{\partial x} = -3/7$, $\dfrac{\partial z}{\partial y} = -2/7$, $\dfrac{\partial x}{\partial z} = -7/3$, and $\dfrac{\partial x}{\partial y} = -2/3$.

7 (a) Let $d(x, y, z) = (x - 1)^2 + (y - 2)^2 + (z - 3)^2$. Then d is at a minimum at the point (5/3, 10/3, 5/3), where

$$d = \sqrt{(4/9 + 16/9 + 16/9)} = 2.$$

(b) $D = \dfrac{|1 + 2(2) - 2(3) - 5|}{\sqrt{1 + 4 + 4}} = 2.$

9 $\dfrac{dy}{dt} = -2\dfrac{dx}{dt}$ and $\dfrac{dz}{dt} = -\dfrac{2}{3}\dfrac{dx}{dt}$, so the tangent has direction $(1, -2, -2/3)$.

11 We have $2yz + xz = 1000$ and $V(x, y, z) = xyz$. There is no maximum volume, but large volumes result when x and y are large and z is small.

---

chapter 4

page 130

4.2

1 (a) 1, (b) 29/12, (c) $-31/30$.

3 (a) $-5/3$, (b) 28/3.

5 (a) 7/3, (b) -4.

7 55/8.

9
$$\int_0^1 dx \int_0^x dy \int_x^y (xy + yz)\, dz$$
$$= \tfrac{1}{2}\int_0^1 dx \int_0^x (2xy^2 + y^3 - 3x^2y)\, dy$$
$$= -\tfrac{1}{2}\int_0^1 \tfrac{7}{12}x^4\, dx$$
$$= -7/120.$$

11 (a) The equation of the intersection of the surface and the positive XY-plane is $x = \sqrt{y + 1}$. For a fixed y, the area under the curve $z = y + 1 - x^2$ from $x = 0$ to $x = \sqrt{y + 1}$ is

$$A(y) = \int_0^{\sqrt{y+1}} z\, dx = \int_0^{\sqrt{y+1}} (y + 1 - x^2)\, dx.$$

(b)

$$A(y) = \{(y + 1)x - \tfrac{1}{3}x^3\}\Big|_{x=0}^{x=\sqrt{y+1}} = \tfrac{2}{3}(y + 1)\sqrt{y + 1}.$$

By Cavalieri's principle,

$$V = \int_0^3 A(y)\,dy = \int_0^3 \tfrac{2}{3}(y+1)\sqrt{y+1}\,dy.$$

13 $\pi/4 + 1/2$.

15 $1/6$.

---

page 138

4.3

1 Consideration of Figure 4–17 shows that $(a+b)^2 = 4A(R) + (a-b)^2$, or $a^2 + 2ab + b^2 = 4A(R) + a^2 - 2ab + b^2$. Hence $4ab = 4A(R)$, or $A(R) = ab$.

3 $A(D_1) + A(D_2)$.

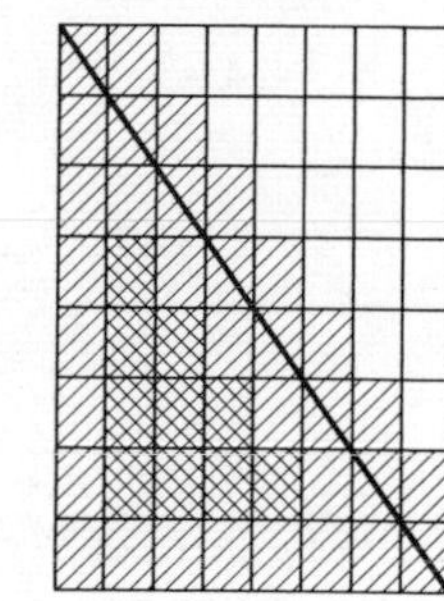

5 If the containing rectangle is divided into N^2 small rectangles (see the figure), there will be $1 + 2 + \cdots + (N-4)$ that lie entirely inside the triangle, so that

$$s(N) = \left(\frac{1}{N}\right)\left(\frac{2}{N}\right)(1 + 2 + \cdots + N - 4) = \left(\frac{2}{N^2}\right)\frac{1}{2}(N-4)(N-3)$$

and $\lim_{N\to\infty} s(N) = 1$. Similarly, there are $2 + 3 + \cdots + N + N$ small rectangles in the outer sum, so that

$$S(N) = \left(\frac{2}{N^2}\right)(1 + 2 + \cdots + (N+1)) = \frac{2}{N^2}(N+1)(N+2)$$

and again, $\lim S(N) = 1$. Thus the area of the triangle exists and is 1, as expected.

9 Let D be a bounded region in 3-space and suppose $D \subseteq R$, where R is a parallelepiped. Partition R into N^3 congruent blocks, and let $s(N)$ (the inner approximation) be the sum of the volumes of those blocks which are wholly contained in D. Let $S(N)$ (the outer approximation) be the sum of the volumes of those blocks which intersect D. The set D is said to have volume α if each of the sequences $\{s(N)\}$ and $\{S(N)\}$ converge to α as $N \to \infty$. In this case we write $V(D) = \alpha$. A region has volume if and only if its boundary has volume 0.

---

page 150

4.4

1 (a) For a small rectangle R_{ij} the maximum point for f over R_{ij} is the lower right corner; the minimum value is attained at the upper left corner. Also, $A(R_{ij}) = 2/25$.

$$\text{URS} = 4, \quad \text{LRS} = 2.$$

$$\int_0^2 dx \int_0^1 (2x - y)\, dy = 3.$$

(b) For a small rectangle R_{ij}, the minimum point for f is the lower left corner and the maximum point is the upper right corner.

$$\text{URS} = 86/25, \quad \text{LRS} = 66/25,$$

$$\int_0^2 dx \int_0^1 (1 + xy)\, dy = 3.$$

3 (a) $-21/4$, (b) $3(e - 1)$.

5 $A(R) = 6$ and $\iint_R f = 17$. Hence the average temperature of f on R is $17/6$.

7 Partition R into 25 small squares, each of area $1/25$, and form the Riemann sum obtained by choosing the point at the lower left corner of each small square. This sum is

$$\sum_{i=0}^{4} \sum_{j=0}^{4} \frac{1}{25} f\left(\frac{i}{5}, \frac{j}{5}\right) = .2066,$$

which is approximately $\iint_R f\, dx\, dy$.

9 Let $I = I(y) = \displaystyle\int_0^1 \frac{x + y}{1 + xy}\, dx$. Then,

$$\begin{aligned} I &= \frac{1}{y^2} \int_0^1 \frac{xy}{1 + xy}\, y\, dx + \int_0^1 \frac{y}{1 + xy}\, dx \\ &= \frac{1}{y^2}[y - \log(1 + y)] + \log(1 + y). \end{aligned}$$

Then, using integration by parts, we have

$$\begin{aligned} \int_0^1 I(y)\, dy &= \left\{-1 + \frac{\log(1 + y)}{y} + 2\log(1 + y) + y\log(1 + y) - y\right\}\Bigg|_0^1 \\ &= 4\log 2 - 2. \end{aligned}$$

11 $V = 7/3$.

---

page 159

4.5

1 (a) We must show that $g_{D_1} \cdot g_{D_2}(p) = 1$ whenever $p \in D_1 \cap D_2$, and $g_{D_1} \cdot g_{D_2}(p) = 0$ whenever $p \notin D_1 \cap D_2$. If $p \in D_1 \cap D_2$, then

$p \in D_1$ and $p \in D_2$, so $g_{D_1}(p) = 1$ and $g_{D_2}(p) = 1$ and hence $g_{D_1} \cdot g_{D_2}(p) = 1$. If $p \notin D_1 \cap D_2$, then p is not in D_1 or p is not in D_2. Thus $g_{D_1}(p) = 0$ or $g_{D_2}(p) = 0$ so that $g_{D_1} \cdot g_{D_2}(p) = 0$.

(b) $g_{D_1 \cup D_2} = g_{D_1} + g_{D_2} - g_{D_1} \cdot g_{D_2}$.

3 For any grid N and choice of points $\{p_{ij}\}$ we have $m \leq f(p_{ij}) \leq M$. Hence $mA(R_{ij}) \leq f(p_{ij})A(R_{ij}) \leq MA(R_{ij})$. Adding these inequalities over all i and j, we see that the Riemann sum for this grid and choice of points satisfies

$$m \cdot A(R) \leq \sum f(p_{ij})A(R_{ij}) \leq M \cdot A(R).$$

Since f is integrable over R, we know the middle term in the inequality converges to $\iint_R f$. Hence

$$m \cdot A(R) \leq \iint_R f \leq M \cdot A(R).$$

5 Apply Exercise 4 to the function $g - f$.

---

page 165

4.6

2 (a) 1/6, (b) 1/8, (c) 1/24.

3 $$\iint_D (2x + 8y)\, dx\, dy = \int_0^1 dy \int_{2y-1}^{\sqrt{y}} (2x + 8y)\, dx = \tfrac{61}{30}.$$

5 Since D_1 and D_2 are disjoint, $g_{D_1 \cup D_2} = g_{D_1} + g_{D_2}$. Let R be a rectangle containing $D_1 \cup D_2$. Then

$$\iint_{D_1} F + \iint_{D_2} F = \iint_R F g_{D_1} + \iint_R F g_{D_2}$$
$$= \iint_R F(g_{D_1} + g_{D_2}) = \iint_R F g_{D_1 \cup D_2} = \iint_{D_1 \cup D_2} F.$$

7 15/8.

9 (a) $\frac{8}{3}L^4$, (b) $\frac{1}{3}$, (c) $\frac{1}{2}\pi R^4$.

11 (a) The double integral is the volume of the solid with vertical sides and base D which is bounded above by the surface $z = f(x, y)$.

(b)

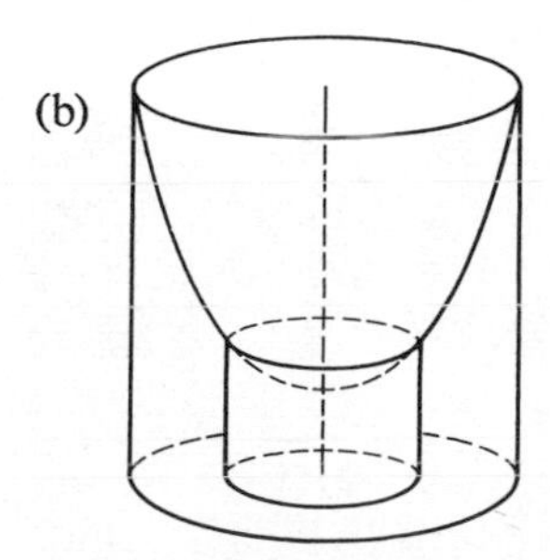

(c) $$V = 4\left[\int_0^1 dx \int_{\sqrt{1-x^2}}^{\sqrt{2-x^2}} (1 + x^2 + y^2)\, dy + \int_1^{\sqrt{2}} dx \int_0^{\sqrt{2-x^2}} (1 + x^2 + y^2)\, dy\right].$$

13 (a) The equivalent integral is

$$\int_{-2}^{0} dy \int_{-\sqrt{2y+4}}^{\sqrt{2y+4}} f\,dx + \int_{0}^{6} dy \int_{y-2}^{\sqrt{2y+4}} f\,dx.$$

(b) 18.

15 32/3.

17 (a) Choose a coordinate system with X-axis along the top edge of the dam and Y-axis through the center of the dam. The total force on the dam is $(62.5)(2/3)(25)^3$.

(b) The total force on the lower 18 feet of the dam is $(62.5)(2/3)(24)^3$.

19 $I_l = \displaystyle\int_0^1 dx \int_0^x (1-x)^2\,dy = \tfrac{1}{12}.$

---

page 177

4.7

1 5/24.

3 Place the sphere of radius R with center at the origin. The density function is $\kappa(x, y, z) = c\sqrt{x^2 + y^2 + z^2}$. The mass is then

$$M = \iiint_D \kappa(x, y, z) = \int_{-R}^{R} dx \int_{-\sqrt{R^2-x^2}}^{\sqrt{R^2-x^2}} dy \int_{-\sqrt{R^2-x^2-y^2}}^{\sqrt{R^2-x^2-y^2}} c\sqrt{x^2+y^2+z^2}\,dz.$$

Using the symmetry of the sphere and of the special function $\kappa(x, y, z)$, this can also be written as

$$8\int_0^R dx \int_0^{\sqrt{R^2-x^2}} dy \int_0^{\sqrt{R^2-x^2-y^2}} c\sqrt{x^2+y^2+z^2}\,dz.$$

5 (a)

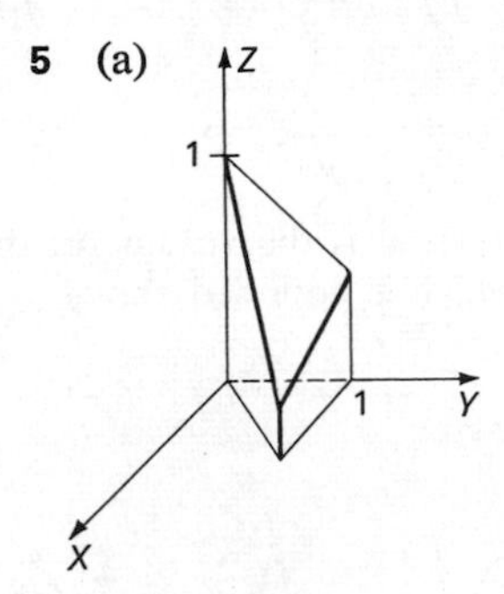

(b) $I_Z = \frac{3}{20}\rho.$

7 $M = \displaystyle\iiint_D z\,dx\,dy\,dz = \tfrac{1}{5}.$

9 The two surfaces intersect in a circle of radius 3 in the plane $z = 9$. The volume of the solid is

$$V = \int_{-3}^{3} dx \int_{\sqrt{9-x^2}}^{\sqrt{9-x^2}} dy \int_{x^2+y^2}^{27-2x^2-2y^2} dz$$
$$= 243\pi/2.$$

11 $M = \frac{176}{45}a^5$.

13 $I_X = \frac{5}{16}\kappa\pi$.

15 By symmetry, $\bar{x} = 0$. Observe that $\bar{y} = \dfrac{4}{3\pi}a$. Hence the centroid of the region is $(0, 4a/3\pi)$.

17 $\bar{x} = \bar{y} = 0$, $\bar{z} = 3/4$.

19 $$\iiint_D \delta_{\mathcal{P}}(x, y, z)\kappa(x, y, z)\, dx\, dy\, dz$$
$$= \frac{1}{\sqrt{A^2 + B^2 + C^2}} \iiint_D Ax\kappa + \iiint_D By\kappa + \iiint_D Cz\kappa + \iiint_D D\kappa$$
$$= \frac{1}{\sqrt{A^2 + B^2 + C^2}} (MA\bar{x} + MB\bar{y} + MC\bar{z} + MD)$$
$$= \left(\frac{A\bar{x} + B\bar{y} + C\bar{z} + D}{\sqrt{A^2 + B^2 + C^2}}\right) M$$
$$= \delta_{\mathcal{P}}(\bar{x}, \bar{y}, \bar{z}) \iiint_D \kappa(x, y, z)\, dx\, dy\, dz.$$

---

page 183

4.8

1 (b) $I = \iint_D x\, dx\, dy = \pi$.

3 $I_Z = \int_0^h dz \int_0^{2\pi} d\theta \int_0^{Rz/h} r^3\, dr = \frac{1}{10}\pi R^4 h$.

5 $I = 15/8$.

7 $M = 350\pi$.

9 (a) $$I_Z = \iiint_D (x^2 + y^2)\, dx\, dy\, dz$$
$$= 2\int_{-R}^{R} dx \int_{-\sqrt{R^2-x^2}}^{\sqrt{R^2-x^2}} dy \int_0^{\sqrt{R^2-x^2-y^2}} (x^2 + y^2)\, dz$$

(rectangular)

$$= \int_0^{2\pi} d\theta \int_0^{\pi} d\phi \int_0^{R} (\rho\,|\sin\phi|)^2 \rho^2 \sin\phi \, d\rho \text{ (spherical)}$$

$$= 2\int_0^{2\pi} d\theta \int_0^{R} dr \int_0^{\sqrt{R^2-r^2}} r^3 \, dz \text{ (cylindrical).}$$

(b) $V = (8/15)\pi R^5$.

11 $V = \pi a^3$.

13 $M = Ca^4\pi(2 - \sqrt{3})/4$.
By symmetry $\bar{x} = \bar{y} = 0$, $\bar{z} = a/(5(2 - \sqrt{3}))$.

15 The sphere and paraboloid intersect when $z = 1$ in a circle of radius 1. By symmetry,

$$\bar{x} = \bar{y} = 0,$$

$$M = 2\pi \int_0^1 r(\sqrt{2 - r^2} - r^2)\, dr = \frac{(8\sqrt{2} - 7)}{6}\pi,$$

$$\bar{z} = \frac{1}{M}\frac{(2\pi)}{2}\int_0^1 (2r - r^3 - r^5)\, dr$$

$$= \frac{7\pi}{12M} = \frac{7}{2(8\sqrt{2} - 7)}.$$

---

chapter 5

page 195

5.1

1 For small values of θ, $\sin\theta \approx \theta$ (in radians). Thus $\sin(\theta_0/2) \approx \theta_0/2$, so that V_0 in (5.8) is about $\sqrt{gL}\,\theta_0$, which is (5.12).

3 When $\theta_0 = \pi/2$, (5.8) yields $V_0 = \sqrt{2}\sqrt{gL}$ and (5.12) yields $V_0 = (\pi/2)\sqrt{gL}$. The relative difference in values is

$$\frac{(\pi/2)\sqrt{gL} - \sqrt{2}\sqrt{gL}}{\sqrt{2}\sqrt{gL}} \approx 11\%.$$

If $\theta_0 = \pi$, then the corresponding relative difference is

$$\frac{\pi\sqrt{gL} - 2\sqrt{gL}}{2\sqrt{gL}} = \frac{\pi - 2}{2} \approx 70\%.$$

5 The easiest general approach to complex pendulum problems is to analyze energy relations. Let L_0 be the distance from the pivot to the center of gravity of the pair of weights, and $M = M_1 + M_2$. If the pendulum is displaced to an angle θ, the potential energy of the system is $MgL_0(1 - \cos\theta)$. The kinetic energy of the moving system is $(1/2)I(d\theta/dt)^2$ where I is the moment of inertia of the system about the pivot. Using the principle of Conservation of Energy, $MgL_0(1 - \cos\theta) + (1/2)I(d\theta/dt)^2 =$ const. Differentiating this, we obtain

$$\frac{d^2\theta}{dt^2} = -\frac{MgL_0}{I}\sin\theta.$$

Comparing this with (5.1), we see that this compound pendulum is equivalent to a simple pendulum with a different length, namely $L = I/ML_0$. For the double pendulum, $L_0 = (L_2M_2 - L_1M_1)/M$ where $M = M_1 + M_2$, and $I = M_1(L_1)^2 + M_2(L_2)^2$. (Note the special case when $L_2M_2 = L_1M_1$. The pendulum balances and does not swing.)

7 Suppose the chemical enters the lake at the rate of s lb./day and pure water enters at the rate of r gal./day. Assume the lake contains V gallons of water and that V is large. Let $x = \varphi(t)$ be the amount of pollutant in the lake at time t, in pounds, and let $A = \varphi(0)$. Then in a small time interval h, rh gallons of pure water enter the lake, sh pounds of pollutant enter the lake, and rh gallons of water leave the lake with pollution concentration x/V lb./gal. Thus at time $t + h$ the amount of pollutant in the lake is

$$x(t+h) = x(t) + sh - rh\frac{x}{V}$$

and

$$\frac{x(t+h) - x(t)}{h} = s - \frac{r}{V}x,$$

so $\dfrac{dx}{dt} = s - \dfrac{r}{V}x$. Thus

$$\varphi'(t) = s - \frac{r}{V}\varphi(t), \quad \varphi(0) = A.$$

9 Let $A(t)$ be the amount of substance A at time t (measured in number of atoms, say) and let $B(t)$ be the amount of substance B. Assume $A(0) = A_0$ and $B(0) = 0$. By hypothesis, in a short time interval h, $kA(t)h$ atoms of A will disintegrate into substance B, where k is a positive constant. Thus at time $t + h$ we have

$$A(t+h) = A(t) - kA(t)h, \quad B(t+h) = B(t) + kA(t)h.$$

This leads to the differential equations

$$A'(t) = -kA(t), \quad B'(t) = kA(t).$$

In carbon dating, scientists use knowledge of the rate of disintegration of radioactive carbon 14, found in plant and animal remains, to determine the age of these remains. In this case, $A(t)$ is the amount of carbon 14 remaining at time t, and t_0 is the time of the organism's death. Since carbon 14 disintegrates at a rate proportional to the amount remaining, A satisfies the equation $A'(t) = -kA(t)$. The constant k can be determined by using the half-life of carbon 14, 5760 years.

11 (a)

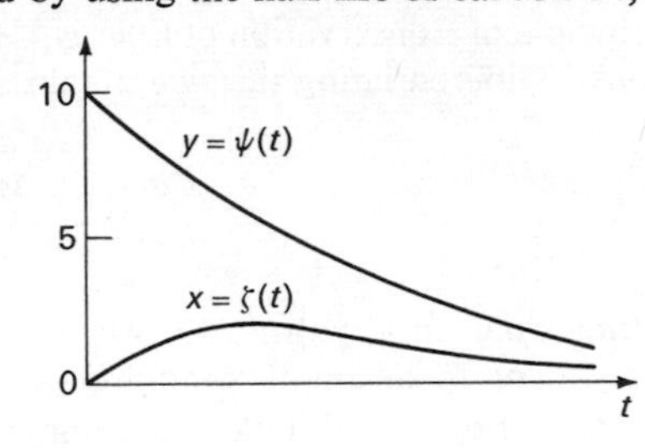

(b)

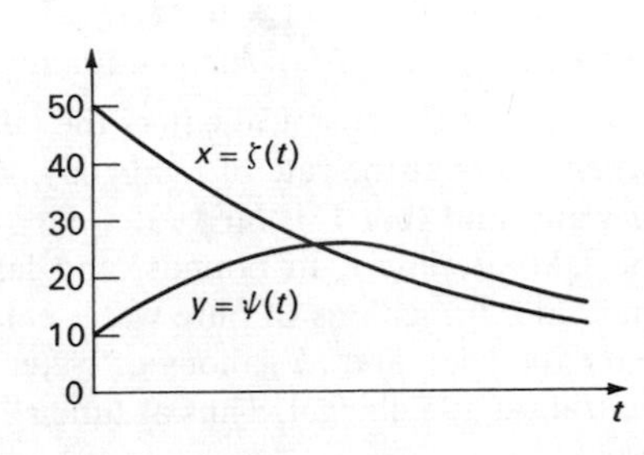

(c) The mixing is much faster, and the water in both tanks becomes purer much faster.

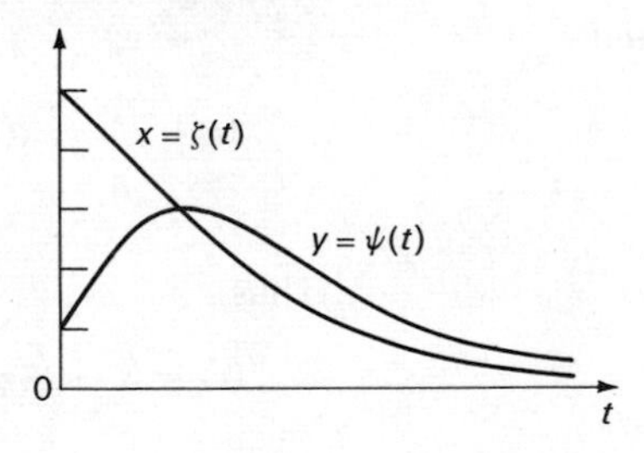

---

page 201

5.2

1 $\dfrac{dy}{dt} + \dfrac{t}{y} = \dfrac{1}{2}(C - t^2)^{-1/2}(-2t) + \dfrac{t}{\sqrt{C - t^2}} = 0$, so $y = \sqrt{C - t^2}$ is a solution.

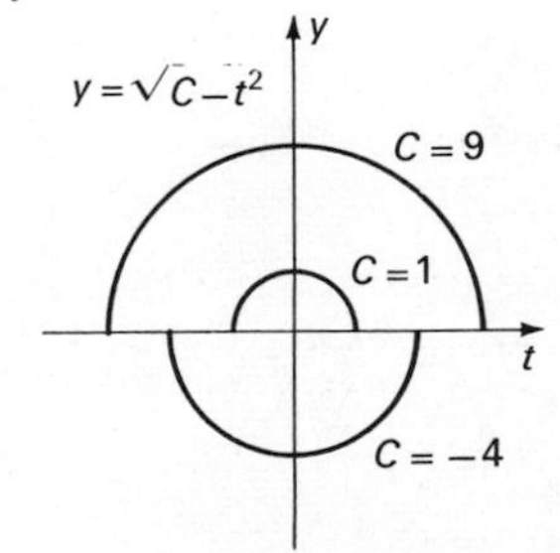

3 $\dfrac{dy}{dt} = \dfrac{2Ce^t}{(1 - Ce^t)^2}, \quad \dfrac{y^2 - 1}{2} = \dfrac{2Ce^t}{(1 - Ce^t)^2}.$

Note that $1 - Ce^t \neq 0$ excludes intervals containing $-\log C$ when $C > 0$.

5

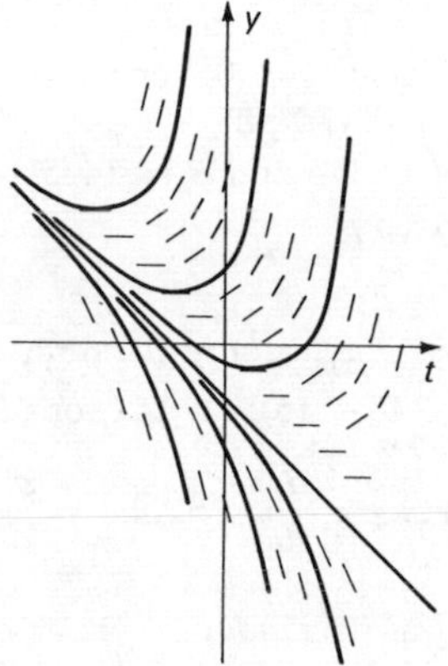

7 Careful sketches will reveal that solution curves seem to have vertical asymptotes to the right of the origin.

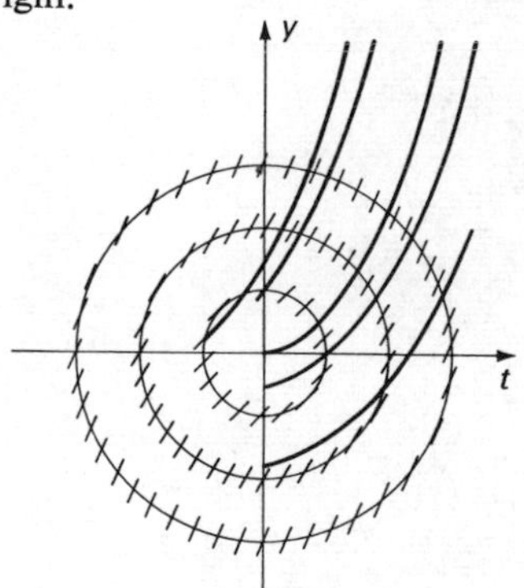

9 $C = 2/(3e)$, $C = 1/3$.

11 (a) $5y - x = 4e^{4t} - 6e^{-6t} = \dfrac{dx}{dt}$; $5x - y = 4e^{4t} + 6e^{-6t} = \dfrac{dy}{dt}.$

13 The given solutions cannot be the only solutions of the system. Others can be found by observing that $\dfrac{dy}{dx} = \dfrac{3}{x^2y^2}.$

page 205

5.3

1 (a) $24y - 2\dfrac{dy}{dt} = 24Ae^{4t} + 24Be^{-6t} - 8Ae^{4t} + 12Be^{-6t}$

$$= 16Ae^{4t} + 36Be^{-6t}$$

$$= \frac{d}{dt}(4Ae^{4t} - 6Be^{-6t}) = \frac{d^2y}{dt^2}.$$

(b) $A = 15/10 = 3/2$ and $B = 1/2$.

3 (a) If $y(t) = t^\alpha$, then $y'(t) = \alpha t^{\alpha-1}$ and $y''(t) = \alpha(\alpha - 1)t^{\alpha-2}$. If $y'' = \dfrac{5}{2t}y' - \dfrac{3}{2t^2}y$ then $\alpha(\alpha - 1)t^{\alpha-2} = (\frac{5}{2}\alpha - \frac{3}{2})t^{\alpha-2}$, $2\alpha^2 - 7\alpha + 3 = 0$, or $\alpha = 3$, $\alpha = \frac{1}{2}$. Hence let $\varphi(t) = t^3$, $\psi(t) = t^{1/2}$ $(t \geq 0)$.

(b) Thus $A = 1/5$ and $B = 14/5$.

5 $A = -3/2 = B$.

8 $(D - 3)x = 2y$.

9 $(D + 5)(D - 3)x = (D + 5)(2y) = 2(D + 5)y = 2(7x) = 14x$. Hence $(D^2 + 2D - 15)x = 14x$, or $(D^2 + 2D - 29)x = 0$.

11 Set $u_1 = y$, $u_2 = \dfrac{dy}{dt}, \ldots, u_n = \dfrac{d^{n-1}y}{dt^{n-1}}$. Then,

$$\frac{du_1}{dt} = u_2$$

$$\frac{du_2}{dt} = u_3$$

$$\vdots$$

$$\frac{du_{n-1}}{dt} = u_n$$

$$\frac{du_n}{dt} = F(t, u_1, u_2, \ldots, u_n).$$

---

page 210

5.4

1 We can take the region D to be the TY-plane. Any point may be used as an initial point.

3 If $-\infty < t < b$, then $dy/dt = 0 = 3y^{2/3}$; if $b \leq t < \infty$, then $dy/dt = 3(t - b)^2 = 3((t - b)^3)^{2/3} = 3y^{2/3}$.

5 With $f(t, y, y') = \dfrac{y}{(4 - (y')^2)}$,

$$f_2 = \frac{1}{(4 - (y')^2)} \quad \text{and} \quad f_3 = \frac{-2yy'}{(4 - (y')^2)}.$$

Each is continuous at all points (t, y, y') with $y' \neq \pm 2$. The only excluded initial conditions are those with $y' = +2$.

7 If the equation does not have a solution, this would indicate the reaction cannot take place. Our confidence in this conclusion should be only as great as our belief that the model is an accurate description of reality. If the reaction does indeed take place, we should conclude that the model is not a useful one.

If the equation has a solution, this does not guarantee that the reaction does take place, for the model may not be an accurate description of physical reality. For example, in the equations for the motion of a pendulum we saw that the model predicted the pendulum would never stop swinging, which certainly is not an accurate description of reality.

---

page 225

5.5

1 $y = 2e^{\frac{1}{3}t^3 - \frac{1}{3}}$.

3 $\varphi(t) = \sqrt{t^2 + 8}$, $\varphi(t) = \sqrt{t^2 + 1}$, $\varphi(t) = -\sqrt{t^2 + 8}$.

5 $y = \dfrac{1}{1 - (1/2)t}$, $-\infty < t < 2$, and $y = 1$, $2 \leq t < \infty$.

7 $\varphi(t) = 1 + 2e^{(t^3-1)/3}$, $\varphi(t) = 1 + e^{(t^3-1)/3}$, $\varphi(t) = 1 - e^{(t^3-1)/3}$, $\varphi(t) = 1 + e^{t^3/3}$.

9 Write $y' + 2y = e^t$. Multiplying each side by e^{2t} gives $e^{2t}y' + 2e^{2t}y = e^{3t}$. Hence $\dfrac{d}{dt}(e^{2t}y) = e^{3t}$, $\varphi(t) = \frac{1}{3}e^t + \frac{2}{3}e^{-2t}$ is the solution through $(0, 1)$, defined for all t.

11 $\varphi(t) = (t + 1)e^t$.

13 $\varphi(t) = 1/t$ is a solution through $(1, 1)$ for $0 < t < \infty$, $\varphi(t) = 1/t - 1/t^2$ is a solution through $(1, 0)$ for $0 < t < \infty$.

15 $y = -3t \pm \sqrt{A + 10t^2}$.

17 $y = (Ce^{t^2} - \frac{1}{2})^{-1}$. To these we must adjoin the extra solution $y = 0$.

19 $\varphi(t) = (t^3 + Ct)^2$ and $\varphi(t) \equiv 0$ are the solutions.

21 $y = \frac{1}{18}t^6 + \frac{1}{3}C_1t^3 + C_2$.

23 Put $u = \dfrac{d\theta}{dt}$ and $\dfrac{d^2\theta}{dt^2} = \dfrac{du}{dt} = \dfrac{du}{d\theta}\dfrac{d\theta}{dt} = u\dfrac{du}{d\theta}$.

Then the equation becomes $u\dfrac{du}{d\theta} = -\dfrac{g}{L}\theta$, which is separable. Solving this with the initial condition $\theta = \theta_0$, $u = 0$, we have $u^2 = \dfrac{g}{L}(\theta_0^2 - \theta^2)$.

For $t > 0$ but small, $\theta < \theta_0$ and thus $u = \dfrac{d\theta}{dt} < 0$. Hence

$$\frac{d\theta}{dt} = -\sqrt{\frac{g}{L}}\,(\theta_0^2 - \theta^2)^{1/2}.$$

This is also separable. Put $\theta = \theta_0 \cos\phi$, obtaining $\dfrac{d\phi}{dt} = \sqrt{\dfrac{g}{L}}$. Since $\phi = 0$ for $t = 0$, we find $\phi = \sqrt{\dfrac{g}{L}}\, t$ and the solution $\theta = \theta_0 \cos\left(\sqrt{\dfrac{g}{L}}\, t\right)$, which describes the motion of the pendulum for $t \geq 0$, for this model.

25 (5.26) becomes $u\dfrac{du}{d\theta} = -\dfrac{g}{L}\sin\theta - Cu^2$. Putting $u = \sqrt{v}$ changes this to $\dfrac{1}{2}\dfrac{dv}{d\theta} = -Cv - \dfrac{g}{L}\sin\theta$, which is linear.

---

page 235

5.6

1 $\varphi(1) \approx .3487$. The correct value $= .36788$.

3 (a) With $h = .2$, we estimate $\varphi(1) = .2548$. With $h = .1$, we estimate $\varphi(1) = .2976$.
(b) $\varphi(1) = 3/8 = .375$. The direction field for this equation makes it very difficult to find the desired solution, since errors are quickly magnified.

5 $\varphi(1) = 1,\quad \varphi'(1) = 2,\quad \varphi''(1) = 1 + 2 + 2 = 5,\quad \varphi'''(1) = 11$, $\varphi^{(4)}(1) = 26$, $\varphi(t) = 1 + 2(t-1) + \frac{5}{2}(t-1)^2 + \frac{11}{6}(t-1)^3 + \frac{13}{12}(t-1)^4 + \frac{7}{12}(t-1)^5$.

7 $\varphi''(0) = 1$, $\varphi'''(0) = 2(-1)2 = -4$, $\varphi^{(4)}(0) = 2(4) - 2 = 6$, $\varphi^{(5)}(0) = 4(2) + 6(2) = 20$,

$$\varphi(t) = -1 + 2t + \tfrac{1}{2}t^2 - \tfrac{2}{3}t^3 + \tfrac{1}{4}t^4 + \tfrac{1}{6}t^5.$$

9
$$y = 1 + t + \frac{1}{2}t^2 + 3\left(\frac{1}{3!}t^3 + \frac{1}{4!}t^4 + \cdots + \frac{1}{k!}t^k + \cdots\right)$$
$$= 3e^t - 2 - 2t - t^2.$$

11
$$y = 1 + \frac{1}{3\cdot 4}t^4 + \frac{1}{3\cdot 4\cdot 7\cdot 8}t^8 + \frac{1}{3\cdot 4\cdot 7\cdot 8\cdot 11\cdot 12}t^{12} + \cdots.$$

---

page 251

5.7

2 (a) $T_1(3t+1) = 15t - 1$; $T_3(3t+1) = 3t + 2 - 9t^2$.

3 (a) $T_1(t^2 - t) = 2t^3 - 3t^2 + t$, $T_3(t^2 - t) = t^4 - t^2$.
(b) T_2 and T_3 are linear.

5 (b) The null space of D^n is all polynomials of degree at most $n - 1$.

7 (a) Show that each of $1, t, t^2$ can be written as a linear combination of the three given functions. For example, $t^2 = \frac{2}{3}(t+1) + \frac{2}{3}(t^2 - t) - \frac{1}{3}(2t^2 - 1)$.

(b) Show that some polynomial of degree at most 2 is not a linear combination of the three given functions.

9 Suppose $Ae^{\gamma t} + Bte^{\gamma t} = 0$ for all t. Setting $t = 0$ we see that $A = 0$. Setting $t = 1$ (any nonzero t will do), we see that $Be^{\gamma} = 0$ or $B = 0$. Therefore the two functions are linearly independent.

11 $P(s) = s^3 + 3s^2 + 3s + 1 = (s + 1)^3$.

13 (a) $C_1e^{(1+\sqrt{2})t} + C_2e^{(1-\sqrt{2})t}$,
(b) $C_1e^{-t}\cos\sqrt{2}\,t + C_2e^{-t}\sin\sqrt{2}\,t$,
(c) $C_1e^{-3t/2}\cos\dfrac{3\sqrt{7}}{2}t + C_2e^{-3t/2}\sin\dfrac{3\sqrt{7}}{2}t$.

15 $y = (-2/5)\cos t + (4/5)\sin t + (2/5)e^{-2t}$.

17 $\varphi(t) = t - t^2 + e^{-t}$.

19 Try a solution of the form $y = A\sin 2t + B\cos 2t$. The general solution is $y = Ce^{-2t} + (1/4)\sin 2t - (1/4)\cos 2t$.

21 (a) $\varphi(t) = (1/2)e^{-2t} + C_1e^{-t} + C_2$.
(b) $\varphi(t) = (1/2)e^{-2t} + e^{-t} + \frac{1}{2}$.

23 (a) $(D^2 - 2D - 15)y = 0$; thus $y = C_1e^{5t} + C_2e^{-3t}$.
$x = (1/8)(D - 1)y = (1/2)(C_1e^{5t} - C_2e^{-3t})$.
(b) $x = e^{5t}$, $y = 2e^{5t}$.

27 $y = C_1e^{t} + C_2e^{2t} + te^{2t}$.

---

page 263

5.8

1 Newton's "law of cooling" becomes $\dfrac{du}{dt} = -k(u - T)$; 42.5°.

3 The differential equation suggested is $dP/dt = kP(M - P)$. The particular model is very poor and should not be used as a basis for discussion of human population problems.

5 9.375 gm/gal.

7 Analyzing the flow of red paint, we arrive at the system

$$\frac{dx_1}{dt} = 5\frac{x_3}{40} - 5\frac{x_1}{20},$$

$$\frac{dx_2}{dt} = 5\frac{x_1}{20} - 5\frac{x_2}{30},$$

$$\frac{dx_3}{dt} = 5\frac{x_2}{30} - 5\frac{x_3}{40}.$$

index

ABCDEFGHIJ— H —76543210